NATO ASI Series

Advanced Science Institutes Series

A series presenting the results of activities sponsored by the NATO Science Committee, which aims at the dissemination of advanced scientific and technological knowledge, with a view to strengthening links between scientific communities.

The Series is published by an international board of publishers in conjunction with the NATO Scientific Affairs Division

A Life Sciences **B Physics**	Plenum Publishing Corporation London and New York
C Mathematical and Physical Sciences **D Behavioural and Social Sciences** **E Applied Sciences**	Kluwer Academic Publishers Dordrecht, Boston and London
F Computer and Systems Sciences **G Ecological Sciences** **H Cell Biology** **I Global Environmental Change**	Springer-Verlag Berlin Heidelberg New York London Paris Tokyo Hong Kong Barcelona Budapest

PARTNERSHIP SUB-SERIES

1. Disarmament Technologies	Kluwer Academic Publishers
2. Environment	Springer-Verlag
3. High Technology	Kluwer Academic Publishers
4. Science and Technology Policy	Kluwer Academic Publishers
5. Computer Networking	Kluwer Academic Publishers

The Partnership Sub-Series incorporates activities undertaken in collaboration with NATO's Cooperation Partners, the countries of the CIS and Central and Eastern Europe, in Priority Areas of concern to those countries.

NATO-PCO DATABASE

The electronic index to the NATO ASI Series provides full bibliographical references (with keywords and/or abstracts) to about 50000 contributions from international scientists published in all sections of the NATO ASI Series. Access to the NATO-PCO DATABASE compiled by the NATO Publication Coordination Office is possible in two ways:

- via online FILE 128 (NATO-PCO DATABASE) hosted by ESRIN, Via Galileo Galilei, I-00044 Frascati, Italy.
- via CD-ROM "NATO Science & Technology Disk" with user-friendly retrieval software in English, French and German (© WTV GmbH and DATAWARE Technologies Inc. 1992).

The CD-ROM can be ordered through any member of the Board of Publishers or through NATO-PCO, Overijse, Belgium.

2. Environment – Vol. 3

Springer-Verlag Berlin Heidelberg GmbH

Remediation and Management of Degraded River Basins

with Emphasis on Central and Eastern Europe

Edited by

Vladimir Novotny

Department of Civil and Environmental Engineering
Marquette University
Milwaukee, Wisconsin 53201-1881, USA

László Somlyódy

Department of Water and Wastewater Engineering
Budapest University of Technology
Müegyetem rpk. 3
H-1111 Budapest, Hungary

Published in cooperation with NATO Scientific Affairs Division

Proceedings of the NATO Advanced Research Workshop "Remediation and Management of Degraded River Basins with Emphasis on Central and Eastern Europe", held at Laxenburg, Austria, June 13–16, 1994

Library of Congress Cataloging-in-Publication Data

Remediation and mangement of degraded river basins : with emphasis on Central and Eastern Europe / edited by Vladimir Novotny, László Somlyódy.
p. cm. -- (NATO ASI series. 2, Environment ; vol. 3)
Includes bibliographical references and index.
ISBN 978-3-642-63346-1 ISBN 978-3-642-57752-9 (eBook)
DOI 10.1007/978-3-642-57752-9
1. Water quality management--Europe, Central--Congresses. 2. Water quality management--Europe, Eastern--Congresses. 3. Water--Pollution--Europe, Central--Congresses. 4. Water--Pollution--Europe, Eastern--Congresses. 5. Watersheds--Europe, Central--Congresses. 6. Watersheds--Europe, Eastern--Congresses. I. Novotny, Vladimir, 1938- . II. Somlyódy, L. (László) III. Series.
TD255.R46 1995
363.73'946'0943--dc20 95-21496
CIP

ISBN 978-3-642-63346-1

Originally published by Springer-Verlag Berlin Heidelberg New York in 1995
Softcover reprint of the hardcover 1st edition 1995

Typesetting: Camera-ready by authors/editors
SPIN: 10492437 31/3136 – 5 4 3 2 1 0 – Printed on acid-free paper

PREFACE AND ACKNOWLEDGEMENT

In June of 1994, the Advanced Research Workshop (ARW) sponsored by the North Atlantic Treaty Organization (NATO), Brussels, Belgium was held in Laxenburg, Austria. The main focus of the ARW was identification of water quality and pollution problems in Central and Eastern Europe (CEE), which also includes some countries of the former Soviet Union, and outlining methodologies of abatement. The ARW used a format of both key note presentations by invited speakers and short communications from the participants with discussions, to outline the scope of the problem and current as well as planned approaches to solutions.

The objective of the NATO Advanced Research Workshop (ARW) was to address in an investigative fashion the tools, strategies and policies of watershed and water quality restoration and management both in Western and CEE countries, focusing on applicability of Western experience to the CEE countries and to develop new innovative strategies which would reflect ongoing and emerging political and economic realities.

The ARW, attended by thirty-one delegates from seventeen countries (4 NATO, 9 CEE, 4 others), provided means for accelerated transfer of knowledge to the CEE countries. Most of these countries have the scientific base to assimilate the Western experience and know-how in an expedited fashion. Fifteen speakers from seven countries (4 NATO, 3 CEE) presented key-note lectures.

The long term indirect benefit of the workshop is improved water quality and use of water resources for economic and social development, and improvement of public health. Also the scientists and professionals from NATO countries had an opportunity to interact with their CEE counterparts and initiate cooperative efforts, resulting in the development of feasible remediation technologies, strategies, and basin management, which will also benefit Western countries. The new ideas being developed could be prototype tested in CEE countries for the benefit of both the East and West.

This edited treatise is a product of the NATO workshop which was held at the International Institute for Applied Systems Analysis (IIASA) in Laxenburg, Austria. The workshop was co-sponsored by IIASA and the International Association on Water Quality. Dr. Vladimir Novotny, Professor of Environmental and Water Resources Engineering at Marquette University, Milwaukee, Wisconsin (USA) and Dr. László Somlyódy, Professor of Water and Waste Water Engineering of the Budapest University of Technology, were co-directors of the workshop.

IIASA is a non-governmental, interdisciplinary research institute which focuses on global environmental change. Integrated river basin management on various scales is one of the major research themes of the IIASA program. It includes global vulnerability of water resources and the management of degraded river basins in the region of Central and Eastern Europe as well as development of related methodologies.

The directors of the ARW and editors of this treatise would like to thank Dr Louis Veiga da Cunha, director of the NATO environmental programme, for his support during preparation of the ARW and active leadership participation during the workshop. Dr Petr E. de Janosi, director of IIASA, was most helpful by providing facilities, logistical and administrative support and resources of the Institute before and during the workshop.

TABLE OF CONTENT

CHAPTER 1

WATER QUALITY MANAGEMENT: WESTERN EXPERIENCES AND CHALLENGES FOR CENTRAL AND EASTERN EUROPEAN COUNTRIES

Vladimir Novotny[1] and László Somlyódy[2]

1. INTRODUCTION AND BACKGROUND

Central and Eastern European (CEE) countries are undergoing fundamental political, economic, and social changes. The outcome of this transition is not yet known, and it will take two to three decades to fully realize in which direction individual CEE countries will go. As far as the situation of the environmental "degradation" in general, and water pollution in the CEE countries in particular, is concerned the heritage of the past is serious. The situation in the mid 1990s is characterized by a high level of contamination and by a multiplicity of problems caused by traditional and toxic pollutants originating from both point and nonpoint sources. The problems are both local and regional, the latter including high level of pollution of the Black Triangle, Black and Baltic Seas, and of several international rivers such as the Danube, Odra, Elbe and others. Additional problems are caused by past contamination of soils, sediments, and groundwater which may be difficult to remedy.

Most surface waters in CEE countries have deteriorated water quality and often, as it is the case of Poland, Slovak Republic and Bulgaria, more than half of the total monitored length of their rivers is classified as having the poorest water quality according to established water quality classification[3]. The

---

1 Marquette University, Milwaukee, Wisconsin, USA

2 Technical University of Budapest, Hungary

3 In the CEE countries, water quality is classified into four or five water quality classes using a number of indicators such as BOD_5, Dissolved Oxygen, temperature, nutrients, and toxic compounds measured or projected to low flow summer conditions. This system is different from the US system based on designated water uses which emphasizes Dissolved Oxygen concentrations and concentrations of priority pollutants extrapolated to very rare probabilities of excedences.

NATO ASI Series, Partnership Sub-Series, 2. Environment – Vol. 3
Remediation and Management of Degraded River Basins
Edited by V. Novotny and L. Somlyódy

most serious problems include high BOD_5 levels which cause low DO concentrations, bacterial contamination, high ammonia and nitrate concentrations threatening the use of water resources for water supply, high phosphorus levels which along with very high nitrogen levels turn many reservoirs into hypereutrophic water bodies with diminished beneficial uses, high salinity of major rivers draining coal mining areas (e.g., the Odra River) due to the very high salt content of mine drainage waters, and serious toxic metal contamination of industrial areas. Many rivers have been impounded for navigation and the pools are now full with contaminated deposits, however, the extent of contamination is largely unknown or measured only recently. To illustrate the problem, it was determined that the average low-flow BOD_5 of receiving rivers downstream of larger urban centers (population greater than 25,000) in five CEE countries (Poland, Czech Republic, Slovakia, Hungary and Bulgaria) was 20 mg/l (level of treated sewage) with extremes reaching 200 mg/l (level of untreated sewage).

Nitrate contamination of grou .dwater and surface water bodies, even though following a worldwide trend, has reached levels that made many water bodies unfit as a source of potable water. For example, nitrate content in some tributaries of the Želivka reservoir, which is a primary source of potable water for the capital city of the Czech Republic, Prague, has reached levels exceeding on occasions 100 mg NO_3^-/l.

In spite of the aforementioned state of "degradation", water pollution of the CEE region should not be considered unique and to some degree it resembles similar situations in some Western countries twenty to forty years ago. For example, in the early 1950s and before the River Thames in London was heavily polluted and often anoxic during summer periods. Similar pollution episodes related to low DO were common in many important U.S. water bodies such as the Delaware and Potomac Rivers and estuaries. Some Great Lakes were "dying" due to excessive inputs of nutrients. At the same time toxicological catastrophes due to metal pollution of water and fish (the so-called Minamata and Itai-Itai diseases) took place in Japan (Krenkel and Novotny, 1980).

Serious efforts to combat pollution in most Western countries have occurred only since early 1970. The Thames River became "alive" again and fish are caught there now. Water quality is good in the Delaware and Potomac systems and eutrophication of the Great Lakes has been to some degree reversed. Meanwhile, after accomplishing most of the point source pollution clean-up in some advanced Western countries[4], the attention has shifted from traditional pollutants to toxic and carcinogenic ones and diffuse sources (combined sewer overflows, urban, agricultural and silvicultural runoff, discharges from abandoned mines, construction runoff pollution control, etc.)

Thus, tools, methodologies and experiences of clean-up are available. However, due to the unique political and economical situation, near future development and efforts of environmental management in the CEE countries may and most likely will not repeat past developments in the Western developed world.

The primary difference is the state of the economies and nature of economic developments. Unlike presently in the CEE region, thirty years ago the economy of the Western countries was healthy, expanding, and the population enjoyed a relatively high affluence. Environmental awareness was initiated by grass root movements of environmental organizations and citizens. Far reaching environmental legislation in some western countries was passed in the early 1970s (e.g., the U.S. Clean Water Act in 1972) and, subsequently, large expenditures have been spent on clean-up. Initially, the federal government of the United States provided substantial pollution abatement grants to install pollution control technology for municipal dischargers (amounting to hundreds billion dollars over the period of twenty years), however, a large portion of the cost of abatement has been borne by the dischargers[5]. No grants were provided for clean-up of industrial sources for which the "polluter pays" principle was fully applied.

Environmental clean-up in Western countries brought about many successes. Yet, in spite of large expenditures, these countries spent a small fraction of their GNP which had no (or little) impact on the national economies and, the fiscal burden on the population was minimal. For example, in Milwaukee, Wisconsin (USA) taxes and fees paid annually by an average family for pollution control by one of the most ambitious and costly pollution abatement[6] programs, amounts to about 1 percent of the family income. Similar figures apply to several Western European countries.

---

[4] There are great disparities of the levels of environmental protection even among the Western countries of the EC, e.g., the status of pollution control in Italy, Spain and Portugal is probably not as advanced as in the Czech Republic.

[5] It should be pointed out that grants to municipalities which initially amounted to up to 75 percent of the capital cost, have been phased out in the late 1980s and today no large grants are provided.

[6] The total present worth of the cost of the Milwaukee Pollution Abatement Program is about $ 2.5 billion. The population of the metropolitan area is about 1.4 million.

The political, economic and social environment is drastically different in CEE countries. On the resource side, the per capita GDP is one fifth to one tenth of the developed Western countries (and one third to one fifth of that in the west about twenty years ago). This is associated with growing unemployment, high inflation, increasing price levels (close to Western ones) and bread-and-butter worries. Thus, the path to environmental clean-up in the CEE countries is initiated under completely differed conditions. What is a small fraction of the GDP in the advanced Western countries could be a significant portion of the GDP in the CEE countries if the same expenditures on abatement are implemented.

The above comparison clearly indicates that :

- CEE countries may not be willing nor able to cover comparable costs of environmental management in the immediate future as was the case in developed Western countries;

- it is perceived that the time horizon of any comprehensive country-wide clean-up efforts may require a rather long time span, i.e., several decades;

- setting and scheduling water quality priorities and goals are critical tasks.

Somlyódy (1993) has pointed out that a feasible strategy of pollution abatement in the CEE countries should be based on cost-effectiveness and affordability in the short run, leaving the possibilities for gradual extension in the long run. Any new strategies should be initiated under the condition of the still ongoing transition. Some of the key elements of this process are (Somlyódy, 1993):

- Centralized strategic and financial planning by governments has been almost completely abandoned, and there is a strong movement towards decentralization and privatization;

- The institutional system is undergoing significant change, leading to an increased role for local governments, creation of new river basin authorities (for example in Poland) and environmental agencies, the disappearance of state owned regional water and wastewater companies and division of water quality management functions into separate water and environmental ministries and district authorities (e.g., in Hungary and in Russia)[7];

- Responsibility for water supply and treatment and ownership of the infrastructure is being transferred to municipalities;

- The decision process has been largely decentralized which can lead to rather peculiar institutional and decision schemes;

- New environmental legislation is being gradually introduced;

- Foreign companies and capital have entered the market in different ways which have both positive and negative consequences. Transfer of knowledge in certain fields (management, operation, environmental impact assessment, auditing, etc.) and providing well proven technologies are beneficial. On the other hand, selling of outdated and unreliable methods can have a long-term negative impact.

The economic transition influences pollution from various sectors differently. The decline in industrial production and plant shutdowns (and the collapse of military industries in some countries) have caused some heavily contaminated rivers (e.g., the Sajó in Hungary) to improve by one or two water quality classes. Decreased use of fertilizers in agriculture has reduced nutrients (nitrogen and phosphorus) loads, although to a smaller extent, due to earlier accumulation of nutrient in the topsoil. The longer-term decline of pollution is also anticipated due to decreasing demand of products as a result of price adjustment and market mechanisms. The present transitional situation provides an opportunity to correct earlier economic discrepancies and make producers economically responsible for the cost of abatement and/or damage, i.e., to combine economic restructuring and implementation of new environmental policies.

Unlike industrial and agricultural sources that have been strongly affected by the economic transition, municipal emissions in CEE countries have not been significantly altered (although industrial

---

7 It may be pointed out that similar division into quantity water management and environmental protection is also administered in the United States. The quantity water management, mostly practiced in the western United States is administered and carried out primarily by the Bureau of Reclamation, which is an agency of the Department of Interior while water quality issues are handled on a regulatory and enforcement side by the Federal Environmental Protection Agency and corresponding state agencies.

changes and water use are declining due to properly set water prices). The investment needed to improve existing systems, at a level far below the standards in Western Europe, would be an enormous burden even for stable economies. Thus, the dilemma is how to develop a strategy which supports the economic transition without exacerbating existing social problems caused primarily by the low per capita GDP (see Somlyódy, 1993 for details).

The objective of this chapter is to give an overview on major principles of water quality management, to outline how legislation works in the U.S. as one of the Western countries having the richest experiences in the field and to discuss major challenges of CEE countries particularly in the short run, during the ongoing transition.

The structure of the material is as follows. First we discuss major imperatives of pollution control mostly as they are defined in the United States (but the basic principles are similar also in other advanced Western countries). This is followed by a discussion on effluent and ambient water quality standard systems, and why both should be applied. The next section outlines main elements of the U.S. regulation and its changes. Subsequently, we summarize some of the challenges which the CEE countries face nowadays. The chapter is completed by conclusions.

2. POLLUTION CONTROL IMPERATIVES

Pollution control in a broad sense is driven by the need to handle the so called externalities. Solow (1971) defined pollution externality as follows : "One persons's use of a natural resource can inflict damage on other people who have no way of securing compensation, and who may not even know that they are being damaged." Overcoming the externality problem and incorporating the cost of damage caused by pollution in the economic thinking of those causing it are the major objectives of abatement policies. Again quoting Solow: "We would like to insure that each resource is allocated to that use in which its net social value is highest," which is called by political economists *Pareto optimality*.

Ignoring externality has serious consequences as it did in the CEE countries where the previous regimes disregarded economic principles. For instance, if acid rain damage to lakes and soils is not included in tariffs of electricity, electric power is then too "cheap" to consumers. This then leads to overproduction and more pollution damage. In this way society will subsidize those using a lot of electricity and cause increased environmental problems. The same reasoning applies to industrial and agricultural production.

In several CEE countries a significant portion of population is at or below the poverty level and price increases to compensate for pollution damages and abatement could be, and very often are, considered detrimental. However, the cost of unabated pollution is real and even more detrimental. Using previous regimes in the CEE countries as an example, "free" health care and sending children from affected highly polluted areas to health resorts was a substitution for pollution abatement, yet, the total cost of these substitutes and the cost of damage to human health damages and lost resources have certainly surpassed the cost of pollution abatement. Unabated pollution also has other costs such as declining life expectancy, lost working hours due to illness, loss of soil fertility, loss of water resources for drinking which must then be replaced by more costly distant sources or more expensive treatment, etc.

The existence of externalities causes so called *market failure* in water pollution abatement. Market forces alone can not determine acceptable levels of pollution because upstream polluters would have a distinct advantage (no restrictions and less cost of treatment) over downstream users who would have to spend more funds for treatment or for whom the use of water resources would be diminished without a compensation from upstream polluters (these are reasons why the practice of water pollution control differs from that of air pollution). Overcoming the externality problem requires regulation and use of economic and legal tools. Rogers and Rosenthal (1988) defined several policy imperatives[8] for control of pollution which though are debatable reflects well the views in the United States (and in several other Western countries). These imperatives are as follows:

Equity. No group of individuals in society should bear a disproportionate cost of meeting environmental quality requirements. The levels of environmental quality chosen should be such that no additional benefits can be derived without making one group of individuals worse off (Pareto optimality). The equity imperative dictates that polluters are made responsible for damages. Implementation of the equity criterion also implies that dischargers (polluters) are treated more or less equally, i.e., no one should be exempt from abatement and abatement should bring about the same burden on the polluters. Furthermore the cost of abatement should not exceed the cost of damage.

---

[8] Imperatives in pollution control can be technological (i.e., it is not technically feasible to remove all pollution form the source without completely ceasing the discharge), economic (a project will fail if its cost exceeds the benefits) or political (taxes are unpopular) .

This imperative is reflected in several aspects of environmental policy that have been implemented or are under consideration in Western countries.

Irreversible impact. No actions may be permitted that would irreversibly harm the environment and/or natural resources. This concern has to be guarded against by society at large (intergeneration impact). This imperative is also known and is now being implemented as a requirement for *sustainable economic development.*

Regulation and statutes. Due to the failure of the general market (see the preceding discussion on externalities) to control the quality of the environment as well as to protect resources, there is a need for legislation which must be clear and not difficult to carry out.

Acceptance. There must be concurrence on the part of the people and groups being regulated that they will, by and large, obey regulations.

Discharger pays principle (DPP). Dischargers bear the primary responsibility for pollution and its abatement. The implementation of this principle requires application of economic instruments (see e.g. OECD, 1991).

Integrated approach. The pollution problem must be resolved in an integrated manner, whereby causes of pollution, all sources, and the combined environmental impact are considered, and resulting combined solutions are therefore most equitable and efficient. The need for an integrated strategy emerges increasingly nowadays (see, e.g., the ongoing revision process of the U.S. Clean Water Act) learning from pitfalls of past practices.

Pollution prevention. Environmental policies will promote economic developments that will anticipate potential pollution problems and react to them by political and technological means before they occur.

Other imperatives have also been included for instance in U.S. environmental legislation. Two of them are fundamental to the CEE countries:

Reasonableness and/or avoidance of widespread social and economic hardship. For instance the U.S.

Clean Water Act and regulations derived from it recognize the fact that implementation of pollution control should not bring about an undue widespread economic and social hardship on the population. This imperative will be elaborated more extensively later on and is one of the tests for formulation, implementation and *enforcement* of environmental standards. Enforcement is particularly stressed in the CEE context since past legislation looked quite all right in theory, but not in practice and setting of unrealistic goals and standards now may lead to a similar failure.

Antidegradation. No action should bring about worsening of water quality in water bodies which presently meet environmental standards even if the action may not result in water quality that would violate the standards.

It should be pointed out that implementation of the above nine policy imperatives may not necessarily lead to "optimal" and "efficient" solutions (surprisingly, "efficiency" is not listed as an imperative of Western environmental policies). Consequently, in some aspects they lead to policies far from being "least-cost" (it is sufficient to refer to uniform treatment dictated by the principle of equity). In another example of contradiction of these imperatives, the antidegradation imperative can make future economic development very difficult if not impossible in situations where proposed economic development is situated on water resources that have good water quality and have available waste assimilative capacity. The economic efficiency criterion would call for allowing additional dischargers and for using the available WAC. It should be noted that the policy imperatives are political and legal instruments and in some cases evolved from legal doctrines and their interpretations by courts.

The dilemma of CEE countries is at what extent the above imperatives can be used during the transition period. Or stated differently, which principle(s) should be the focus, i.e., which should be made primary and followed immediately, and which should be secondary ? Or again differently, for instance, how should conflicting equity and efficiency imperatives be reconciled ? These are extremely important issues which should be addressed as a first step of developing detailed environmental regulations in the CEE countries.

Environmental damages of mostly intangible resources foregone are often related to the living standard and, indirectly, to the GDP. To follow the Pareto optimality axiom it would mean that the affordable cost of abatement in the CEE countries is less than that in the West and the equilibrium between the cost of abatement and damage (and the value of resources forgone) will be perceived as being lower[9] until the GDP catches up with the West. There are many in Western countries and even many environmentally conscious people in the CEE countries who will argue for full immediate implementation of Western effluent limitations but it is quite evident that implementation of such measures would put an undue burden on the CEE societies and economies (see the imperative of reasonableness).

What also differentiates CEE countries from developed Western countries is the notion and comprehension of damages by pollution. Estimating the cost of pollution abatement is relatively straightforward (they are not much different from those in the West). It is the comprehension of the cost of damage and of the resources forgone (often hard to quantify) that may be perceived lower under economic scarcity than the economic necessities of providing food and employment. Consequently, it appears logical that the damage cost in the CEE region is not the same as that in the West[10] and the notion of efficiency should play a more important role during the transition than in industrialized countries.

3. STANDARDS FOR POLLUTION CONTROL

Two sets of standards are generally in force in environmental management. The first set is designed to protect human health as well as aquatic fauna and flora. These standards are called receiving water standards. They are typically expressed in terms of maximal allowable concentrations during certain specified adverse conditions (such as low flow), where often a certain level of specified violations are accepted (Krenkel and Novotny, 1980; US EPA, 1983, 1988).

---

[9] Putting value on damage function and value of resources forgone is a very sensitive issue. As it can be seen in many developing countries and in some CEE countries which overuse resources and do not abate pollution discharge, satisfying bare living needs overshadows the notion of environmental damages, i.e., the perceived value of the environmental damage is not as high even though great damage is occurring. Many environmental economists use the notion of "willingness to pay" for environmental benefits as a substitution for the damage function.

[10] To overcome this dilemma of different values of notions of environmental damage and " willingness to pay" for environmental benefits the concept of side payments may be a solution. However, this concept will work only under a situation in which a wealthier country affected by pollution from a "poorer " country provides an economic incentive for abatement which theoretically should equal to the difference between the value of damage of the wealthier and poorer country. Examples of such side payments are numerous but obviously there is a danger of misuse as it is in any use of subsidies and side payment. Typically, if side payments are a possibility the polluter may do nothing until the side payment is received.

The second set is not directly related to receiving water quality. It enforces, as a tool of legislation, directly or indirectly certain minimal abatement technologies for all sources (within the same source category, for example, an industrial or municipal category). Typically for industrial sources, the limitations are expressed in terms of allowable kg of the pollutant discharge per unit of the product (for example, kg of BOD_5/ton of pulp) or for municipal sources as mg/l in the effluent or grams of pollutant discharge per capita. In practice, most of these standards are based and enforced on equity and DPP imperatives - that is, all dischargers within the same category should reduce "uniformly" their waste load (regardless the actual damage caused). Subcategories of sources can also be used, for example, different standards can be used for "large" and "small" municipal dischargers and/or for "old" and "new" sources. Other standards may be based for instance on the avoidance of irreversible harm to future generations by overuse of the resource for waste disposal. Since most of these standards require mandatory application of certain management technologies they are known as *technology based.* In this category *effluent standards* are used for point source controls while *performance standards* can be used for nonpoint source controls.

3.1 Why Two Sets of Standards ?

Water uses are specified in terms of ambient criteria and for this reason they form an important element of legislation in many countries. However, monitoring and enforcing ambient water quality standards are difficult, which then calls for the usage of effluent limits. Other reasons for using "safe" effluent standards include difficulties in defining benefits of receiving water quality improvements and the negative attitude by some to employ water quality models for evaluation of the impact of emissions and their reductions on ambient quality (if such models were accepted as reliable they could be easily used to translate ambient standards to effluent ones or *vice versa*). While development of a least-cost, "efficient" river basin policy cannot be done without application of receiving water quality standards and models, a strategy based on the principle of equity (as it is the case in many Western countries) suggests to rely solely upon uniform emission reductions and effluent limits.

From this brief explanation it is clear that legislation of most countries apply both sets of standards, where priorities depend on the underlying philosophy, tradition, experiences collected and others. For instance several CEE countries used in the past mostly ambient criteria, while the U.S. and Western

European countries relied, in the past almost exclusively, on effluent limits. In some countries, such as Hungary or France, regional variable standards are operational (expressing differences in "vulnerability" and the "assimilative capacity" of receiving waters). Many systems have advantages and disadvantages, but perhaps the most significant failures have resulted when standards have been set too lenient or if they are set so stringent that they can not be afforded for economic reasons (see imperative of Reasonableness earlier). The final outcome in most CEE countries is no control for both cases.

In the past, CEE countries tried to use both lenient effluent or ambient standards and sometimes very strict ones which, however, were not enforced at all. The present tendency is to introduce Western effluent standards "immediately", often driven by political pressures without really thinking about the possibilities and limitations of implementation. Those, who advocate this strategy tend to forget that the introduction of presently accepted standards and legislation in the industrialized world has been a very long (two decades or so) and difficult process. It is also often overlooked that regulations in Western countries undergo significant alterations as experience is gathered, hence, today most countries apply a combination of effluent and ambient criteria with priority being given to effluent limits.

As mentioned before, sole reliance on ambient receiving water standards may violate the equity criterion and make enforcement difficult. The immediate adoption of, for instance, the rather strict ambient criteria and regulations of the European Community may also fail since they may not be "*economically achievable.*" Thus, actually the "reasonableness" imperative and the economic Pareto optimality rule according to which the cost of abatement should approximately be equal to the cost of environmental damage[11], which are now considered seriously in the Western world, would be also violated (see later the discussion on the U.S. Clean Water Act).

The sole reliance on effluent standards has also other drawbacks. Many water bodies have limited

---

[11] This relationship between the cost of abatement and environmental damage is presently greatly distorted by the currency relations between the western and CEE countries. The cost of abatement is more related to hard currencies while the damage function is expressed in local, mostly undervalued currencies.

However, there are difficulties with application of the Pareto optimality rule in Western countries as well. For example, past efforts in the US to equate the cost of environmental damage to the market value of fish in the receiving water body were rejected by courts. To overcome the problem the value of a healthy environment is perceived as being very high, in excess of any cost of abatement. The laws require to reach a state of an healthy and sustainable environment which would not be impairing the integrity of the resource as long as achieving the environmental goals does not result in a "wide spread adverse socio-economic impact" on a large segment of population (CFR 40, 1991.

sources, the receiving water bodies may still have poor water quality and may not be usable for some important beneficial uses (see the "integration" imperative). Hence, expenditures would have been spent inefficiently or even wasted. The other extreme is an adoption of effluent standards that would be excessively stringent. For example, requiring nitrification-denitrification for every municipal point source not only increases significantly the cost but in many cases has no major environmental "benefit".

Some countries now accept cleverly a gradual adoption of standards with a final goal of reaching EC levels of protection. For example, in the Czech and Slovak Republics the municipal emission standards are progressively put in force. By the end of 2004, discharges (above 100 000 P.E.) must comply with effluent limitations of 30 mg/l of BOD_5, 10 mg/l of ammonia - nitrogen, and 3 mg/l of total phosphorus, respectively, beginning with 2005, limits are tightened to 25 mg/l, 5 mg/l, and 1.5 mg/l, respectively (standards are less stringent for smaller communities). In addition to effluent standards ambient ones are also set although it is not clear how this combined policy will work in practice (see subsequent Chapter by Grau).

In summary, usage of dual standards offers an opportunity to find a compromise between different imperatives, including economic affordability, equity, DPP, sustainability and others. Where to put the focus during the transition period requires careful analyses. It appears to us that at the beginning "efficiency" and "sustainability" should play a more important role than in industrialized countries nowadays with a gradual shift towards "equity" phasing of which should be also reflected in defining the combination of different standard systems and their actual levels (see also subsequent discussion).

3.2 Effluent Limited and Water Quality Limited Water Bodies

The use of dual standards can lead to the following categorization of receiving water bodies (adopted in the US as a useful tool for planning and water quality management):

1. **Water bodies where mandated effluent limitation will suffice.** If after implementation of the mandated (point) source abatement specified by nationwide effluent standards the water quality goals are achieved, the receiving water body can be classified as *effluent limited.*

2. **Water bodies where uniform mandatory effluent control will not achieve water quality goals.** Consequently more stringent controls of regulated sources and other actions extending controls

to other unregulated sources, water body restoration and waste assimilative capacity improvement will be necessary. Such water bodies are classified as *water quality limited.*

At the onset of the analysis and water quality planning process, before most of the mandated abatement is implemented, it is usually not known whether the water body is effluent or water quality limited. The process which will lead to such classification is depicted on Figure 1 taken from Novotny *et al.* (1995). An *impaired water body* is a subcategory of the water quality limited water body classification used in the U.S. for water bodies where the water quality problem is caused primarily by (unregulated) nonpoint sources. This categorization certainly depends on setting the levels of ambient and effleunt standards by each country, i.e., on the actual water pollution control legislatioon.

Achieving water quality goals in effluent limited water bodies is straightforward and in most cases requires enforcement of permits which are issued to dischargers based on nationwide categorized effluent limitations. The cost, in most cases, is borne by the dischargers according to the Discharger Pays Principle. Water Quality Limited water bodies require a combination of incentives, more strict enforcement, adequate and effective institutional arrangement, and other management components that, in overall, would be classified as *water quality management.* Although the above definitions are useful for planning the real issue in the CEE countries, at least during the transition period, is the actual level of the two sets of standards and their priority.

Subsequently, we discuss water quality management in the U.S. to illustrate the role of different principles, the rather complex process of legislation, the type of analyses necessitated and observed trends in the system.

4. WATER POLLUTION CONTROL IN THE U.S.

In the United States, according to the Clean Water Act (1972) which authorized a nationwide strategy for surface water quality control, water quality goals and standards are defined in terms of the designated use (such as protecting human health, supporting aquatic life, contact recreation, drinking water supply etc.) of the water body. According to the law the state water quality standards[12] must

[12] In the United States it is the states' responsibility to issue and enforce water quality standards which, however, must comply with federal guidelines issued by the U.S. Environmental Protection Agency, following the intent and mandate of the Clean Water Act.

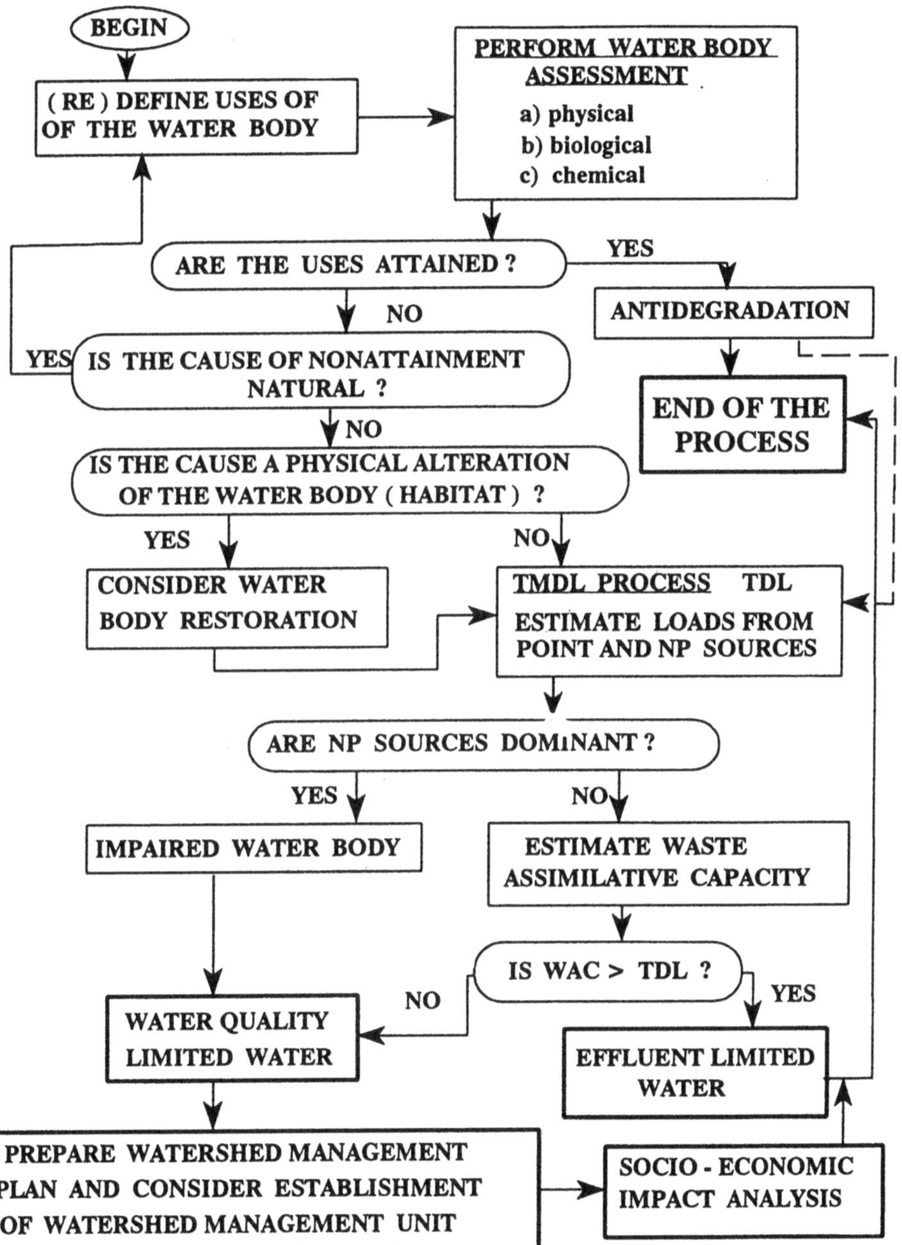

Figure 1 Process of classification of water bodies into water quality and effluent limited (Novotny et al., 1995).

(1) include provisions for restoring and maintaining chemical, physical, and biological integrity of state waters;

(2) provide wherever attainable, water quality for the protection and propagation of fish, shellfish, and wildlife and recreation in and on the water; and

(3) consider the use and value waters for public water supplies, agricultural and industrial purposes, and navigation.

The Clean Water Act (CWA) is based on basic principles discussed earlier. For instance, when states designate uses, considerations must be given to whether such uses can be attained (see imperative of Reasonableness). If the state does not intend to designate uses that would comply with the goals of the Clean Water Act, the so-called Use Attainability Analysis (UAA) must be performed to justify the "downgrade" of the use. This is a vehicle to allow lower water quality to accommodate important economic or social development in the region and to find a compromise with the antidegradation principle which is another important element of the U.S. environmental policy incorporated into law.

4.1 Effluent Control of Point Sources

Point source control relies on the National Pollution Discharge Elimination System (NPDES) permits that are issued by states to regulated point sources. The distinction between regulated and unregulated sources is important because in the U.S., due to judicial interpretations of the constitution, enforcement is only feasible for point sources discharging into navigable[13] waters. The discharge permit which contains the maximal loads and concentrations of key pollutants and conditions under which limits can be legally exceeded was the major vehicle for relatively successful pollution abatement program in the U.S. during the past twenty years.

The effluent limitations are derived, source category by category(e.g., municipalities, various industrial and some agricultural sources), from an assumption of implementation of the *Best Available Technology (BAT)*. The reasonableness imperative was again incorporated into the guidelines for state effluent limitations because the original essential premise of these limitations was that they were

---

[13] The legal definition of "navigability " in the U.S., though very important, is very loose and includes all surface waters on which a canoe can be floated.

Economically Achievable. Under present economic situation, the point source discharger subjected to the NPDES permit, is in most cases fully responsible for the cost of installing treatment technology. Past practices of subsidizing municipal polluters were eliminated in the mid 1980s.

Time has tested the reasonableness of these limitations. Even though there were many complaints by polluters at the beginning of the NPDES implementation period there are almost no industrial dischargers in the US which would have had to cease or relocate their operation because of effluent limits. Most industries that have relocated from a state or even the nation have done so because of other reasons (cheaper labor, loss of market, etc.). This is due to the fact that the NPDES system treats all polluters within the same category equally and no one would gain an advantage by relocating somewhere else[14].

At the beginning effluent standards were limited to a few key pollutants, presently the list of regulated pollutants is expanding to include toxic (or priority) pollutants. By virtue of the definition of allowable duration of excedence for priority pollutants, the acute toxicity criteria are applicable in the U.S. to a very small mixing zone, essentially to the effluent. The chronic toxicity criteria are applicable only to the receiving water body at the edge of a large mixing zone extending up to 4 days travel distance from the effluent[15]. The NPDES permit does not cover the majority of nonpoint sources because of legal ramifications of the U.S. legal system (actually the lack of integration was the major shortcoming of the

---

[14] This is also an argument against so called "optimization" of wastewater discharges in which different limitations would be imposed according to various cost optimization scenarios and models. A polluter in a "high cost" water body could relocate to a "low cost" watershed thus increasing the load to a "low cost" water body and taking up a part of the waste assimilative capacity and freeing up waste assimilative capacity on the "high cost" water body. Consequence of such action is the requirement for more treatment for the remaining dischargers until the cost of treatment would become equal at the two water bodies. Note also that the *antidegradation* imperative would make such transfers very difficult if not impossible.

However, optimization of waste load discharges may be achieved by economic means such as transferrable discharge permit in which dischargers on a stressed (water quality limited) water body negotiate among themselves the level of treatment subjected to the overall waste load limitations and other constraints. A public (governmental) agency, such as the watershed management agency, would have to overview and approve the final negotiated allocation.

[15] The allowable exceedences of toxicity criteria have been defined by the U.S. EPA as follows:

frequency	- once in three years
duration	- acute toxicity - one hour grab or daily maximum sample
	- chronic toxicity - four consecutive days average concentration

U.S. system which called for a revision to be discussed below). Consequently, most of the nonpoint sources remained unregulated. One of the present trends is to recategorize some nonpoint sources as point sources (for instance CSOs, urban storm water runoff and animal feedlot runoff) for which the NPDES permit is then required.

4.2 Water Quality and Watershed Management: the Need for an Integrated Approach

Even though with a few exceptions the point source control programs defined by the BAT effluent standards have been essentially completed, an appreciable number of surface water bodies and coastal waters are still not meeting water quality goals (US EPA, 1992). This portion may actually "increase" when standards for toxic contaminants in water and sediments are enforced. Many water bodies are impacted by unregulated mostly nonpoint sources of pollution. In a situation where the water quality goals are and/or will not be met after implementation of mandated point source controls, water quality and watershed management and protection should be considered. The watershed management approach is well suited to track holistic cause-and-effect water quality relationships since it can link upstream uses with downstream effects.

Watershed management activities are considered by Congress for authorization in the Water Pollution Prevention and Control Act, which is to amend the Clean Water Act. There are several new amended sections included in the Act (as well as some previous rules in the Clean Water Act) which will make watershed planning and management necessary in many watersheds where water quality goals have not been met. The highlights of the WPPCA related to water quality and nonpoint pollution management are:

- provides new authority for watershed planning,
- expands authority to control nonpoint pollution,
- redefines the scope of watershed monitoring, and
- incorporates antidegradation principles into the Act which until the present was included in only government directives but not in the law.

The goal of watershed planning is to correct water pollution problems in impaired watersheds. Pertinent clauses to watershed water quality planning and management are:

- For all watersheds each state shall periodically (every five years) prepare a plan on attainment and maintenance of water quality.

- The Comprehensive Watershed Management should

  - identify water quality impairment and the pollutants, sources, and activities causing impairments;

  - integrate water quality protection efforts with other natural resource protection efforts;

  - define long-term social, economic and natural resource objectives and the water quality necessary to attain or maintain the objectives;

  - through public participation increase public support for improved water quality;

  - identify the most cost effective measures to achieve objectives of the Act.

- For intrastate watersheds, the Governor of the State, and for interstate basins jointly the Governors of the States, may designate waters (including groundwater) and associated land areas within the State(s) as a *watershed management unit*.

- Each watershed management unit designation must be approved by the Administrator of the U.S. Environmental Protection Agency. The management entity may be an agency of the State governments[16], a substate regional planning organization, a conservation district, lake management district or any other natural resources management agency with the capacity to carry out tasks and responsibilities specified by the Act.

---

[16] Florida's Water Management Districts are examples of a state watershed wide management units. These districts cover Florida's major inland water bodies.

The proposed initiatives are a milestone in water pollution abatement efforts. They promote an *integrated ecosystem approach* (although not evenly and with the same efficiency) differing from past technology driven approaches (see Fig. 2. for five groups of factors which can have an impact on ecosystems indicating the importance of water body integrity). It is not clear whether the proposed initiatives will supersede, replace or overlap present activities authorized under the previous amendment of the Act.

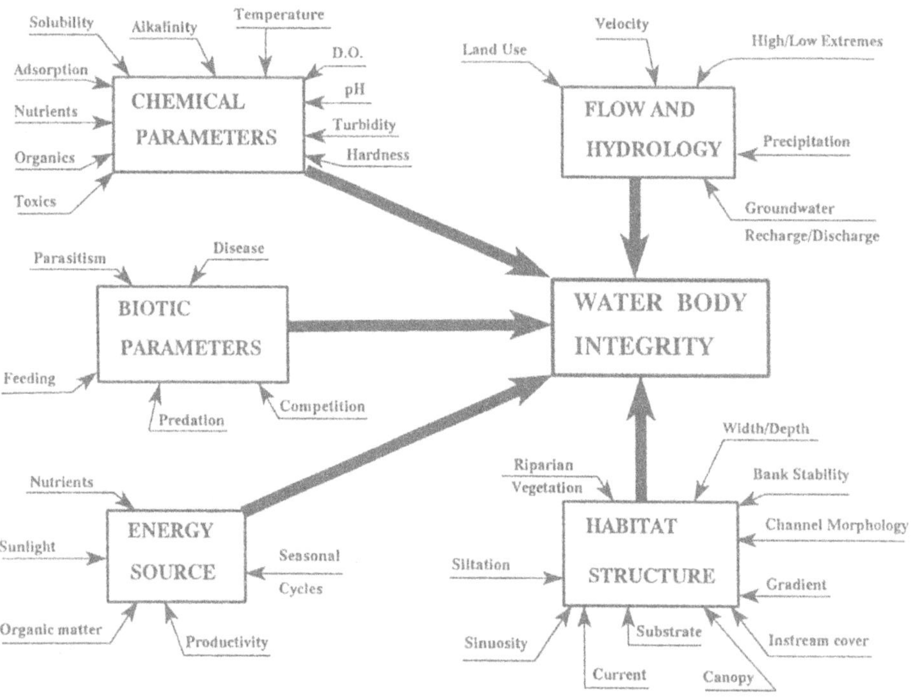

Figure 2 - Concept of water body integrity and affecting factors (after Karr *et.al*, 1986).

4.3 Use Attainability Analysis Planning Process

As noted earlier, if a water body has been classified as partially supporting or not supporting the designated use (i.e., if the actual ambient quality is not fully meeting the criteria set on the basis of water uses) the states or designated agencies perform *a Use Attainability Analysis (UAA)* to determine the proper use. The UAA is a structured assessment (or a planning study) of the factors (physical, chemical, biological, socio-economic and other) affecting attainment. At the same time the output of the UAA is a legal document by which state regulatory agencies justify water quality standards (for water bodies for which the designated use is not attained). Regulated dischargers may use the same approach to justify and even challenge water quality based NPDES effluent limitations, i.e., those which are more stringent than the technology based (BAT) effluent limits (Novotny *et al.*, 1995).

The process of the Use Attainability Analysis is the most important component of an integrated approach to environmental protection of water resources in the United States. Based on the UAA process it is possible to modify or change non-existing designated water use (or establish subcategories within the designated uses) if it is demonstrated that attaining the original use is not feasible for one or more of the subsequent reasons (US EPA, 1983):

(1) Naturally occurring pollutant concentrations prevent attainment of the use;

(2) Natural, intermittent (ephemeral) or low flow conditions prevent attainment (unless these conditions may be compensated for);

(3) Dams, diversions or other hydraulic structures preclude attainment of the use, and it is not feasible to restore the water body to its original conditions (or to operate such modifications in a way that would result in attainment);

(4) Physical conditions associated with the natural features of the water body (such as the lack of proper substrate, cover, flow, depth etc.) unrelated to quality preclude attainment of aquatic life protection;

(5) Effluent controls more stringent rules than those required by the CWA would result in substantial and wide-spread adverse social and economic impact (the "reasonableness" imperative).

There are basically two reasons why the UAA is needed. The first one is to determine what levels of quality are possible to attain by implementation of various feasible point and nonpoint source abatement measures. Typically for water quality abatement of streams and estuaries, EPA and states recognize only source control measures even though rule (3) above offers the concept of water body restoration.

The integrated ecosystem approach referred to earlier represents an improvement over past practices by assessing (in addition to pollution source controls), water body (ecosystem) restoration as well as waste assimilative capacity enhancement (or ecotechnological) measures (Committee on Restoration of Aquatic Ecosystems, 1993). These could include a variety of measures such as in-stream aeration for different purposes, dredging of contaminated sediments, removal of dams which have lost their usefulness, low flow augmentation, restoration of riparian wetlands and others.

The second purpose of the UAA is to determine the most desirable use if the designated one can not be attained. This concept involves a study of the socio-economic impacts of attainment of a specified use and may lead to a formulation of site-specific standards.

The designated uses cannot be removed if:

(1) They are existing uses (antidegradation imperative applies) which means that the present water quality is in compliance with standards; or

(2) Such uses can be attained by implementing mandatory effluent technology based limits of point sources and "cost-effective" and "reasonable" Best Management Practices for nonpoint source control, respectively (i.e. a change of use is not permissible for effluent limited water bodies - see the next Chapter by Novotny and Capodaglio for definition of effluent and water quality limited water bodies).

In the present context, the Use Attainability Analysis process uses physical, biological and chemical criteria (standards) and evaluations (Figure 3) based on the ecological integrity shown on Figure 2.

5. CHALLENGES TO CEE COUNTRIES

The overview of U.S. legislation shows that extensive experience is available. This is particularly true if we also consider the French, German, Dutch and other European regulations. Thus, the question is which portion of these experiences could be utilized under the rather peculiar conditions of the CEE

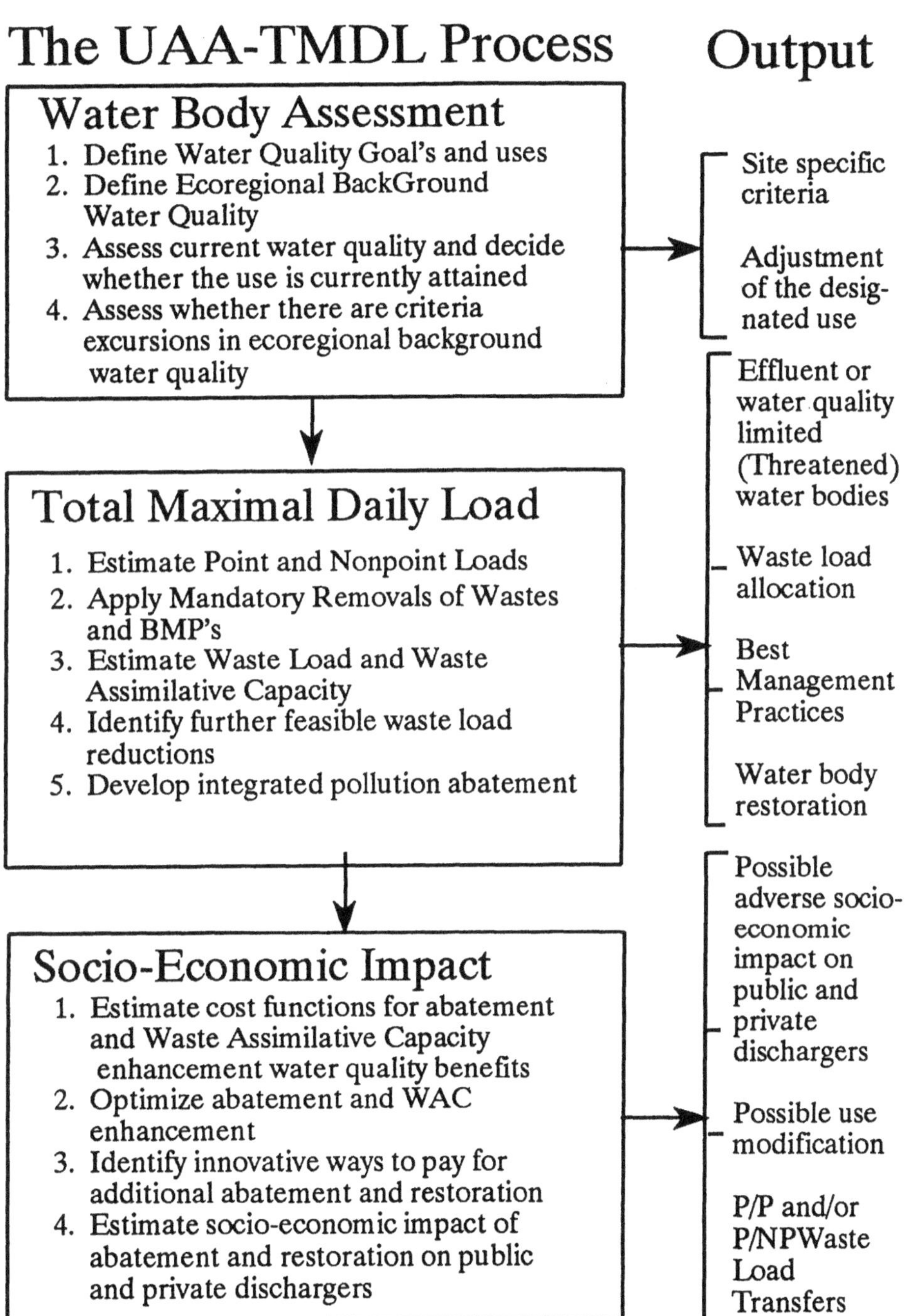

Figure 3 The Use Attainability Analysis Process (Novotny et al., 1995)

countries. Correspondingly, what should be done differently? Which mistakes can be avoided? Subsequently, we will discuss some of the major (and often interrelated) challenges of CEE countries primarily in the light of the U.S. experience.

5.1 Standards and Priority Setting

Environmental laws now in place in the West have evolved over many years during which the early seventies represented only a milestone in most countries. Actually legislative environmental control efforts in the US began shortly after World War II (but the first efforts were made even long before[17]). The comparison of the U.S. and other Western legislation would show that they are similar but not identical. This is one of the arguments why the CEE countries should not exactly copy these laws but rather to build on "successful" provisions.

One of the most important practical questions we address is whether lower standards than those used in the West would be appropriate in the CEE countries. Numerous arguments for and against such criteria can be summarized as follows:

(1) The low per capita GDP and notions of affordability and/or application of "economically achievable" abatement may imply that lower standards, at least for some transitional period, until the economies are brought on a par level with the more advanced Western countries. The process nature of tightening standards should be stressed (which was also the case in the West) in contrast of setting for example the most stringent recommendations of the EC "everywhere" and "immediately" as it was often unrealistically suggested.

(2) Industries and municipalities undergoing dramatic changes may object against DPP environmental polices and more stringent effluent limitations which may lead to bankruptcies and plant closures. However, as it was pointed out, the same claims were used in the U.S. twenty years ago, yet, no wide spread plant closures have materialized. The economic viability of production processes or their failures were due to a number of factors from which implementation of pollution control

[17] The first environmental law was passed by U.S. Congress in 1896. The "Harbor Refuse Act" prevented discharge of refuse from nonmunicipal sources into navigable U.S. waterways. It is interesting to note that this archaic law was used by the U.S. administration in the late 1960's to require permits for industrial wastewater discharges into navigable waterways.

measures was not the most important determinant. The transition nature of the economy and high unemployment in the CEE region certainly makes the issue more difficult than in the West in the past, but a clear policy controlling the consequences of privatization and enhancing the introduction of clean technologies can not be avoided. In terms of toxic pollutants and their risky impacts which is difficult to estimate, it is unlikely that a different strategy can and should be followed than that followed in Western countries.

(3) The loans which might be required to implement more stringent pollution controls based on Western standards would add to the already great indebtness of most CEE countries. Simultaneously, as noted earlier, countries are not yet in the situation to develop clear water infrastructure development programs.

(4) On the negative side, inefficient pollution control of meeting environmental goals will lead to waste of limited funds. Actually, as it will be subsequently shown, this issue raises the question of setting priorities.

(5) Low standards can lead to pollution imports: companies producing pollution (either in the production process or in the product itself) may relocate to countries with less stringent environmental limits. Such practices are common and many less developed countries with lax environmental laws suffer from excessive "imported" pollution caused by foreign industries[18]. For example, deplorable environmental conditions in Northern Mexico and in the developing countries of the Pacific Rim is caused mainly by pollution import as characterized. Frequent examples can also be taken from the press of various CEE countries. Thus, less stringent standards (and policies) under free market conditions will inevitably lead to pollution import.

(6) Another, interrelated important factor is the strong pressure from developed Western countries to push for the same stringent (international) standards in the CEE region (and the desire of these countries to join the European Union which will require an adoption of and an effort to comply with the EC requirements). For instance, this tendency was also clearly manifested during NAFTA

---

[18] Many argue however, that the relocation of industrial processes to less developing countries is mainly due to cheaper labor rather than due to lax environmental laws. Nevertheless, there is no doubt that a majority of the relocating and new industries moving into the less environmentally restricted countries will comply only with the local environmental restrictions.

(North American Free Trade Alliance) negotiations. Overall, more advanced countries today and in the future may not tolerate lax environmental controls in their international trade negotiations.

(7) As pointed out, omitting the cost of environmental damage from costs of producing goods leads to greatly distorted markets and to overproduction of goods produced by the polluting processes. It is noted, that implementation of an environmental control is often an economic stimulus rather than a deterrent. New efficient processes based on "sustainable" use of resources (for example, reuse and recycle) are created instead of processes which produce much residuals and a new branch of the national economy dealing with environmental protection and restoration may be created.

From the above pro's and con's it follows that in the long run CEE countries can hardly avoid introduction of the same or similar ambient and effluent standards as those applied in Western Europe. Thus, actually the question remains how to proceed during the transition.

Our related comments are as follows.

(1) Tightening standards should be scheduled over a longer period of time (as it was done in the Czech and Slovak Republics) to assure affordability and allow dischargers to develop their adjustment plan in a realistic fashion.

(2) Due to a lack of financial resources, the importance of priorities should be recognized. In practice this is equivalent to setting standards regionally and temporally different during the transition (although this fact may not be declared) or to set the intermediate "minimum" level of treatment lower than the BAT. This would allow upgrading of existing primary, biological and half-completed treatment plants (see e.g., Hahn, 1994) or the usage of "natural" treatment systems primarily in rural areas. For the specification of priorities river basin least-cost studies (or UAA type of analyses) are recommended to be performed (see e.g., Somlyódy et al., 1994).

(3) Consequently, it is suggested to put a stronger focus on meeting ambient water quality goals in comparison to effluent criteria during transition. In other words the two sets of standards should be harmonized such that not the more stringent one is selected, but the more cost-effective, in terms of ambient water quality improvement (this would mean by using U.S. terminology that "ambient quality limitation" would be given priority over a stringent "effluent limitation"). The

rationale of this approach is that frequently a small ambient quality improvement can be achieved while minimizing unnecessary expenditures (Somlyódy et al., 1994).

(4) Wastewater phosphorus and particularly nitrogen control requires special care (the latter due to the associated high marginal cost). An automatic focus on effluent standards for these substances without considering contributions of non-sewage (primarily nonpoint) origin of these emissions and their enforcement - as shown by Western experiences - often leads to unnecessarily expensive strategies. For instance, in many watersheds loads are dominated by agricultural sources. By considering these limitations on a case by case basis the CEE countries are in a position to implement improved integrated approaches now.

6. INSTITUTIONS

Successful water quality management requires three or more cooperating but sometimes adversarial institutions. These institutions can be broadly categorized as (1) regulators, (2) enforcers, and (3) managers. In addition, users and dischargers should also organize themselves or water use utilities can be created (by responsible legislative bodies).

The regulating group includes those institutions that translate environmental laws into a form of executable regulations, directives, standards, etc. This agency is responsible for environmental protection. In the U.S. these tasks are carried out by the U.S. Environmental Protection Agency and similar state agencies (e.g., Florida Department of Environmental Regulation, California Environmental Protection Agencies, Wisconsin Department of Natural Resources, etc.). State agencies are also responsible for issuance of permits and often develop water quality management plans that would either support the permits or may lead to other water quality management actions such as waterbody restoration[19].

In the EC and CEE countries these functions are fulfilled by the ministries and agencies belonging to them that deal with environmental protection. The enforcement activities may or may not be carried out by separate agencies. Typically, in the United States, the regulators also enforce regulations, permits

---

[19] In the U.S. the states are responsible for environmental protection on their territories. The U.S. EPA provides guidelines and directives to the states and oversees state compliances with federal laws. The EPA, however, can take over the regulating responsibility if a state does not comply.

and monitoring for compliance. In many European and CEE countries these functions are carried out by water or environmental inspectorates and their regional units. These inspectorates often oversee both effluents and the receiving water bodies and resolve violations. National water quality monitoring networks and compliance systems, and associated laboratories are well developed in the CEE countries[20].

Dischargers can act as individuals (such as individual communal or industrial dischargers or a single farmer or an agricultural cooperative) or be organized with various degrees of formality. In the U.S. (and many CEE countries, see next section) urban centers have their, often regional, wastewater disposal utilities which are responsible for operation and maintenance of wastewater collection and treatment systems. These utilities have the authority to collect fees from users and to borrow funds to finance their activities. Recent trends to privatize wastewater disposal utilities in the U.S. were not as successful as privatization of solid waste disposal which is now provided in the U.S. frequently by private companies[21].

In the CEE countries the regional sewer works had similar responsibilities. However, under centrally planned economies these agencies could not collect fees to cover their expenses, and consequently could not borrow funds for capital improvements. All funds were derived from planned state budgets. After the change of the political system new laws that would enable them to became at least partially financially independent may not have been passed while funding from national central sources has diminished. As indicated in the Introduction, the issue is strongly influenced by the overall tendency of decentralization, privatization and the increasing role of local governments. Several sewerage agencies became insolvent and/or were dissolved. Sometimes the institutional setting is still confusing due to ongoing transformations and the responsibilities are not clear.

---

[20] For example, in Wisconsin the bulk of analyses for compliance monitoring is performed by the Laboratory of Hygiene which is affiliated with the University of Wisconsin. Some dischargers carry out their own compliance monitoring, e.g., the Metropolitan Milwaukee Sewerage Commission, or contract it to private analytical laboratories.

[21] The most important objection against privatization of urban wastewater disposal utilities is the fact that due to the uniqueness of the collection system these agencies would become monopolies and under such situation would be subjected to very strict public oversight vis a vis private power producers or local telephone companies. This is not true for solid waste disposal where, theoretically and practically, several companies can provide the service and compete.

6.1 Basin Wide Management Approaches

Unlike in the United States where basin wide management is a novelty and will be evolving in the future following the impetus of the Water Pollution Prevention and Control Act (see previous discussion), the river basin management agencies have been in operation in many CEE countries for decades(the same is true for several Western countries, *vis a vis* the British Basin Authorities)[22].

In the U.S. it is felt that not every receiving water body requires a river basin approach. If the water quality goals are presently met or if they can be achieved by implementation of mandatory effluent controls (i.e., if the water body is effluent limited), enforcement of the permits and collection of the discharge fees by the water quality inspectorates is all that may be required (following principles of the Use Attainability Analysis, see previous discussion). Establishing large and bureaucratic water basin management agencies for "effluent limited streams" would be wasteful and counterproductive. In turn, for "water quality limited water bodies" (i.e., where mandated point and nonpoint source controls will not achieve water quality goals due to presently unregulated sources and physical alteration of the water body) a water basin management agency is needed because:

(1) Additional treatment will be required for regulated point sources which may need economic incentives. This is especially necessary if the cost of such treatment exceeds the economically achievable levels for the dischargers.

(2) Pollution abatement will be required from unregulated point and nonpoint sources which will also call for incentives.

(3) Water body and flood plain ecosystem restoration and waste assimilative capacity enhancement may be needed and be more economical than excessive effluent controls. The water quality management agency would be responsible to carry out and secure financing of such in-stream abatement practices.

In the CEE countries, water pollution control is in a less advanced stage than in the U.S. It is too early to talk about management of receiving waters where goals are met (in contrast, as already noted,

---

22 There are several successful basin wide water management agencies in the U.S., including Florida's Water Management Districts, Delaware River Basin Commission and Tennessee Valley Authority.

a large part of the receiving waters belong to the poorest water quality class), and there is no distinction between "effluent" and "ambient limited water bodies" (and it is not sure that these notions will be used). More important, however, is that river basin agencies form a traditional element of practice of water management in the CEE region and they can be used for careful planning and implementation of regional strategies. A procedure similar to the UAA process certainly would make sense with differences in criteria of evaluation which stem from particular conditions of the transition. In short, as stressed several times before, among the imperatives outlined a stronger focus should be put on "efficiency" and "cost-effectiveness".

6.2 Issues of Financing and Economic Instruments

Most of the CEE countries face at least two main problems related to environmental investments: (i) financial resources are very scarce and (ii) the inflation rate is high (though variable within the region) which may persist for some time into the future. Both factors are obstacles in the classical scheme of implementing infrastructure developments such as building sewers and treatment plants (which would, e.g., suggest issuing a bond for capital cost and repaying it over a longer period, say thirty years which is a typical amortization period for wastewater facilities).

For instance, for a 100 000 P.E. town - where a collection network exists, which can be considered as a typical case - the capitalized construction (investment) cost of a biological treatment plant would be about US$ 3.98 million US$ (see subsequent Chapter by Henze and Ødegaard), while the operation, maintenance and repair cost is US$ 2.42 million per year. The total cost is then US$ 6.40 million/year, which would correspond to a cost of treatment of US$ 64/cap./year , which then leads - without water supply and wastewater collection expenditures - to about 10% of the annual income of a "typical" family in Hungary or the Czech Republic. It would be much higher in other CEE countries. Such costs in Western developed countries would be considered high, not "affordable" and not "reasonable". If the interest rate is only around its nominal value (say 2 %) then the sewerage fee to be collected - assuming that the municipality is self sustaining - could be reduced to around 3 to 5 % of the family income, perhaps a more realistic value but still much higher than that typical for Western developed countries.

If the affordability of BAT pollution abatement is in question, as it would be in most CEE countries, this would lead logically to the idea of phased investments and borrowing (depending on the interest rate) which can significantly reduce annual payments. From a technological point of view this can be realized by constructing first only a primary, a chemically enhanced primary or a single stage partial biological treatment plant (depending on site-specific conditions such as the type of the receiving water), which is upgraded later on in a multi-stage fashion (see Henze and Ødegaard, 1994; Somlyódy et al., 1994; and subsequent Chapter by Henze and Ødegaard). Although the final goals would be the same, the phase-in, pay-as-you-go approach would dramatically reduce payments and, conceivably, bring them to affordable levels.

The above simple example raises a number of crucial questions. Would and could the state issue bonds or create infrastructure banks offering loans with small interest rates? Would foreign aid banks be willing to support such initiatives? Could environmental and/or water funds of sufficient sizes be created which could be used at least partially for the above purpose? What grants should be given if at all? How to collect money for covering the above expenses? Unfortunately very few positive or certain answers can be given nowadays to the above questions (which then suggests once again the need to stress careful planning, and cost-effective solutions).

The application of the Discharger Pays Principle obviously leads to the application of economic instruments such as sewage charges (in addition to water and collection fees which should be sufficient to cover the operation of existing infrastructures and facilities). The purpose of charges is often unclear in practice which can lead to unrealistically low levels. Charges set on the basis of careful analyses should force polluters to meet set standards (irrespectively whether effluent or ambient ones or their mix are used). They should be related to marginal costs of treatment such that the introduction of sewage treatment or its upgrading would be cheaper than the payment of charges (it is noted that unlike the U.S., CEE countries are in a favorable position to introduce instruments for controlling non-point source pollution in an integrated fashion together with point dischargers). In addition, fines (penalties for violating standards) should also be maintained but in the future they should be collected primarily for spills and unlawful violations of discharge and/or ambient limits. The concept of charging for discharging pollutants, however, has serious drawbacks since its existence gives polluters a "right" for polluting the environment, and for this reason the tendency is to introduce strict legal restrictions.

Due to the particular economic and financial conditions in CEE countries, effluent charges could be progressively scaled. In the initial period, each discharger would be given a reasonable grace period (specified in the discharge permit) to install pollution reduction facilities. During this period fees could be lower. However, to encourage installation of treatment, after the grace period fees should be increased, exceeding the cost of treatment for polluters who did not comply with the permit. To be equitable discharge fees should probably be uniform for a given discharge category (the charges could be collected by inspectorates).

However, a distinction should be made between equitable charges and uniform treatment and cost effective non-uniform strategies. It is also noted that the present situation in the CEE countries is far from being "equitable" thus in this case, a gradual shift is justified.

There are a number of other economic instruments which could be considered. For instance, grants (and other tools) can be given for implementing a well-designed priority policy on the short run, but a discussion on these and other issues of clever financing is beyond the scope of the present paper. At the end of this section we stress that the past practice when funds were derived solely from the state should be abolished. A number of methodologies and schemes are known in advanced Western countries which could be analyzed by CEE countries to decide at what extent they correspond to their conditions and needs. Unfortunately, nature of these transitional measures does not promise easy solutions, but professional skill is available to find and implement good compromises.

7. CLOSING REMARKS

The new economic and political situation in the Central and Eastern European Countries and serious environmental degradation left therein requires a new, innovative management approach. However, none of the observed problems are new since they occurred in the West twenty or thirty years ago. What is different is the co-existence of different issues (point and non-point loads, traditional and toxic contamination, pollution of sediment, soil and groundwater, local and regional impacts, etc.) and the transitional economy and institutions with all of its consequences (low per capita GDP, high prices, inflation, unemployment, high national debts, privatization, decentralization etc). At the same time there are also a number of positive features such as the professional skill, some of the existing institutions (e.g., river basin authorities), the opportunity to introduce an "integrated" approach now and the desire to join the European Union. In addition to these, plenty of experiences are available in water quality management from developed Western countries which should be utilized (according to the specific features of the CEE countries) and past failures can be avoided. Concepts and methodologies are available and it is also clear that - as it occurred in the West - CEE countries face a long and difficult process to solve their environmental problems. The evolution process is well reflected and documented by the ongoing and past significant revisions of the U.S. Clean Water Act since its passage in 1972.

This introductory chapter discusses a number of issues of water quality management, such as main imperatives of pollution abatement (equity, efficiency, reasonableness, the discharger pays principle, etc.), standard setting, details of U.S. regulation, impact of the process of the European integration, the role of river basin planning, the importance of proper financing and economic instruments and others.

Our major conclusion is that in the future (a few decades ahead) the practice of water quality management should be the same as those in Western Europe (at least we do hope it will be the case). So we see converging trends. However the path to reach this stage is extremely difficult in some CEE countries and no clear guidelines exist which could be followed without a problem. The development process may be different from that which has occurred in the advanced Western countries during the past 20-30 years.

Our short recommendation to professionals and politicians in the CEE countries is avoid copying any specific Western systems without critical analysis of its impact. Rather, it is advisable to analyze them carefully, in light of past experiences, local conditions and actual needs. We believe that elements of existing principles, methodologies, technologies etc. can be used, but what will be different during the transition is the combination of different tools and the overall focus. In the latter respect, the role of priority setting and the notion of efficiency should be much more important than they are today in the Western world. Under present economic, social and institutional conditions well-tailored strategies are necessitated and, as said before, local knowledge and skills are there to develop and implement such policies.

REFERENCES

Code of the Federal Register 40 CFR 131 (7-1-1991 Edition).

Committee on Restoration of Aquatic Ecosystems (1992) Restoration of Aquatic Ecosystems, National Academy Press, Washington, DC.

Henze, M. and H. Odengaard (1994) " An analysis of wastewater treatment strategies for Eastern and Central Europe," Paper presented at the 17th International Biennial Conference of IAWQ, Budapest, Hungary, 24-29 July, 1994, also published in Water Sci. & Technol.

Karr, J.R. et al. (1986) "Assessing biological integrity of running waters. A method and its rationale," Illinois Natural History Survey, Spec. Publ. No 5, Champaign, IL.

Krenkel, P. and V. Novotny (1980) Water Quality Management, Academic Press, New York, NY.

Hahn, H.H. (1994) " Upgrading municipal wastewater treatment strategies for Central and Eastern Europe," Paper presented at the 17th International Biennial Conference of IAWQ, Budapest, Hungary, 24-29 July, 1994, also published in Water Sci. & Technol.

Novotny, V. *et al.* (1995) Use Attainability Analysis: Identification and Evaluation of Use Attainability Methodologies for Aquatic Ecosystems, RP91-NPS-1, Water Environment Research Foundation, Alexandria, VA

Organization for Economic Cooperation and Development (1991) The State of the Environment, OECD, Paris, France

Rogers, P., and A. Rosenthal (1988) "The imperatives of nonpoint source pollution control," in Political, Institutional and Fiscal Alternatives for Nonpoint Pollution Abatement Programs (V. Novotny, ed.), Marquette University Press, Milwaukee, WI

Solow, R.M. (1971) " The economist's approach to pollution and its control," Science **173**:497-503

Somlyódy, L. (1993) " Quo vadis water quality management in Central and Eastern Europe," WP-93-68, International Institute for Applied System Analysis, Laxenburg, Austria also a paper presented at the 17th International Biennial Conference of IAWQ, Budapest, Hungary, 24-29 July, 1994, also published in Water Sci. & Technol.

Somlyódy, L., M. Kularathma and I. Masliev (1994) " Development of least-cost water quality control policies for the Nitra River basin in Slovakia," Paper presented at the 17th International Biennial Conference of IAWQ, Budapest, Hungary, 24-29 July, 1994, also published in Water Sci. & Technol.

U.S. Environmental Protection Agency (1983) Water Quality Standards Handbook, Office of Water Regulations and Standards, Washington, DC.

U.S. Environmental Protection Agency (1988a) Introduction to Water Quality Standards, Office of Water, Washington, DC.

U.S. Environmental Protection Agency (1992) The Quality of our Nation's Water: 1990, EPA 841-K-92-001, Washington, D.C.

CHAPTER 2

USE OF WATER QUALITY MODELS

Vladimir Novotny[1] and Andrea Capodaglio[2]

1. WHY AND WHEN WATER QUALITY MODELING IS NEEDED

The use of dual standards in water quality management, i.e., the effluent and water body limitations, leads to two different water quality management situations. In the first scenario the mandatory effluent limitations will be sufficient to achieve the water quality goals, in the second they are not. The former situation is characterized as *effluent limited*. The latter situation in which further actions beyond the mandatory effluent controls specified by the effluent limits are needed to achieve the water quality goals, is called *water quality limited* control. The economic mechanisms and incentives in either case are different. The effluent limited controls rely heavily and often exclusively on the "discharger pays" funding mechanisms while the water quality limited controls may require incentives, benefits transfers and some public financing.

Conceivably, if it is known that water quality goals can be achieved only by mandatory effluent controls, water quality modeling may not be needed. Such situations may arise if the effluent limits are set very stringent and are enforced and implemented. There is no doubt that such stringent effluent limits guaranteeing effluent limited control for every receiving water body would be very costly and in many cases in excess of necessary controls. Furthermore, in the planning stage of the processes it not clear whether the controls will be effluent limited or water quality limited. Modeling can resolve these uncertainties.

In water quality limited situations, however, modeling is essential and should lead to an optimal

[1] Department of Civil and Environmental Engineering, Marquette University, Milwaukee, WI, USA

[2] Department of Environmental and Hydraulic Engineering, University of Pavia, Italy

NATO ASI Series, Partnership Sub-Series, 2. Environment – Vol. 3
Remediation and Management of Degraded River Basins
Edited by V. Novotny and L. Somlyódy

and/or equitable[3] allocation of waste loads among dischargers, both point and nonpoint. The process by which the waste assimilative capacity of the water body is divided among the dischargers is called *the waste load allocation (WLA)* or, in the United States, the Total Maximum Daily Load (TMDL) process.

2. THE WASTE LOAD ALLOCATION PROCESS

The basic concept of the WLA is the relationship between the *loads of pollutants* from all sources (including background) and *concentrations of key contaminants - pollutants* in the receiving water body. Upon comparison of the concentrations resulting from the loads under some specified hydrologic and hydraulic conditions and/or specified probability of excedence with the established water quality standards, the allowable loadings can be determined, which provides the basis for establishing water quality - based controls. These controls should provide the pollution reduction necessary for a water body to meet water quality standards (US EPA, 1991). Even though the input into the WLA procedures are the loads of pollutants, the key decisions are based on concentrations in the receiving water body and their comparison with the ambient standards and criteria. Consequently, the primary emphasis is on concentrations of key contaminants in the receiving water body and their short and long term impact on aquatic biota and human health.

The WLA process focuses primarily on chemical standards and criteria for the simple reason that they are numerically defined and can be tied to numerically defined loads by a water quality model or procedure. Establishing a link between waste loads and biological criteria (such as the Saprobien Index in Europe or various biological indices adapted in western countries) is still being researched. The WLA determines allowable loads and provides a basis for establishing or modifying controls of pollutant sources.

The objective of a WLA process is to allocate allowable loads among different pollutant sources so that the appropriate control action can be taken and water quality standards achieved. The WLA provides an estimate of pollutant loadings from all sources and predicts the resulting pollutant

---

[3] The terms optimal and equitable have two different connotations. The optimal allocation typically means the least costly, however, this requires that different dischargers will receive different waste load allocation which may not be fair. The equitable allocation implies that all dischargers are treated equally.

concentrations (EPA, 1991). The WLA process distributes portions of the waterbody's Waste Assimilative Capacity to various pollution sources - including natural background sources and Margin of Safety - so that the water body achieves the water quality standards.

To protect surface water quality of water quality limited receiving water bodies, the US EPA defines Waste Load Allocation limits based on the WLA for both conventional and toxic pollutants (US EPA, 1991). Waste Load Allocation can be considered as the "inverse function" of Waste Assimilative Capacity: while the latter identifies the residual assimilation potential of a water body, the former describes the amount of pollution that can be discharged into the water body while maintaining the desired water quality characteristics.

The US EPA has recommended 5 steps in the TMDL process which are presented below.

Step 1: Identification of Problem Water Bodies Requiring TMDL Process

Step 2: Priority Ranking and Targeting

Step 3: WLA Development

Step 4: Implementation of Control Actions

Step 5: Assessment of Water Quality-based Control Actions[4]

Step 1 is essentially the Water Body Assessment which also defines the need for the WLA and this also implies the need for water quality modeling.

In some cases, water quality goals cannot be achieved even after implementation of most stringent effluent controls. This occurs when in place contaminants such as pollutants stored in the sediment or past adverse actions by man prevent the attainment of the water quality goals. Typically, impounding rivers

---

[4] This final step is essentially a part of what should be a continuing process of Water Body Assessment which is also monitoring for compliance. A monitoring requirement may be put directly into the discharge permit and be carried out by the permittees or can be carried out by the regulatory-enforcing agencies (water quality inspectorates) as part of their compliance monitoring.

for navigation or electric power production reduces the waste assimilative capacity of the water body and causes contaminated sediments to accumulate in the impoundment. In addition, the organic character of such sediments increases the Sediment Oxygen Demand, which can significantly reduce the Dissolved Oxygen content of water. Modeling in the WLA process should then estimate the limits of the effluent controls on water quality improvement and assess the benefits of water body restoration on water quality.

Box 1 shows the activities of the TMDL process as envisioned by the US EPA (1991).

Box 1 - WLA DEVELOPMENT ACTIVITIES

- Selection of the pollutant to consider.
- Estimation of the waterbody assimilative capacity.
- Estimation of the pollution from all sources to the water body.
- Predictive analysis of pollution in the waterbody and determination of total allowable pollution load.
- Allocation (with a Margin of Safety) of the allowable pollution among different pollution sources in a manner that water quality standards are achieved.

In the US, State water pollution control agencies and EPA use the WLA process for identifying watersheds requiring further water quality management beyond the effluent mandated effluent controls and prioritizing. In most water quality limited situations the WLA process will result in more stringent effluent limitations than those based on technology based effluent standards. In some cases the WLA process may show excess Waste Assimilative Capacity after the technology based standards are implemented and/or considered in the previously issued permits. In this situation, the antidegradation principle will rule and restrict future downgrade of water quality.

2.1 WLA Concept

The logic of including modeling in the planning process can be illustrated using a simplified step by step procedure of the WLA process.

In the first step the Total Daily Load (TDL) is estimated such as

$$TDL = L_p + L_{NP} + BL \tag{1}$$

where

L_p is the load from point sources after the mandated emission limitations are fully implemented

L_{NP} is the load from nonpoint sources (e.g., urban and agricultural runoff) assuming that reasonable and economic Best Management Practices are implemented

BL is the load from background (natural sources) such as pollutant content, e.g., nitrate in base flow

In the second step the maximum concentrations allowed by water quality standards are used to develop loading capacity (maximal load), LC, of the pollutants to the water body. This is in most cases accomplished by established and tested water quality models of various complexity. This can be represented by

$$LC = F^{-1} (WQS) \tag{2}$$

where WQS is the pertinent water quality standard and F^{-1} represents an inversion application of the water quality model (i.e., estimation of the maximal load from the limiting in-stream water quality standards).

The Loading Capacity represents the ability of the receiving water body to accept potential pollutant and background loads without harm to the ecology of the water body and impairment of uses. The TMDL is then the maximal permissible load from all sources, including background sources which are not injurious to the uses of the receiving water body or its ecology. The Waste Assimilative Capacity (WAC) is then the Loading Capacity minus the background load, or

$$WAC = LC - BL = F^{-1}(WQS) - BL \tag{3}$$

In this concept the loading capacity (LC) includes both waste discharges and natural loads while the Waste Assimilative Capacity (WAC) only considers waste (pollution) discharges.

Typically, for some conventional pollutants (BOD_5, DO, nutrients) WAC is calculated for specified low flow conditions. However, in the U.S. the standards for the Priority (Toxic) Pollutants are defined, in addition to magnitude and duration, in terms of their allowable excedences[5] and not in terms of their application to a steady state predetermined flow situation. Using a single value to characterize the loads may not be appropriate. Consequently, the loads (TDL, L_p, L_{NP} and BL) are essentially a time series which should be, at a minimum, characterized by the probability distribution or, at best, by a stochastic (ARMA) time series model. By the same argument, WAC and LC are also stochastic quantities. The methodology of dealing with stochastic time series with various degrees of complexity will be elaborated in the subsequent sections of this chapter. Consequently, it is not possible to specify a certain single flow value for toxic pollutants because wet weather (point and nonpoint) and dry weather sources must be considered and WAC should be determined using either statistically based models (such as Monte Carlo applications) or long term dynamic modeling.

Margin of Safety (MOS). Special attention must be given to determining the MOS when steady-state or quasi-dynamic models are used[6]. Such models typically use averaged parameters and input loads and must be calibrated, using parameters and input values averaged over time periods over which the calibration and verification data were obtained (ranging from a few days to a season). However, average values are of little use when probability of violation of the WQS is in question. In this case conservative assumptions of the model (MOS) must be selected to satisfy the regulatory agency[7]. Typically this is accomplished by the modelers in the four steps outlined as follows (Figure 1):

1. The selected steady-state or quasi-dynamic model is calibrated and verified using average values of input loads and system parameters and variables. Documentation on calibration and verification should be reviewed by the overseeing agency.

2. For WAC determination a critical flow period is selected. For example, if WAC for priority pollutants based on acute toxicity standards is to be determined, the critical period could be one day low-flow with a recurrence interval of once in ten years (US EPA, 1986).

---

[5] See further discussion on flow selection for modeling.

[6] It should be pointed out that most of the water quality models developed and distributed by the US EPA will require a MOS. Such models include the well known and promoted QUAL 2E, which is essentially a steady-state model, and WASP which is a quasi-dynamic model.

[7] Such "typical approach" may under some circumstances result in overprotective Waste Load Allocation. See Novotny *et. al* (1995) for discussions.

Figure 1
The concept of MOS in steady state and quasi-dynamic modelling for WAC determination.

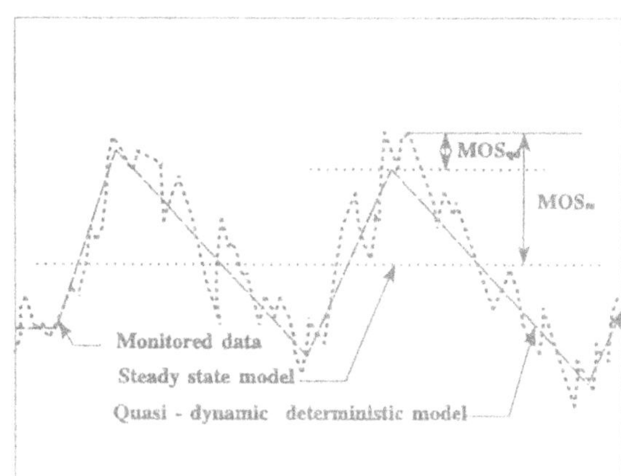

3. Using a log-probability distribution of loads to the system, a representative load with a probability of being less or equal ranging from 90 to 99.9 percent will be selected. No specific guidelines are available from the regulatory agencies, however, such probabilities must be either specified and/or approved by the regulatory agency. Consequently, the selection of the load may be quite arbitrary[8].

4. The model run is executed with the above assumptions. If the simulated concentrations exceed the WQS the loads are decreased[9] until the maximum (minimum in the case of Dissolved Oxygen) of the calculated pollutant concentrations meet the WQS. To satisfy the MOS requirement the maximal total allowable load should be reduced by a fraction of the WAC that will reflect the MOS.

---

[8] The laws and methodologies of joint probability estimation of two or more independent events (e.g., flow and waste water loads) are well known and should be used and presented to the regulatory agency in the documentation explaining the selected MOS.

[9] The loads are decreased uniformly for all sources. The legal ramification in the US do not typically allow to use so called "optimal waste allocation" scenario in which the load reduction would be based on the least cost to the dischargers. This can only be achieved by application of the "transferrable discharge permits" and similar economic instruments.

Such an approach may be adequate if the critical concentrations occur during low - flow periods, implying that only point source pollution may be considered. However, Novotny (1994) has pointed out that this approach may lead to greatly overprotective discharge limitations. Water quality models based on stochastic concepts (e.g., Monte Carlo) may not require consideration of MOS.

2.2 Allowable Waste Load

Both WAC and TDL are expressed in Mass/Time units such as kg/day, however, as pointed out previously, both quantities may be derived from or related to concentrations. In this step the WAC is based on existing conditions of the water body. WAC enhancement is possible but should be considered only for water quality limited water bodies.

If WAC is smaller than the TDL with an appreciable Margin of Safety, then the water body is *water quality limited* or when WAC $\approx$ TDL and/or there is no MOS between the WAC and TDL, then the water body may be considered *threatened.*

According to the EPA, and based on the above concepts, the TMDL equals the WAC plus the BL minus the MOS (or LC minus the MOS). In water quality limited watersheds, the WLA process divides the Waste Assimilative Capacity among dischargers. For point sources this would imply establishing new more stringent limits. For nonpoint sources, new NPS controls (BMPs) will have to implemented.

In the Waste Load Allocation (WLA) process, WAC is converted by the regulatory agency into the allowable loads which includes both allowable point (WLA) and nonpoint (LA_{NP}) sources. According to the EPA (1991) guidelines, Margin of Safety (MOS) is considered and included in the WAC determination in situations where parameters and processes affecting the magnitude of the TMDL are uncertain. Therefore, in water quality limited water bodies

$$TMDL = WLA_P + LA_{NP} + BL = LC - MOS \tag{4a}$$

and

$$WLA_P + LA_{NP} = TMDL - BL = WAC - MOS \tag{4b}$$

where

WLA_P = Waste Load Allocation for point sources and

LA_{NP} = load allocation for nonpoint sources.

Often unregulated nonpoint sources and/or pollutants stored in the water body (e.g., contaminated sediments or sediments with a high Sediment Oxygen Demand) may be the major source of impairment. In such cases, putting the entire burden of remediation on regulated point sources would lead to unequitable and more costly solutions. Furthermore, without considering WAC enhancement and/or water body restoration and by putting the entire burden on remediation of regulated point sources there may be a possibility that the water quality goals become unattainable due to the in-place contaminants and background loads. These considerations should also be investigated by modeling in the WLA process.

Hence, attainment of water quality goals can be achieved both by reducing waste loads by effluent controls and BMPs and by increasing the Waste Assimilative Capacity using water body restoration and WAC enhancement practices (Novotny and Olem, 1994). Waste Assimilative Capacity enhancement has been used and is an accepted remedial alternative in many instances, including:

1. Lake management (aeration, sediment dredging, nutrient inactivation, etc.);

2. Contaminated sediments remediation (dredging, sediment capping, etc.);

3. Aeration enhancement of streams with high in-stream sinks of Dissolved Oxygen;

4. Cases where a desired use of the water body is not attained because of man-made physical past alteration of the water body and its aquatic habitat (water body restoration is needed).

Implementation of nonpoint source control and Waste Assimilative Capacity enhancement (water body restoration) may require innovative financing relying on both private and public resources (see subsequent chapters by Paulsen and Smith) and, possibly, policy changes by agencies.

2.3 The Waste Assimilative Capacity Concept

The concept of Waste Assimilative Capacity (WAC) refers to the ability of a receiving water body to assimilate any discharge of pollutants in it. The concept of WAC was first investigated and proposed following the studies by Streeter and Phelps on the Ohio River in 1914 -15, concerning the relationship between Dissolved Oxygen concentration and wastewater discharges (Streeter and Phelps, 1925).

Cairns (1975) defines assimilative capacity as "*the ability of an aquatic system to assimilate a substance without degrading or damaging its ecological integrity*".

The capacity of a water body to receive waste discharges without ecological damage may have been altered by past actions. For example, streams impounded for navigation and/or power production have lost a part of their reaeration capacity and reduced their ability to safely absorb biodegradable organic wastes. The same impounded streams may become a place of deposition of contaminated sediments with severe ecological consequences while in a free flowing state the sediments would pass downstream to a safer place of final disposal (for example, a sea). A question can be asked whether the entire burden of water pollution should be put on the dischargers discharging into a water body with a reduced WAC or whether the WAC could be restored and increased, approaching pre-alteration conditions, resulting in a less costly and far more equitable plan.

Effect of Temporal Dynamics on WLA. Traditionally, conventional pollutant limitations were given based on critical low flow conditions. WLAs for conventional pollutants from point sources, for example, are still based on steady-state modeling of the stream, using in most cases 7-days low flow with recurrence interval of 10 years (also known as 7Q10[10]) or 1 percent low flow as a standard critical condition. The guidance frequency of exposure of biota to concentrations at toxic levels recommended by the EPA is 3 years, which, in the "best scientific judgment" of EPA scientists, the minimum amount of time needed for an ecological community to recover from an event consisting either of an acute or chronic exposure to a toxic compound (US EPA, 1986).

---

[10] It may be noted that the 7Q10 flow is defined in U.S. EPA (1995) handbook as the average flow during the seven consecutive daily low flows. Hydrologically, however, 7 days excedence low flow would be the highest flow during the 7 consecutive days of low flows.

3. WATER QUALITY MODELS AND THEIR APPLICATION

A state-of-the art review of modeling methodologies has been published by Somlyódy and Varis (1993).

Steady-state water quality models have been widely used in the past to simulate the effect of waste discharges and water bodies, to predict the Waste Assimilative Capacity of rivers and streams, and as a consequence, to develop Waste Load Allocations for wastewater discharges. As a matter of fact, the large majority of older available water quality models can be considered as operating in a steady-state mode (example: EPA's model QUAL-2E).

To fulfill EPA's requirements regarding the duration/frequency criteria on toxic pollutant limitations, daily variations of both receiving water and wastewater discharges should be considered. Steady-state modeling can still be used to describe intrinsically dynamic frequency/duration-related events by incorporating these conditions in an appropriately designed simulation framework. For example, EPA recommends that, when using steady-state modeling, the 1-hour duration of the acute critical event be interpreted as a minimum daily flow, and the 4-day chronic event criterion as a 4-day average flow.

There is no clear indication from the US EPA on how to interpret the 3-year frequency-duration period, yet. The EPA, for example, has determined that the extreme value analysis traditionally used for Waste Load Allocation modeling cannot be considered appropriate when applied in a toxicological framework. Specifically, a 3-year return flow determined by traditional extreme value analysis, would occur too often during droughts to provide the required recovery time that is considered necessary for aquatic organisms (US EPA, 1986). While other methodologies are currently under development, the EPA (1986) recommends that a 1-day in 10-year (1Q10) recurrence flow be used for the acute criterion and the 7-day, 10-year recurrence (7Q10) low-flow see Footnote 10) for the chronic criterion in stressed systems (1Q5 and 7Q5 flows are recommended for unstressed riverine ecosystems). Since these low flows *a priori* presume a period of extended severe drought this approach does not facilitate evaluating water quality impairments caused by stormwater and CSO discharges.

Dynamic simulation or probabilistic (stochastic) techniques are, however, deemed more appropriate because they define a probability distribution of water quality concentrations of the compounds of interest, rather than a single value based on assumed worst-case conditions. There are three dynamic simulation techniques recommended by the EPA for determination of Waste Load Allocations:

- continuous simulation,
- Monte Carlo techniques, and
- log-normal probability modeling.

Continuous simulation can be conducted using a dynamic water quality model (e.g., WASP4 - Ambrose et al., 1991) over an extended period of time. The results are then plotted on probabilistic paper to yield a concentration/probability diagram. The process is illustrated in Figure 2.

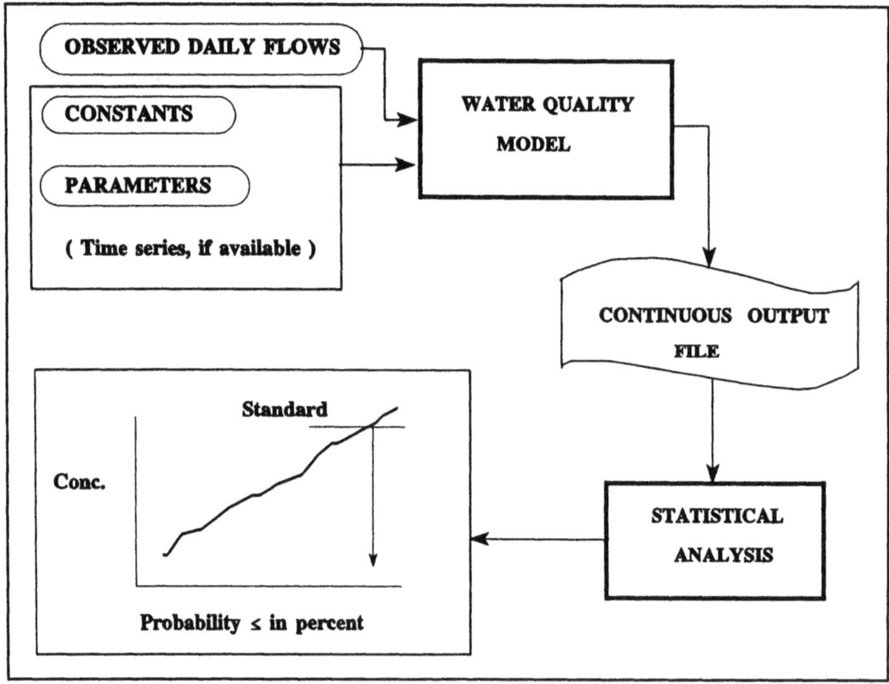

Figure 2 Continuous simulation in the WLA process

In this application, observed continuous series of flow data should be available. In addition, water quality and other parameters should be monitored to provide model calibration and verification data. Although it is not clear how long a continuous simulation should be, it is reasonable to assume that simulation conducted on a daily basis for a period between 10 and 25 years should produce a meaningful statistical distribution of water quality constituent concentrations. Simulation can be conducted using daily flow values observed over a period of time much longer than the prescribed maximum exposure frequency of three years. If a sufficient historical record of observed flows is not available, flow values generated by a stochastic model reflecting distribution of the observed flow series could be used.

Since the response and representativeness of the model may vary according to the hydrodynamic conditions in the water body, parameters (describing environmental conditions) and constants (describing reactions and processes) should be allowed to vary over time during continuous simulation. Ideally, models should be calibrated using independent data sets for different hydrologic and seasonal conditions, however, this process may become too cumbersome and resource-intensive to be applied in most circumstances. As an alternative, a single set of constants and parameters could be used, provided that the chosen set represents a conservative situation.

Monte Carlo techniques derive their name from the gambling analogy: as in the roulette game, random numbers fitting an observed or hypothesized probability distribution are generated (with computer programs in this case) and "shot" as input into a deterministic relationship. If this procedure is repeated many times (in the order of 1000 or greater) the results of the deterministic simulation conducted with random input generates probability distribution of the output domain, which can be used to estimate the required frequency/duration relationship for load allocation. Figure 3 illustrates the principle of application of a Monte Carlo simulation methodology.

A Monte Carlo model is basically constituted by a deterministic portion (the deterministic water quality model), of variable complexity that is used to mathematically represent the system under observation, and a probabilistic portion, constituted by the probability distributions of both the deterministic model parameters (if available) and the observed variables (conditions). These should be derived from a representative observation data base, by calculating the necessary moments and comparing the obtained distribution (normal, log-normal or other appropriate one) to the observed data. The Monte Carlo approach requires that the inputs of the model be independent. Therefore, in-stream flows and pollutant concentrations should be independent in order to achieve maximum accuracy. The presence or

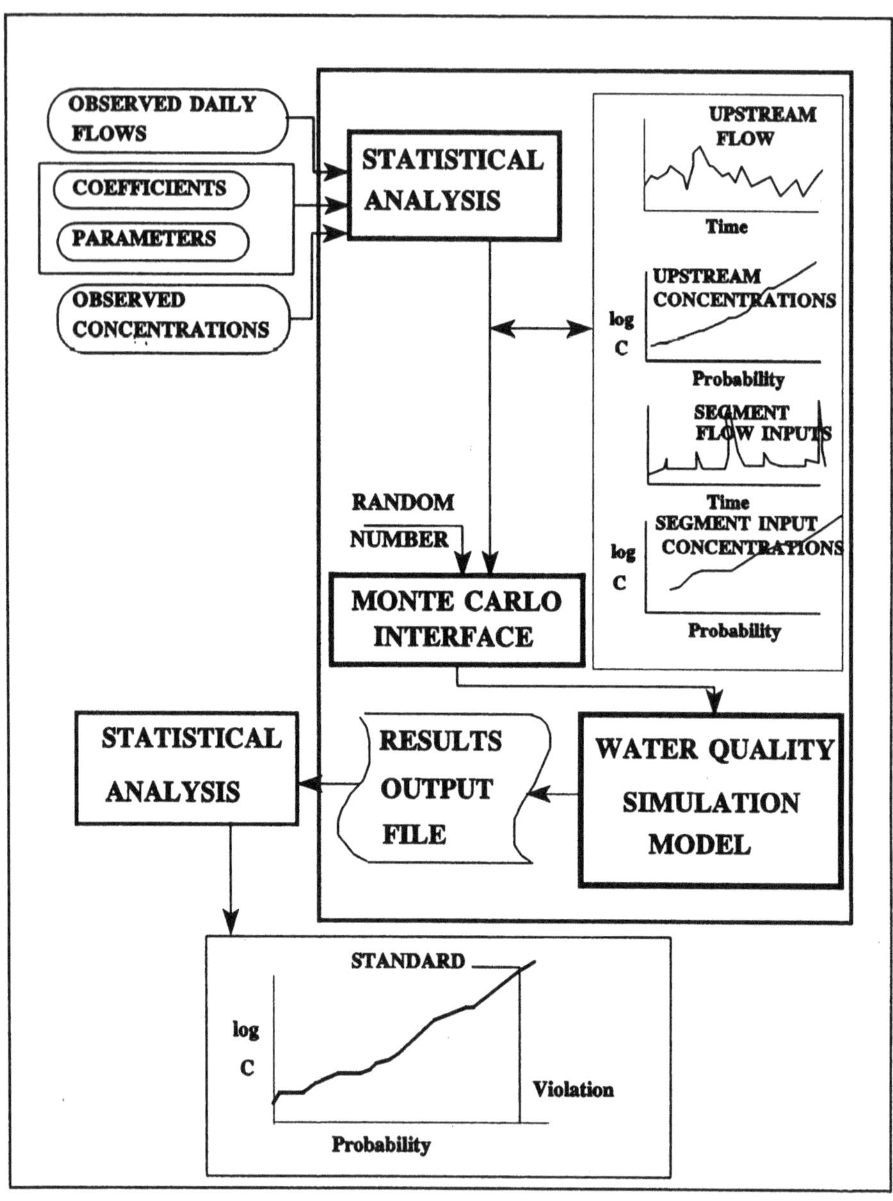

Figure 3 Monte Carlo simulation concept

absence of cross-correlation between input parameters should be verified by appropriate statistical methods. If two or more input parameters are cross-correlated then one of them is considered as the primary independent parameter and the other parameters are dependent variables estimated by the same Monte Carlo principle using a multi-regression connective model.

Advantages of the Monte Carlo approach are:

- the ability to handle uncertainty and variability associated with model coefficients and forcing functions,

- the fact that it can potentially be applied to any deterministic modeling structure,

- flexibility with respect to the types of probability distributions that can be used to characterize the inputs of the model, and

- the fact that it is conceptually simple.

Disadvantages of Monte Carlo approaches are: the very large number of iterations required for calculation of the final probability distribution, and the fact that cross-correlations between different types of probability distributions are difficult to simulate.

The random number, generated by a suitable computer program, is transformed into an excedence probability value, which is then applied to the probability distribution of the parameter(s) of interest, thus obtaining a value that is used as an input to the deterministic water quality model. This process must be repeated a very large number of times (in the order of thousands). In this fashion, an equal number of data points becomes available, that is subsequently subject to statistical evaluation. Halm *et al.* (1993) and Novotny *et al.* (1994) presented examples of Monte Carlo modeling of toxic compounds for a stream.

The log-normal probability modeling approach is based on the assumption that stream flows, and effluent flows and concentrations are log-normally distributed and uncorrelated. The log-normal modeling approach is also known as the probabilistic dilution model (PDM).

3.1 Fate of Conventional and Toxic Pollutants in Aquatic Environments

Traditionally, the class of "conventional" pollutants (BOD, nutrients) has received most of the attention from water quality planners, although in recent years growing concern has been focused on the class of toxic pollutants. Concentrations of conventional pollutants are usually expressed in mg/l, toxic pollutants are expressed in μg/l or smaller units.

Conventional pollutants are mostly transported in dissolved or colloidal form. A fraction of conventional pollutants may be adsorbed to sediment particles (e.g., phosphorus or ammonia) and is usually not available to the resident biota. In slow moving receiving water bodies such as impoundments or estuaries, settleable conventional pollutants (e.g., organic solids) associated with point sources usually deposit within a short distance of the outfall and may cause water quality problems by resuspension (during high flow events) or by exerting a Sediment Oxygen Demand. Most conventional organic pollutants are partially degraded biologically into harmless substances (CO_2, CH_4, NH_4^+).

Toxic pollutants may be strongly sorbed to suspended and benthic layer sediments. They may also undergo biochemical transformations, but the final products of such transformations may still be toxic (or even more toxic than the original compounds). The residence time in the system of these compounds may be in the order of years, depending on the hydrodynamic characteristics of the water body and their chemical properties. It has been determined that sorption processes on sediments are important for 60 out of 129 priority pollutants and negligible for 28, while there is still uncertainty for the others; volatilization processes are important for 52 pollutants, not important for 44 and uncertain for the remaining ones. The EPA (Mills *et al.*, 1985) suggests in fact that the lab-determined half-life values are usually shorter than in most natural systems.

Processes Affecting the Fate of Pollutants in Water Bodies. Water bodies can be dominated by mixing and transport processes that can be quite different. For example, rivers are dominated by one-dimensional flow that can be described by advective transport (so called plug flow regime) while in some lakes the dispersion process may be predominant. In impoundments and estuaries the effects of stratification are the main driving force in water quality issues. The structure and approach methodology of the model applicable in each case can be very different.

Usually, complex surface water quality models are constituted by at least two complementary units: a hydrodynamic model, that is first used to determine flow conditions in the water body, and a water quality model, that will simulate the processes affecting water quality constituents according to the flow and velocity conditions in the water body. For modeling priority pollutants the sediment movement component must be included. In simpler cases, the hydrodynamic modeling part of the simulation can be substituted by knowledge of the appropriate flow and velocity fields in the water body in critical conditions (for example, from statistical analysis of historical records).

The processes that affect the fate of pollutants, both conventional and toxic, in the aquatic environment can be classified into the following four main categories:

1. **Transport processes** which include the following mechanisms:
   a. *Advection*. Pollutants, in dissolved, suspended or sorbed (to suspended sediments) form are transported by flow.

   b. *Sedimentation* affects suspended pollutants and pollutants sorbed to suspended sediments. In addition, sediment deposition rates can affect dissolved pollutant concentrations via adsorption/desorption equilibria.

   c. *Volatilization*. Volatile pollutants may enter the atmosphere from the water surface, with an overall effect of reducing aquatic concentrations. The type of flow in the water body may affect significantly the volatilization rates of certain pollutants.

   d. *Precipitation/dissolution* may be considered a transport process inasmuch as it causes a change in the phase of the pollutant. It may affect the availability of the pollutant to other transport and transformation processes.

2. **Transformation processes**, including:
   a. *Biodegradation* is carried on by microbial metabolism. In the case of toxic organic compounds, microbial metabolism may alter the original toxicity of a compound.

   b. *Hydrolysis*. Compounds may react with water, yielding simpler (usually less harmful, or more easily biodegradable) products.

c. *Photolysis*. Compounds may be degraded or chemically modified by adsorption of sunlight energy. Changes due to photolysis may affect toxicity of organic compounds.

d. *Oxidation/reduction*. Organic pollutants and metals may undergo redox reactions in the aqueous and benthic phases. These may considerably influence their environmental-related properties. Almost all redox reaction occurring for organic compounds are microorganisms-mediated.

3. **Speciation processes** which include:

a. *Acid-base equilibria* determine the ionic states of acid and bases, and may influence volatility and precipitation/dissolution processes.

b. *Sorption*. Hydrophobic organic compounds are sorbed to suspended sediments according to sorption equilibria. Fate of sorbed matter is then correlated to the fate of suspended sediment.

4. **Bioaccumulation**. Bioaccumulation refers to the intrusion of nondegradable, nonassimilable (usually toxic) compounds in the food chain. In this context, bioconcentration refers to the primary uptake of such compounds (i.e., through the gills of fish), while biomagnification refers to their uptake as part of a prey/predator food chain scheme (i.e., consumption of "contaminated" food or lower order predators). Bioaccumulation is the final step in the pathway of pollutants in the aquatic environment. It is not considered as part of water quality/Waste Assimilative Capacity determinations; it is rather a starting point to determine allowable pollutant concentrations in the water.

Prediction of the fate of pollutants requires the modeler:

a) to know which processes act on which pollutant;

b) to define the environmental compartments between which the processes occur;

c) to select (or develop) a model that simulates a process of interest; and

d) to estimate the input parameters and/or coefficients for that model.

Assessment of pollutant fate. In conducting Waste Assimilative Capacity evaluations it is important to determine whether a given pollutant is likely to cause problems in the water body, i.e., if the assimilative capacity of that water body is limited for that particular pollutant, and if so, to what level.

a. the standard can be met, or

b. the standard can not be met under those conditions.

To proceed with this task efficiently EPA (Mills *et al.*, 1985) suggests a three-tiered approach to assess the fate of pollutants in water bodies:

1. first, pollutants are treated as conservative substances

   Under this scenario, decay, degradation and other removal processes are neglected. Simple dilution models will calculate pollutant concentrations that are most likely higher (or WAC's that are lower) than those that will be actually observed in the water body. If these concentrations do not cause violation of standards, or other problems, then it can be safely assumed that the criteria for that water body can be upheld.

2. then, transportation and speciation processes are also considered, and

   Under the second option, chemical speciation and pollutant transport out of the aquatic environment are considered. These include; acid-base equilibria, solubility, removal to bottom sediments (sorption) and compound volatilization.

3. transportation, transformation and speciation processes are considered.

   Under the third option, transformation (degradation) processes are added to speciation and transport. The transformation processes include biodegradation, photolysis and hydrolysis.

The important processes affecting the aquatic fate of organic priority pollutants are shown in Table 1.

Table 1 Relative importance of processes influencing aquatic fate or organic priority pollutants (after Callahan *et al.*, 1979)

COMPOUND	PROCESS					
	Sorption	Volatilization	Biodegradation	Photolysis	Hydrolysis	Bioaccumulation
PESTICIDES						
Acrolein	-	+	+	+	-	-
Aldrin	+	+	?	-	-	+
Chlordane	+	+	?	-	-	+
DDD	+	+	-	-	-	+
DDE	+	+	-	+	-	+
DDT	+	+	-	-	+	+
Dieldrin	+	+	-	+	-	+
Endosulfan and Endosulfan Sulfate	+	+	+	?	+	-
Endrin and Endrin Aldehyde	?	?	?	+	-	+
Heptachlor	+	+	-	?	++	+
Heptachlor Epoxide	+	-	?	?	-	+
Hexachlorocyclohexane (α,β,δ isomers)	+	?	+	-	-	-
Hexachlorocyclohexane (Lindane)	+	-	+	-	-	-
Isophorene	-	-	?	+	-	-
TCDD	+	-	-	?	-	+
Toxaphene	+	+	+	-	-	+
PCBs and RELATED COMPOUNDS						
Polychlorinated Biphenyls	+	+	+[a]	?	-	+
2-Chloronapthalene	-	?	+	+	-	-
HALOGENATED ALIPHATIC HYDROCARBONS						
Chloromethane (methyl chloride)	-	+	-	-	-	-
Dichloromethane (methylene chloride)	-	+	?	-	-	-
Thrichlomethane (chloroform)	-	+	?	-	-	-
Tetrachloromethane	?	+	-	-	-	?
Chloroethane (ethyl chloride)	-	+	?	-	+	-
1,1-Dichloroethane	-	+	?	-	-	-
1,2-Dichloroethane	-	+	?	-	-	-
1,1,1-Trichloroethane	-	+	-	-	-	-
1,1,2-Trichloroethane	?	+	-	-	-	?
1,1,2,2-Tetrachloroethane	?	+	-	-	-	?
Hexachloroethane	?	?	?	?	?	+
Chloroethene (vinyl chloride)	+	-	-	-	-	-
1,1-Dichloroethene	?	+	?	-	-	?
1,2-*trans*-Dichloroethene	-	+	?	-	-	-
Trichloroethene	-	+	?	-	-	-
Tetrachloroethene	-	+	+	-	-	-
1,2-Dichloropropane	?	+	-	?	+	?
Hexachlorobutadiene	+	+	?	-	?	+
Hexachlorocyclopentadiene	+	+	-	+	+	+
Bromomethane (methyl bromide)	-	+	-	-	+	-
Bromodichloromethane	?	?	?	?	-	+
Dibromochloromethane	?	+	?	?	-	+
Tribromomethane (bromoform)	?	+	?	?	-	+
Dichlorodifluoromethane	?	+	-	?	-	?
Trichlorofluoromethane	?	+	-	-	-	?

Table 7.1. Cont.

COMPOUND	PROCESS					
	Sorption	Volatilization	Biodegradation	Photolysis	Hydrolysis	Bioaccumulation
<u>HALOGENATED ETHERS</u>						
Bis(chloromethyl) ether	-	-	?	-	++	-
Bis(2-chloroethyl) ether	-	+	-	-	-	?
Bis(2-chloroisopropyl) ether	-	+	-	-	-	?
2-Chloroethyl vinyl ether	-	+	?	-	+	-
4-Chloroethyl phenyl ether	+	?	?	+	-	+
4-Bromophenyl phenyl ether	+	?	?	+	-	+
Bis(2-chloroethoxy) methane	-	-	?	-	+	?
<u>MONOCYCLIC AROMATICS</u>						
Benzene	+	+	-	-	-	-
Chlorobenzene	+	+	-	-	-	+
1,2,-Dichlorobenzene	+	+	-	?	-	+
1,3-Dichlorobenzene	+	+	?	?	?	+
1,4-Dichlorobenzene	+	+	-	?	-	+
1,2,4-Trichlorobenzene	+	+	-	?	-	+
Hexachlorobenzene	+	-	-	-	-	-
Ethylbenzene	?	+	?	-	-	-
Nitrobenzene	+	-	-	+	-	-
Toluene	+	+	?	-	-	-
2,4-Dinitrotoluene	+	-	-	+	-	?
2,6-Dinitrotoluene	+	-	-	+	?	?
Phenol	-	+	+	-	-	-
2-Chlorophenol	-	-	?	+	-	-
2,4-Dichlorophenol	-	-	++	-	-	-
2,4,6-Trichlorophenol	?	-	?	?	-	-
Pentachlorophenol	+	-	+	+	-	+
2-Nitrophenol	-	-	-	++[b]	-	-
4-Nitrophenol	+	-	-	++	-	-
2,4-Dinitrophenol	+	-	-	++[b]	-	-
2,4-Dimethyl phenol (2,4-xylenol)	-	-	?	+	-	-
p-chloro-m-cresol	-	-	?	++	-	-
4,6-Dinitro-<u>o</u>-cresol	+	-	-	++	?	?
<u>PHTHALATE ESTERS</u>						
Dimethyl phthalate	+	-	+	-	-	+
Diethyl phthalate	+	-	+	-	-	+
Di-n-butyl phthalate	+	-	+	-	-	+
Di-n-octyl phthalate	+	-	+	-	-	+
Bis(2-ethylhexyl) phthalate	+	-	+	-	-	+
Butyl benzyl phthalate	+	-	+	-	-	+

Table 7.1 Cont.

COMPOUND	PROCESS					
	Sorption	Volatilization	Biodegradation	Photolysis	Hydrolysis	Bioaccumulation
POLYCYCLIC AROMATIC HYDROCARBONS (PAHs)						
Acenaphthene[c]	+	-	+	+	-	-
Acenaphthylene[c]	+	-	+	+	-	-
Fluorene	+	-	+	+	-	-
Naphthalene	+	-	+	+	-	-
Anthracene	+	+	+	+	-	-
Fluoranthene[c]	+	+	+	+	-	-
Phenanthrene[c]	+	+	+	+	-	-
Benzo(a)anthracene	+	+	+	+	-	-
Benzo(b)fluoranthene[c]	+	-	+	+	-	-
Benzo(k)fluoranthene[c]	+	-	+	+	-	-
Chrysene[c]	+	-	+	+	-	-
Pyrene[c]	+	-	+	+	-	-
Benzo(ghi)perylene[c]	+	-	+	+	-	-
Benzo(a)pyrene	+	+	+	+	-	-
Dibenzo(a,h)anthracene[c]	+	-	+	+	-	-
Ideno(1,2,3-cd)pyrene[c]	+	-	+	+	-	-
NITROSAMINES AND MISCELLANEOUS COMPOUNDS						
Dimethylnitrosamine	-	-	-	++	-	-
Diphenylnitrosamine	+	-	?	+	-	?
Di-n-propyl nitrosamine	=	-	-	++	-	-
Benzidine	+	-	?	+	-	-
3,3'-Dichlorobenzidine	++	-	-	+	-	-
1,2-Diphenylhydrazine (Hydrazobenzene)	+	-	?	+	-	+
Acrylonitrile	-	+	?	-	-	+

KEY TO SYMBOLS

++	Predominate fate determining process	-	Not likely to be an important process
+	Could be an important fate process	?	Importance of process uncertain or unknown

Notes

a Biodegradation is the only known process to transform polychlorinated biphenyls under environmental conditions, and only the lighter compounds are measurable biodegraded. There is an experimental evidence that heavier polychlorinated biphenyls (five chlorine atoms or more per molecule) can be photolyzed by ultraviolet light, but there are no data to indicate that this process is operative in the environment.

b Based on information for 4-nitrophenol

c Based on information for PAHs as a group. Little or no information for these compounds exists.

4. SELECTION AND APPLICATION CRITERIA FOR WATER FLOW AND QUALITY MODELS

The selection and application of a model should follow these general steps:

1. *Clearly identify the water quality problem (or problems) and their possible causes. The type of pollutants of interest should be identified as accurately as possible. Almost all water quality models have the capability to simulate simple conventional pollutant and nutrient dynamics.*

Some models have the capability to include algal growth and eutrophication. A few models include algorithms to account for volatilization, hydrolysis and sorption of organic compounds. Fate of metals is usually difficult to simulate, but some programs can simulate their speciation and interaction with particulate matter.

Sources of pollutants are almost as important. Simpler models usually accept constant point source inputs at multiple locations. If diffuse pollution is an issue, the model should be able to accept distributed loads. Complex models may accept both point and distributed time-variable input loads. Deterministic steady-state or quasi-dynamic models require determination of a Margin of Safety that should be reviewed and approved by the EPA or state regulatory agency. A Monte Carlo or stochastic model may not need MOS but this should first be approved by the overseeing regulatory agency.

2. *Define objectives and constraints of the study, including the degree of accuracy required, an assessment of the available resources and applicable tools, and the time-schedule of the project.*

All complex models require an extended set-up period and calibration and verification efforts. If a rapid assessment of water quality conditions is desired, a simplified procedure is used instead.

Also, an estimate of the financial resources available for the study may be the deciding factor for the selection of a specific modeling approach at this stage of the project.

3. *Review all available data on the site. The available database should match the objectives of the study.*

For example, if biota exposure to toxics is of concern, data on sediment classification and analysis should be available.

Assess the opportunity to complement the available data base with field monitoring, taking into account the resources mentioned in item (1), and the project time-schedule. The climatic characteristics of the site should also be taken into account, since in colder climates some types of sampling may be possible only in a few summer months. The availability of data also will have to be considered in the selection of a specific model.

4. *Identify those processes that are most important for the water body and the pollutants of interest.*

For example, if heavy metals are of concern for the water quality of a river, it is important that the candidate models be able to simulate their speciation and interaction with sediments. In this case, sediment dynamics should also be simulated if biota exposure to metals is an issue. Some models are able to simulate certain processes at different degrees of complexity, i.e., taking into account interactions that occur at different levels in the global process.

Usually, increasingly complex interaction models require a larger number of parameters for the definition of the individual process rates. Since an error is usually associated with the estimation of these parameters, the global simulation error may increase rather than decrease when going from a simpler to a more complex model.

5. *Model selection. A model is selected from a group of models that are deemed compatible with the resources and constraints identified under items 1 to 4.*

Availability of the models and adequate documentation should be evaluated at this point. Literature reported applications of a specific model to similar problems can also be searched and evaluated. Previous experiences with the model, or access to a user support center are important factors to consider, especially in the case of more complex models. Preference should be given to models that have been thoroughly tested.

4.1 Selection of a Hydraulic Model

The strategies available for the definition of flow characteristics in a water body include:

- analysis of historical records,
- extensive field measurements,
- measurements coupled with hydrodynamic model simulation.

Traditionally, steady-state simulation for a set of critical conditions (usually the so-called 7Q10, or the 7-day low flow occurring once every 10 years, or 1, 2 or 3 percent flows) were used for the evaluation of the impact of conventional pollutants in water bodies. New US regulations regarding toxics, however, address the issue of dose/frequency exposure of biota to these compounds. This introduces a probabilistic aspect in the analysis of the problem. To select an appropriate flow model, it is first necessary to characterize flow patterns within the water body.

This preliminary analysis will determine the proper transport and mixing processes that are active. A model with appropriate dimensional capabilities can thus be selected. The dimensional classification characterizes the water bodies and models as zero-, one-, two- or three-dimensional. For example, a zero-dimensional (or more often called *completely mixed*) water body would be a nonstratified circular lake with long retention times, a narrow shallow river would be one-dimensional, while estuaries with saline wedge intrusions would require at least a two-dimensional model. Use of three-dimensional models is very rare due to complexities, data requirements and model instability involved in the use of such models.

The choice of model capabilities with respect to dimensionality of the flow field may also depend on the degree of accuracy with which sources of pollution can be identified. For rivers, one-dimensional modeling is usually adequate, except in close proximity of the mixing zone of a point-source discharge (complete lateral and vertical mixing assumptions are not satisfied), or for very large rivers (width > 500 meters) because of lateral mixing assumptions.

For nonstratified lakes with long residence times, frequently a one-dimensional vertical model is used, ignoring wind-induced circulation patterns. One-dimensional models with two or more water column layers can also be used, although they do not account for effects of horizontal circulation. It should be noted that two- and three-dimensional circulation models are available, however, usually they are not operative models, but rather experimental or research-oriented. In estuaries, two- and three-dimensional models can be applied, however, they are extremely complex both from data input and numerical processing aspects and thus require advanced computer capabilities and qualified expertise to be applied and calibrated properly. This makes them extremely expensive to use. Two-dimensional models are also common for modeling mixing zones. Table 2 lists the available principal public-domain surface flow models and their capabilities.

Table 2 Summary of selected surface flow models

CHARACTERISTICS	MODEL NAME						
	HEC-2	HEC-6	HSPF	DYNHYD5	SEDONE	DWOPER	EXPLORE-I
Applications							
River	X	X	X	X	X	X	X
Lake				X	X		X
Estuary				X	X		X
Dimensions							
1-D	L	L	L	L	L	L	L
2-D			Q				Q
3-D							
Mode							
Steady-state	X						
Dynamic		X	X	X	X	X	X
Processes							
Sedimentation		X	X	X	X		
Water quality capabilities			X	X			
Others							
Autocalibration						X	
Support/ distribution							
HEC	X	X					
EPA			X	X			ND
NOAA					X		
Other/Unknown						OR	

L = Longitudinal
Q = Quasi-two-dimensional
ND = Model is developed by EPA, but not currently distributed or supported
OR = Model was developed by Oak Ridge National Laboratory. Distribution/support sponsorship unknown

Thomann (US EPA, 1992) pointed out that a mathematical model of hydrodynamics is needed only when:

- the transport regime is complex in space and time and can not be specified *a priori*,
- the transport regime will be changed under some future WLA conditions which may be changed, for example, by deepening, straightening or restoration of rivers or construction of tidal barriers,
- the absence of a hydrodynamic model would weaken water quality models credibility in the eyes of peer scientific review.

In most situations, hydraulic parameters can be calculated by the model from flow using morphological relationships such as:

Velocity	$v = a\,Q^b$
Depth	$H = c\,Q^d$
Width	$B = e\,Q^f$

From flow continuity it follows that $a \times c \times e = b + c + f = 1$.

4.2 Selection of a Water Quality Model

Water quality models are available in different degrees of complexity. They can range from screening procedures to full-fledged transient models with interactions between pollutants.

The simplest "model" available is perhaps the US EPA screening procedure (Mills *et al.*, 1985) which is constituted by a collection of procedures of data analysis and algorithms with the purpose of conducting preliminary or even detailed assessments of toxic and conventional pollution in surface water bodies. The procedures contained in the report can be applied at different degrees of detail, based on the influence of individual processes on the overall fate of pollutants. Procedures are identified for applicability to different types of water bodies, such as streams and rivers, lakes and impoundments and estuaries. A section covers also groundwater fate of contaminants. The procedures listed in the report are implementable on a hand-held calculator, although they may be programmed into more sophisticated computer models. The following section discusses briefly the main processes that influence aquatic quality in surface water bodies. Table 3 contains a list of the principal public-domain water quality models available.

Table 3 Summary of selected water quality models

CHARACTERISTICS	MODEL NAME						
	WQAM	QUAL-2E	HSPF	WASTOX	WASP4	FETRA	EXAMS
Applications							
River	X	X	X	X	X	X	X
Lake	X	X	X	X	X		X
Estuary	X		X	X	X	X	
Dimensions							
1-D	A	X	X	C	X		C
2-D			C		Q	H	C
3-D			C				C
Mode							
Steady-state	X	X			X		X
Dynamic			X	X	X	X	
Processes							
Settling	X	X	X	X	X		
Scour/depos.			X	X	X	X	
Exchanging bed			X	X	X	X	X
Transport			X	X	X	X	
1st order decay	X	X	X	X	X	X	X
2nd order decay					X		
Photolysis			X	X	X		X
Hydrolysis			X	X	X		X
Oxidation			X	X	X		X
Biodegradation		X	X	X	X		X
Eutrophication		X	X	X	X		X
Volatilization			X	X	X		X
Sorption			X	X	X	X	X
By-products			X		X	X	X
Support/ distribution							
EPA	X	X	X		X		X
Other/Unknown				1		2	

A = Analytical model consists of a set of equations applicable to diverse processes (see Mills *et al.*, 1985)
C = Compartment model
Q = Quasi-two-dimensional
H = Horizontal two-dimensional
1 = Public domain model developed at Manhattan College, supported in the past by the U.S. EPA
2 = Program developed at Battelle Pacific Northwest Laboratories

The public domain models distributed by the US EPA can be characterized as deterministic steady-state or quasi-dynamic models. As pointed out in the preceding chapter, these models are not compatible with the water quality standards that have been defined by the U.S. EPA and states in terms of probability of excedences of concentrations. A vaguely defined Margin of Safety was introduced by the EPA into the TMDL process in an effort to overcome this incompatibility. But almost in any case, the modeler can not provide an accurate estimate of the probability of excedence of the WQS unless a special extensive evaluation of errors of the model output (an error in this context means a deviation of the model output from a corresponding real concentration that was or would be measured in the receiving water body) using probabilistic -stochastic concepts is performed.

Monte Carlo and other stochastic models could overcome this drawback. At this time, the US EPA does not sponsor any research activity in this area, and in the near future modeling efforts anticipated by the EPA in the TMDL will most likely rely almost exclusively on previously developed deterministic models. The users of the models and permitees affected by the outcome of such models should be aware of the drawbacks of deterministic modeling approaches and carefully evaluate the reliability, outcome and MOS associated with the deterministic steady-state or quasi-dynamic models. A Monte Carlo water quality model for streams has been developed by Marquette University and AquaNova International, Ltd. (Novotny *et al.*, 1994). A Monte Carlo approach is not a model *per se* it is more-or-less a methodology of a model application.

Thomann evaluated and critiqued estuary models for the US EPA (1992). In his assessment he provided some basic rules on the selection of water quality models.

Rule # 1 - *The best models are often the simplest.*

Complex models require experienced modelers, expensive data bases for calibration and verification and large resources for executing the modeling study. Unnecessarily complex models sometimes tend to obscure uncertainty behind a facade of "reality". A significant part of the variations is random and cannot be represented by any deterministic model. Simple models may turn out to be entirely sufficient for the TMDL - WLA process.

Rule # 2 - *Analytical solutions should be used to compute initial response and test numerical model computations.*

Analytical models such as steady-state solutions of differential equations describing the movement and transformation of pollutants, should be used to evaluate the magnitude of the problem and to test more complex numerical models.

Rule # 3 - Temporal-spatial scales begin with steady-state larger space scales, followed by seasonal, more detailed spatial definition, then followed by intra-tidal-fine scale models.

Thomann suggests to define initially the temporal scale of most riverine and estuarine models with a steady-state to determine the overall relationship between input loads and water quality.

Following steady-state modeling, dynamic modeling may be necessary when

- for fresh water impoundments and large estuaries to simulate phytoplankton dynamics and its response to nutrient inputs,

- to simulate impact of dynamic time variable inputs such as storm water discharges,

- to simulate complex hydrodynamics of estuaries, if required by the circumstances.

The fact that an estuary has a tidal oscillation is not in itself justification for constructing an intratidal model. The same reasoning is valid for diurnal oscillation of DO in water bodies affected by algal photosynthesis or daily variations of waste loads.

Rule # 4 - Keep state variables to a minimum; model only those for which data exists; always include those state variables which will be impacted by a WLA.

The modeler must make a decision whether, for example, to model or input state variables such as pollutant fluxes from the sediment, including SOD, photosynthesis and respiration of algae, water temperature, etc. Inability to calibrate the state variables severely limits the utility of the models.

5. MODELING SCENARIOS IN THE WLA (TMDL) PROCESSES

5.1. A Simple Point Source Steady Discharge into Streams and One Dimensional Nonstratified Estuaries.

Model selection: calibrated and verified steady-state deterministic mass balance model

Modeled hydrological period: 7-Q-10 for Dissolved Oxygen and chronic toxicity
1-Q-10 for acute toxicity
30-Q-10 for ammonia in situations involving POTW's design with low effluent variability

Model parameters: typical mean values for the low flow period derived form calibration

Pollutant load characterization: average load[11] or source concentration

Margin of Safety: a) incorporated in the standard
b) incorporated in selection of low flow period.

5.2. A Variable Point Source Input with Deterministic Time Series of Nonpoint Source Inputs[12]. (Streams and Estuaries).

Model selection: a quasi-dynamic model, calibrated and verified

Modeled hydrological period: 10 to 30 year hydrological period

---

[11] The rationale behind selecting the mean concentration (or pollutant load from sources) is the fact that the probability of the concentration (load) being greater during the low flow period is the same as being smaller. Consequently, the frequency of violations over a long period may be about the same or less as if the source concentration (load) was constant, equaling the mean.

The use of the mean load decreases the risk of the standards being exceeded. For example, if the probability of the 7 days consecutive flow in any year of being less or equal the magnitude of 7-Q-10 is $p_Q = 0.1$ and the probability of the load being less or equal the mean is $p_L = 0.5$ then the joint probability of excedence of the concentration in the receiving water body calculated from these two parameters is $p_{Q-L} = 0.1 \times 0.5$, which means that the frequency of possible violations decreases to once in twenty years.

[12] This condition means that concentrations of each event are constant or known for each event. Using statistically characterized Event Mean Concentrations has been shown to be valid in the U.S. EPA (1983) NURP studies.

Pollutant load characterization:

average (mean) or linearly changing point source load during the modeled subperiods,

or

a time series of nonpoint loads

Model parameters: averaged for selected simulation subperiods

Margin of Safety: same as #5.1.

5.3. Variable Point and Nonpoint Sources (Streams and Estuaries)

Model selection: Monte Carlo interface with a deterministic mass balance water quality model[13]

Modeled hydrological period:

10 years or more which includes a critical low flow period.

Pollutant load characterization:

flow - deterministic (measured) or a generated stochastic flow series

concentration - log-normal distribution or ARMA stochastic model

Model parameters: mean directly entered or represented by mean and standard deviation and generated by the Monte Carlo interface.

Margin of Safety: incorporated in the standards

5.4. Impoundment System with a Longer Residence Time Impacted by Nutrients Primarily from Allochthonous (External) Sources.

Model selection: steady-state mass balance models

Pollutant load characterization: once in three years annual load expressed per unit area of the impoundment.

Model parameters: averages

Margin of Safety: typically, criteria for nutrients in impoundments are site-specific. MOS should be considered when developing the criteria.

---

[13] It is advisable to use simpler mass balance models and concepts to evaluate feasibility and need for complex and expensive modeling efforts. In some cases, simpler models may be adequate and far less expensive.

5.5. Impoundment System Significantly Impacted by Pollutants from Internal (Autochthonous) Sources.

Model selection: dynamic lake models with sediment interaction. Ideally, the model should be capable of simulating thermal regimes of the impoundment and pollutant release from the sediment under different oxidation - reduction potentials.

Pollutant load characterization: seasonal, quasi-steady-state. Once in 3 years loads.

Modeled hydrological period: several years containing critical hydrologic period.

Model parameters: average for the seasons.

Margin of Safety: see note for Case 5.4.

5.6. Large Estuary with Higher Flushing Capability.

Model selection: a quasi-dynamic two dimensional model, calibrated and verified.

Modeled hydrological period: an average and once in three year hydrological years.

Pollutant characterization: time series of point source loads (weakly averages)
time series of nonpoint loads

Model parameters: averaged for selected simulation subperiods (see Footnote)

Margin of Safety: same as Case 1.

5.7 Calibration and Verification of Models

Both deterministic and Monte Carlo water quality models require calibration and verification of the model parameters during the design and model identification process. Many mathematical equations and formulas used in the deterministic flow-quality models are of an empirical or semiempirical nature requiring knowledge of a large number of coefficients and reaction rates. The model manuals and other literature sources, provide guidelines for a rough estimation of the most important parameters, however, the ranges are commonly quite wide. Using the WASP model as an example, more than forty parameters must be input for each subunit of a complex model. Under these circumstances, the only way to arrive at a set of coefficients adequately describing the water quality system is by calibrating the model against measured data.

The question of calibration and verification is crucial in the TMDL process since modelers must document to the regulatory agency that the model used in the WLA is "satisfactorily calibrated and verified."

Calibration means varying the coefficients of the designed model within acceptable ranges until a satisfactory agreement between measured and computed output values is achieved. Trying to adjust all twenty or more coefficients for each segment at the same time is tedious and often impossible. Consequently, calibration is a subjective process requiring experience and an evaluation of what constitutes a calibrated model is purely judgmental. Orlob (US EPA, 1992) calls for formalizing calibration and verification procedures. Thomann in the same report provided a definition such as "a model is considered representative of the real estuary when the key model state variables reproduce the observed data over a range of expected conditions and within expected statistical variability. "However, no quantification was given on what is a measure of "expected statistical variability." One has to conclude that the measure of what is satisfactory calibration and verification has to be negotiated between the modeler and the reviewing regulatory agency.

Once the model has been calibrated, that is, once a satisfactory (according to the judgement of the modeler and the reviewing agency) fit of computed and simulated data has been achieved for one calibration storm or time series, the model must be verified. **Verification** is accomplished by running the model with the coefficients established during calibration and with inputs corresponding to another (verification) water quality series. If a satisfactory fit of computed and measured data is obtained, calibration and verification are accomplished. Very often this is not the case, and calibration and verification must be repeated until a satisfactory, often compromising fit is found.

Caveats. Often the models structure may have an excessive number of "degrees of freedom" which is illustrated by the parameter determination requirements for the WASP model. For example, Dissolved Oxygen concentrations can be adjusted by modifying parameters of any oxygen source and sink processes. There is no methodology available to distinguish between the parameters, yet, selection of a wrong magnitude of the parameter may lead to erroneous conclusions on WLA. Therefore, the modeler must be experienced and familiar with the water body system. A reconnaissance tour of the watershed by the modeler, with taking photographs and notes, should always be a part of the model design, calibration and verification process.

The difficulty with achieving satisfactory calibration and verification of complex deterministic estuarine models was documented in the US EPA (1992) report. Four estuarine systems were modeled which included (1) a box dynamic model of Saginaw Bay; (2) Dynamic model of Potomac Estuary (this model was included into the WASP system and constitutes a major part of EUTRO); (3) The MIT intratidal model of Manasquan Estuary; and (4) the RECEIV[14] model of the Calcasieu River Estuary. The modeling efforts and success were then critiqued by four independent experts. In spite of the complexity of the models and enormous expenses associated with the development and execution of the models, all experts had reservations as to whether the models were satisfactorily calibrated and verified and even pointed out that some models were inappropriately and unsatisfactorily calibrated and verified.

[14] The RECEIV - II model is a dynamic deterministic water quality model that is a component of the US EPA Storm Water Management Model (SWMM). This model was developed in the 1970s and can be run independently of the other components of SWMM.

REFERENCES

Ambrose, R.B. Jr., T.A. Wool, J.L. Martin, J.P. Connolly and Schanz R.W. (1991) WASP5.x, A Hydrodynamic and Water Quality Model - Model Theory, User's Manual and Programmer's Guide, Environmental Research Laboratory, U.S. Environmental Protection Agency, Athens, GA.

Cairns, J. Jr. (1975) "Quantification of biological integrity," pp. 171 - 185 In: The Integrity of Water, L.K. Ballantine and L.J. Guerrie, eds., U.S. Environmental Protection Agency, Washington, D.C.

Callahan, M.A. *et al.* (1979) Water-related Environmental Fate of 129 Priority Pollutants, Vols. I and II. US EPA, Washington, D.C.

Halm, M.J., V. Novotny and T.W. O'Brien (1993) "A simple backward finite difference model for toxic substance response and its application on the Milwaukee River," paper pres. at the 1st Int'l. IAWQ Conf. on Contaminated Aquatic Sediments, June 14-16, 1993, Milwaukee, WI; also published in Water Science Technol..

Mills, W.B. *et al.* (1985) Water Quality Assessment: A Screening Procedure for Toxic and Conventional Pollutants in Surface and Ground Water - Part I and II, EPA/600/6-85-002a,b, US EPA, Environmental Research Lab., Athens, GA.

Novotny, V. (1994) "Evidence of designated use attainment: Priority pollutants," Proc. Surface Water Quality & Ecology, pp.575-586, 67th Annual Conference and Exp., Water Environ. Federation, Chicago, IL, October 15-19

Novotny, V. and H. Olem (1994) Water Quality: Identification, Prevention and Management of Diffuse Pollution. Van Nostrand-Reinhold Publ., New York, NY.

Novotny, V. *et al.* (1995) Identification and Evaluation of Use Attainability Methodologies for Aquatic Ecosystems. Project 91-NPS-1, Water Environment research Foundation, Alexandria, VA

Novotny, V., L. Lai Feizhou, and W.G. Wawrzyn (1994) " Monte Carlo modeling of water and sediment contamination by toxic metals at the North Avenue Dam, Milwaukee, WI, USA," paper presented at the 17th Bienial Conference of the IAWQ, Budapest, Hungary, July 24-29, 1994 also published in Wat. Sci. & Technol.

Somlyódy, L. and O. Varis (1993) " Modeling the quality of rivers and lakes," In Environmental Modeling: Computer Methods and Software for Simulating Environmental Pollution and its Adverse Effects (P. Zanetti, ed.), Vol. 1, CMP, Southampton

Streeter, H.W. and E.B. Phelps (1925) A Study of the Pollution and Natural Purification of the Ohio River," U.S. Public Health Service, Pub. Health Bull. No. 146, reprinted by U.S., DHEW, PHA, 1958.

US Environmental Protection Agency (1983) Results of the Nationwide Urban Runoff Program, Water Planning Division, Washington, DC

US Environmental Protection Agency (1986) Technical Guidance Manual for Performing Wasteload Allocations. Book VI: Design Conditions, Office of Water Regulations and Standards, Washington, D.C.

US Environmental Protection Agency (1991) Guidance for Water Quality-based Decisions: The TMDL Process, EPA 440/4-91-001, Office of Water, Washington, D.C.

US Environmental Protection Agency(1992) Technical Guidance Manual for Performing Waste Load Allocations. Book III Estuaries. Part 4 Critical Review of Coastal Embayment and Estuarine Waste Load Allocation Modeling. Office of Water, Washington, D.C.

U.S. Environmental Protection Agency (1994) Water Quality Standards Handbook: Second Edition. EPA-823-B-94-006, Office of Water, Washington, DC

CHAPTER 3

MODELS FOR RESERVOIRS, LAKES AND WETLANDS

Milan Straškraba[1]

1. INTRODUCTION

Management of reservoirs, lakes and wetlands is closely coupled with management of respective watersheds; however, some watershed/water quality problems might become serious only for watersheds that include a lake or a reservoir. The problem of phosphorus and eutrophication serves as an example inasmuch as excessive amounts of this critical nutrient creates far more serious problems in standing than in flowing waters. As far as water quality is concerned very often only one side of the problem is considered , i.e., how to remedy, how to clean our waters, how to control the pollution. The other side, the one that is much less costly and more farsighted, i.e., how to *prevent* pollution, is usually neglected.

This chapter will first outline consequences of principles of ecotechnology relevant to watershed management, then it will focus on models dealing with different alternatives relevant for lake, reservoir and wetland management. Three groups of alternatives will be distinguished: (a) those used in the watershed, (b) those used in the water body, and (c) those used at the outflows. Two management aspects have to be considered, i.e., water quantity and water quality, however, this chapter deals mainly with water quality aspects and addresses water quantity aspects only if they are closely related to water quality. Emphasis will be mainly on approaches and models relevant for *integrated watershed management*.

Also more weight will be given herein to procedures used in the watersheds, one reason being that a detailed review of reservoir water quality models and in-lake modeling methods has recently been published by the author (Straškraba, 1994). Global effectiveness of in-lake approaches will be compared with methods used within watersheds and at the outflows. Subsequently a vision for integrated management of a lake, reservoir or wetland, i.e., one considering the whole watershed will be outlined. A lack of adequate algorithms for ecotechnological - economic evaluations and other research needs will be stressed.

---

[1] Biomathematical Laboratory, Institute of Entomology, Č. Budějovice, Czech Republic

NATO ASI Series, Partnership Sub-Series, 2. Environment – Vol. 3
Remediation and Management of Degraded River Basins
Edited by V. Novotny and L. Somlyódy

The review of models in this chapter is not an in-depth treatment of the scientific background of water quality modeling; this was recently summarized in a comprehensive article by Somlyódy and Varis (1992). Instead a listing of models relevant for different problems will be provided. It has become apparent that even modelers can profit from recognizing the fact that they need not to start from scratch if faced with a task to develop and/or use a model for a particular application.

2. ECO-TECHNOLOGICAL ASPECTS OF WATERSHED MANAGEMENT

The respective water bodies are *ecosystems*, elements of nature, usually heavily affected by human activities and therefore in need of management. Ecosystems are dynamic, elastic systems with variable structural systems - in the terminology of Ruberti and Mohler (1975)- which are ruled by principles outlined in Jørgensen *et al.* (1992). This makes their modeling and management complicated. The ecosystem management methodology is based on *ecotechnological principles* (Uhlmann, 1983; Straškraba, 1984, 1993; Mitsch and Jørgensen, 1989). Ecotechnology in this sense is defined as application of technological means for ecosystem management, using knowledge of principles on which natural ecological systems are based, and transfer this knowledge into management so that the cost and damages caused to the whole environment be minimized. This definition focuses on the use of ecological theory for deriving sound environmental management directed to long-term and global sustainability. The common practice of sound use of technologies considering only local benefits may be very different from the evaluation respecting global consequences. Both long-term aspects of management alternatives for the watershed in question and for its inhabitants as well as for the broader national and world aspects of the decisions and the sustainability of the environment and satisfying instantaneous needs have to be considered. The rules presented in the next section represent an extension of Straškraba (1994).

2.1 Practical Rules for Lake, Reservoir and Wetland Management

CONSIDER RESERVOIR ECOSYSTEM DYNAMICS

The high degree of dynamics of the reservoir ecosystem can be seen, for example, when the outlet level of water is changed. If a deep impoundment is created stratification in the water body creates, in a short time, profound changes of water quality. Reproduction of fish species can change not only the population of organisms representing the food base of fishes (either by the direct effect of this fish species' feeding or indirectly through changing populations of other fish species) but also the phytoplankton and other water quality characteristics. Without respecting such kind of dynamics, the changes intended to improve water quality may instead lead to its deterioration.

Dynamic changes are both temporal and spatial. The temporal changes of water quality are well known - seasonal effects. Spatial variability is less understood but plays an important role in many environments, e.g., in lakes between bays, shores and open water, in reservoirs gradients in the direction from inflow to the dam site change in - depth distribution of different water quality variables.

A typical feature of ecosystem dynamics is the feedback effects between its components. This feature makes the ecosystem and environmental studies very difficult, as it is mostly impossible to distinguish clearly between cause and effect. The consequences of one causative variable affecting another one are again the cause for changes of the original variable. The effect of phytoplankton on the heat budget of the system and, therefore, its temperature and mixing depths and the concurrent effect of temperature and mixing on phytoplankton growth is a recently discovered yet well established example of such feedback.

RETAIN NATURAL STRUCTURES - SHORES, FORESTS, INDIVIDUAL OR GROUPS OF TREES, LANDSCAPE HETEROGENEITY

It is necessary to take steps to protect and improve spatial heterogeneity by ensuring the existence of riparian forests and a natural vegetation diversity. An important aspect of ensuring forest recovery, qualitatively and quantitatively, is planting native species of trees rather than quick growing, non-indigenous species. The riparian forest can function as a biological filter - buffer strip which can remove phosphorus and nitrogen from the inflow as well as retain sediment and suspended material thus decreasing inputs to the reservoir.

Natural shorelines function better than concrete lined or paved shores, even when shoreline erosion in wind-exposed sites might be high. Protection by vegetation (bioremedation) combined with protective wire mesh represents near-natural technology (ecotechnology). Methodology of riparian zone and habitat restoration has been extensively discussed by Novotny and Olem (1994) and by the Committee on Restoration of Aquatic Ecosystems (1994).

RETAIN BIODIVERSITY

Taking steps to maintain, protect and encourage recovery of areas of natural riparian wetlands enhances biological diversity and encourages denitrification. Wetland areas near the impoundment and within the watershed form significant buffer zones at the land/water interface that behaves as a conservation zone for native aquatic species. During reservoir construction they shelter seed species for colonizing the reservoir. The wetlands become epicenters for the recolonization of the reservoir biota.

Colonization with non-indigenous species is a dangerous practice, as the relationships to native fauna and flora are not known *a priori*. Due to the different environment into which they are introduced species behave quite differently from original localities. Many examples have proven that species favorable in one environment become noxious in other environments to which they have been introduced. Detailed studies are needed before such introduction is made.

CONSIDER RESERVOIR SENSITIVITY TO INPUTS

Reservoir design has to consider not only the inputs of water, but also the input of solar radiation which also affects evaporation. Neglecting this basic principle as it happened, for example, by the construction of the Aswan Reservoir on the Nile River at the geographical latitude where evaporation highly exceeds precipitation, lead to disasters in downstream agriculture, public health and fishery of the Nile Delta all the way to the shores of the Mediterranean Sea.

If a safe use of water for drinking purposes is to be guaranteed, all point sources of nutrient pollution must be controlled (treated and/or diverted to a section downstream from the impoundment) .

The following practices may be used to reduce inputs of sediment, nutrients and toxic chemicals from non-point sources: (i) changing agricultural practices within the watershed (crops selection; nature and manner of fertilizer application, that is, when and how the fertilizer is applied), or (ii) protection of the reservoir shore-line with vegetated buffer strips including tree species which can withstand flooding and/or floating, submergent of emergent macrophytes (Novotny and Olem, 1994).

Using headwater (inlet) settling reservoirs on tributaries is an effective technique for reducing nutrient and sediment inputs, allowing simultaneously other uses of the pre-impoundments.

SELECT MANAGEMENT ALTERNATIVES BASED ON MUTUAL INTERACTIONS BETWEEN BIOTIC COMPONENTS AND THE INTERRELATIONS OF BIOTA WITH THE SURROUNDINGS

From the ecotechnological point of view, most useful are those technologies which use nature capabilities. Procedures like wetland recovery, biomanipulation or epilimnetic mixing are not only very cheap, but also limit additional environmental deterioration due to use of energy and chemicals, extensive transportation and others. However, a drawback is the need for a high degree of knowledge to successfully use such options.

RESPECT SUSTAINABILITY OF DEVELOPMENT

A question to be asked during water quality management of an impoundment is how long the present way and degree of use can be guaranteed. The reservoir siltation with sediment is an obvious one, pointing to the need to consider measures for reducing sediment load. Less obvious is, for example, the question of the ever increasing water demand: not only immediate satisfaction of the needs of society must be guaranteed but also measures to save water for the future must be considered and incorporated in the design. The very question of reconsidering the need for construction of new reservoirs is urgent. In using any management option the question arises, how long can the supply of necessary energy sources, chemicals and so on be satisfied, and at what environmental cost.

MANAGE THE RESERVOIR AS A COMPONENT OF THE WATERSHED SYSTEM

Any change in the watershed may have considerable effects on the reservoir. Particularly negative effects result from changes in agricultural practices. A relationship between watershed activities and the reservoir is very tight not only in respect to water quality, but also to its quantity. Decreased flow due to massive irrigation has changed, for example, a very large lake, the Aral Sea in the former USSR to a dust bowl with fishery trawlers rusting in the sand.

A considerable feedback effect between the reservoir and the watershed becomes apparent particularly during reservoir construction. Complete social change of the entire territory occurs.

EVALUATE LONG-TERM EFFECTS OF ALTERNATIVES

Effects of management alternatives that are positive in the short term can lead to long-term negative effects. A typical example is the use of copper containing algicide to decrease algal blooms. Poisoning with Cu^{2+} is an effective means of control which acts almost instantaneously. However, after several applications of the algicide accumulation of Cu in sediments and its possible release to water become harmful to water use for man and animals (Hanson and Stefan, 1984). Usage of other chemicals, e.g., alum can have similar long-term negative consequences either directly endangering human health, or indirectly by producing secondary changes in water chemistry.

EVALUATE GLOBAL ENVIRONMENTAL EFFECTS OF MANAGEMENT OPTIONS

The use of heavy-force technology, e.g., heavy machinery and tons of chemicals for environmental management have to be evaluated in respect to its global environmental effects. One adverse effect is the environmental cost of the energy use during operation of the machinery and the necessary construction of access roads. More important from the environmental deterioration viewpoint may be damages due to such activities as mining the materials and chemicals used during construction and operation, to treating the materials and construction machinery. These are usually carried out in other locations which are remote from the site and, therefore, they are not obvious to the local user. However, it is meaningless to improve the environmental situation locally by deteriorating it globally.

CONFRONT CONFLICTING USES

Dynamic interactions exist not only in the natural domain, but also in the social sphere. One use of water may preclude or at least restrict another use. Not only should this be considered during the planning stage, it is also necessary to find sound compromises between conflicting incentives.

MANAGE RESERVOIR ECOSYSTEMS AS INTERCONNECTED SETS OF SUBSYSTEMS

Any management option directed to one component of the ecosystem, e.g., phytoplankton has to take into account the other components. Mass destruction of algae by chemicals can lead to poisoning of fish populations. Simultaneously, zooplankton which controls phytoplankton growth in the subsequent period may also be poisoned and/or its numbers decrease due to lack of food. Consequently, algae start to grow rapidly after the poisoning effect is over as they are not predated by zooplankton. Zooplankton development is much slower than that of phytoplankton.

The recognized predomination of indirect effects over direct ones in the aquatic ecosystems warrants consideration during management. Most evaluations of management alternatives use direct effects as a tool, without respect to indirect effects.

DETERMINE ASSIMILATIVE CAPACITY OF WATER BODIES FOR VARIOUS POLLUTANTS AND DO NOT EXCEED IT

Nature, including water bodies, is capable to assimilate a certain amount of pollutants of different kinds without harm. However, this amount is limited and, after exceeding the limits, a considerable and harmful increase of the respective concentrations will result and the ecosystem structure may break down. Moreover, the capacity for various pollutants greatly varies and depends also on the physical nature and chemistry of the environment, as well as on the ecosystem composition. It is for these reasons that the determination of the assimilation capacity of a water body is not simple. A further complication is due to the synergistic effect of various chemical components, which is very little known. Nirmalakhandan *et al.* (1994), however, found that for 50 tested chemical compounds the effects are simply additive. The development of the corresponding theory is now only in the beginning stage.

3. MATHEMATICAL MODELS FOR RESERVOIR WATER QUALITY MANAGEMENT

One way to address the above requirements for sound environmental management of water resources in general and of reservoirs in particular is to apply mathematical models of various types. The complexity of the processes and interactions within the system which an individual has difficulties to adequately comprehend is digested and translated into the language of models that enable more comprehensive applications. The models contain collective knowledge accumulated through time. However, the conclusions of any model are to be used with caution, considering the limitations of the model, possible inadequacies of its formulation and the input data gaps.

The kinds of models available for use in management can be classified according to different criteria, as shown in Table 1.

Each model can belong to several categories according to different given classification criteria. For instance a model can be a deterministic continuous simulation model or a static stochastic operational model, etc.

The following discussion describes details about the respective categories and their use in management:.

TABLE 1. Classification of models for use in management according to different criteria.

CRITERION	MODEL TYPES[a]	
MODE OF USE IN MANAGEMENT	for prediction	for direct decisions
	PRESCRIPTIVE	OPTIMIZATION
TIME DEPENDENCE	neglected	present
	STATIC	DYNAMIC
BASED ON	observation	theory
	EMPIRICAL	THEORETICAL, SIMULATION
STOCHASTIC VARIABILITY	neglected	covered
	DETERMINISTIC	STOCHASTIC
TIME HORIZON	long	short
	OFF-LINE	ON-LINE
CLASS OF DYNAMICS	kinematics	higher order
	FIXED STRUCTURE	SELFADAPTATION, SELF-ORGANIZATION
USER FRIENDLINESS	not covered	elaborated
	STANDARD	METAMODELS (EXPERT SYSTEMS, DECISION SUPPORT SYSTEMS)
INFROMATION USED	quantitative	qualitative
	STANDARD	FUZZY LOGIC-INFROMATIONAL

[a] The first row always defines the models, the second in capitals their names.

(1) Prescriptive models do not calculate directly the management options appropriate for a given situation. By means of a *scenario analysis* simulating the outcome of different management alternatives and different possible situations of the system, they can be used to indicate satisfactory management possibilities. The model WASP4 (Ambrose *et al.*, 1988) is an example in this category.

(2) Management or optimization models incorporate selection procedures for choosing the best suitable management option according to a set of criteria appropriate for the situation. Major components of a water quality management model are:

(i) A management objective which determines the critical water quality variable(s) and their critical values. The processes decisive for changes of the critical variables have to be specified and variables that can be manipulated (decision variables) have to be distinguished from those that cannot.

(ii) Goal (objective) function and constraints, to which the goal function is subjected. The most common formulation of the goal function in water quality management is minimization of costs of achieving and/or preserving a given water quality. Typically, water quality is characterized in the model by prescribed levels of certain variables such as Dissolved Oxygen or algal biomass. One basic constraint is due to the relations described by the model, others are due to physical limits and capabilities of different management options (see also the chapters by Hahn and Müller in this volume).

iii) Realization costs for each management alternative. For those alternatives that assume continuous or stepwise application the estimate of unit costs is necessary. Capital and operation/maintenance costs are distinguished or they can be combined into a Present Value, using an appropriate discount rate and amortization period (life time of the project) . The costs can also be expressed in non-monetary (intangible) values.

iv) An optimization algorithm for selecting among the various parameter combinations leading to the "optimal" solution in terms given by the goal function and constraints. Such procedure selects only among the possibilities included in the model and is limited by the validity of the model, its assumptions, formulations and the imposed constraints. Optimization can be instantaneous, made at each step independently, or dynamic, considering the future development as predicted by the model. Multiobjective formulations aim at satisfying simultaneously different objectives and multiparameter solutions select among several management alternatives.

An example of this model type is described in Kalčeva *et al.* (1982) and Rathke *et al.* (1987) which describes a multiobjective decision model.

(3) Static or empirical models are based on the black-box approach, using different kinds of analysis of the data describing a certain set of observations. Usually different statistical methods are used for their construction. The advantage over simulation models is that they are more accurate when used for analyses under the circumstances included into model construction, but may fail completely at the margin or out of the observed data range.

Typical models of this type are the Vollenweider (1968 and seq., e.g., 1975, 1976) eutrophication models.

(4) Dynamic (theoretical simulation) models are based on processes governing the problem in question and, therefore, are more general than empirical models. However, they are less precise for use in a specific situation. Optimal simulation models will be those based on parameters exactly determined from theory. This is now achievable just for physical processes where the models can be considered fixed parameter, fixed structure models. For the more complex ecological models, characterized by variable parameters and variable structure (see below) one has to rely on indirect estimates of the parameters corresponding to a particular situation. The models can be considered either *continuous*, described usually by differential equations (partial or ordinary, depending if the space structure is considered or not) or *discontinuous*, for which usually the object oriented methodology is used. The calculation for a dynamic model can be either *time dependent*, dynamic, or *steady state*, time independent.

Literature data are often used for parameters of simulation models. A comprehensive summary is included in the treatise by Jørgensen *et al.* (1991). Steady state solution of a continuous dynamic model is included, for example, in Imboden and Gächter (1978), however, most models are solved numerically as time-dependent. Discontinuous models are less frequent, they have begun to gain importance during the last decade because of their ability to cover the complex spatial structure of the system. However, no such models have been used in water quality management so far.

(5) The standard dynamic models may be called models with simple kinetics, constant model parameters. For the biotic components of aquatic ecosystems, newer developments attempt to include the changes of model parameters to reflect the adaptation of the organisms present in the system to the effect of changed inputs. These models have been called self-adaptation and self-organization (structural) models. An example is the model SELFOPT (Radtke and Straškraba, 1982), where the organisms are automatically selected by the model based on the assumption that some goal function is followed by the selection procedure (Darwinian selection). The models by Jørgensen (1986b) use thermodynamic criteria for determination of the corresponding parameter values. Self-adaptive models are those with parameters changing according to some rules in dependence on the state of the system and its environment. Self-organization models change their structure (i.e., the composition of model entities and relations between them) as a consequence of changes of the state of the system.

(6) Deterministic models use average values of parameters and neglect the stochastic variability of events in nature. Their calculation is much less time-consuming but only the average behavior of the system can be modeled.

(7) Stochastic models predict the confidence band, within which the state of the system is to be expected. For static models based on statistics it is relatively easy (with more problems for nonlinear models) to determine the confidence intervals. For dynamic models the only way to accomplish this is to repeat the simulation many times (Monte Carlo models). The type of probabilistic distribution of different parameter values is determined and random numbers generator input modified by the probability distribution of the variable creates then confidence bands of the time trajectory of the system state under various conditions. This gives more realistic representation of the expected events (assuming the range and type of distribution of parameters can be estimated with some confidence) at the cost of computer time.

Stochastic models are gaining importance in management for their ability to indicate confidence intervals of model predictions (e.g., Reckhow, 1994).

(8) Off line, long horizon prediction models. Most of the present models are of this type. They are intended for an *a priori* knowledge of different situations that can occur within the system in question under various circumstances. They are used for planning of feasible management alternatives when the system is subjected to various environmental inputs such as different pollution loads, different management strategies and others.

(9) Operational models are intended for on-line (real time), automatic controls of the system performance. Operational models are much less developed for water quality management, they represent the future. They use the feedback of information from the system. The assumption is that the model is being constantly updated based on measurements of the actual system state and short-term prediction of input values.

A summary of operational water quality modeling is written by Beck (1981).

(10) Expert systems use qualitative and quantitative expressions for guiding the user toward relevant answers to complex questions. Expert systems consist of two parts: the expert system shell and the knowledge base. The former is the software product that contains the computer code handling the knowledge base. The knowledge base covers a set of rules for a specific problem that, in the judgement of experts, serve to distinguish between different alternatives. Expert systems can cover both qualitative characteristics and complex decision rules. For application of expert systems to water quality see the summary in Henderson-Sellers (1992).

(11) Decision support systems represent an extension of expert systems. In addition to expert systems they incorporate computer software relevant for solving a decision problem, such as empirical and simulation models, data bases for parameter values, explanations of management options available and others. An integral part of a decision support system can be a graphical package generating explanatory drawings and texts. Inclusion of a Geographical Information System (GIS) is common. For a watershed a GIS can generate from input data or model results maps of distribution of different variables like pollution sources. All parts of a decision support system are driven automatically based on questions by the computer and answers by the user so that the decision process is maximally computer supported and is user friendly (Henderson-Sellers, 1992; Somlyódy and Varis, 1992).

4. MODEL CATEGORIZATION ACCORDING TO WATER QUALITY PROBLEMS AND MANAGEMENT OPTIONS FOR THEIR SOLUTION

The most common water quality problems are listed in Table 2.

TABLE 2. Common water quality problems in lakes, reservoirs and wetlands.

A. CLASSIC ORGANIC MATTER POLLUTION.

B. BACTERIAL AND VIRAL CONTAMINATION.

C. WATER-BORNE DISEASES.

D. CONTAMINATION BY NITRATES. Health problems created.

E. EUTROPHICATION: excessive organic matter production within the reservoir due to high nutrient (particularly phosphorus) inputs.

F. ACIDIFICATION: decrease of pH due to mass transfer of atmospheric gases and acid rain, associated with leaching of metals.

G. TURBIDITY problems caused by siltation.

H. SALINIZATION due to excessive fertilizer application on land or due to soil salinization in connection with irrigation in arid and semi-arid regions.

I. HEAVY METAL POLLUTION.

J. AGRO-CHEMICALS AND OTHER TOXIC CHEMICALS. Accumulation in sediments and bio-accumulation in organisms.

Some water quality problems are specific for certain types of water bodies. For reservoirs the processes resulting in specific management problems is the reservoir "aging" or "trophic upsurge". This will be dealt with separately. Topics B and C in Table 2 have more public health - medical than typical water quality problem meanings and will not be discussed in this chapter. Topics I and J are treated in this volume by Jørgensen. For certain topics there are close relationships to the subjects covered in this volume by Novotny and Capodaglio.

The models to be categorized according to the following two criteria:
a) water quality problems as listed in Table 2,
b) different management options.

The following groupings of management alternatives will be distinguished:

A) Alternatives to be implemented in the watershed. They are identical whether the watersheds are those of lakes, reservoirs or wetlands. They are not necessarily identical with management alternatives used for watersheds without standing waters; the difference is that some types of pollution can be more damaging to standing waters than to rivers. An example is the concentration of phosphorus, which is less damaging in streams than in standing waters where it causes eutrophication. In the following sections models simulating sources of the respective pollutant will be distinguished from models dealing with fate of pollutant discharge and the consequences of pollution. The reason is that this somewhat unusual approach focuses more on preventive, rather than on curative measures. However, for some models dealing with both sources and consequences, it is difficult to make this distinction and the respective model will be included under the source category.

B) Management options implemented in the body of the lake or reservoir (in-lake or in-reservoir options). These management alternatives will not be covered here in detail since a paper dealing with water quality models relevant for simulation of these alternatives has been recently published (Straškraba, 1994).

C) Management alternative implemented at the outflow(s) of a lake or a reservoir. The goal of management is to preserve adequate water quality for downstream water uses. Most alternatives in this category are used for reservoirs rather than for lakes. For drinking water intakes from lakes and reservoirs these alternatives include drinking water treatment in the corresponding plant, which will not be discussed here.

D) Management alternatives specific for wetlands. These will be directed mainly to natural wetlands. Extensive concepts of constructed wetlands have been appearing in the new journal *Ecological Engineering*.

According to the opinion of the author, insufficient attention is being paid in general and by the water agencies in particular to pollution prevention. From the water quality point of view prevention is far less expensive and therefore more important than curative methods of water purification. However, present approaches concentrate on water purification which often implies extremely costly removal of polluting substances which were previously carelessly introduced into water by man. In a factory it is often much easier and less costly to extract a specific substance from water or prevent it to be lost to water rather than to extract the substance from a mixed effluent. These substances may be valuable resources that, if diluted with large quantities of water, cannot be extracted any more and, therefore, contributes to general source emissions.

Also, restoration projects for water bodies are always very expensive. The pollutants are accumulating in the environment and the expenses of their treatment, which have to be paid anyhow, are higher than expenses for conventional treatment. If treatment is postponed to get temporal cost saving more funding will be needed later and the restoration may become very expensive. Therefore, it is always cheaper to treat and remediate as early as possible.

According to this view an attempt will be made in the following sections to distinguish systematically the models for the sources of pollution, which can estimate the priorities and possibilities of prevention, from models devoted to the fates and effects of pollution to lakes, reservoirs and wetlands. Three categories of models are distinguished:

a) sources of pollution due to human activities,

b) fates of pollutants in the environment before they reach water and their behavior in groundwater, rivers and streams, and

c) fate of pollutants within the respective lake, reservoir or wetland.

Because groundwater and stream pollution is the subject of other chapters of this book this chapter will mainly deal with items a) and c).

In the following listings the intent is to indicate models that have been or can be used to assist in identification and/or solution of different management problems. The starting year for inclusion into these tables is 1980, earlier references being given only exceptionally. The selection of 1980 is arbitrary and omits the importance of history. It is hoped that earlier models which had impact on further development would reappear in more recent papers by their authors or in other works.

5. MODELS FOR POLLUTION SOURCES

Management models of lake, reservoir and wetland water quality must consider human activities within the watershed. Such activities result in the disposal of domestic waste water, agricultural waste water, runoff of nutrients, organic and toxic compounds such as pesticides and herbicides used in agriculture and forestry and xenobiotics used as industrial catalysts.

A quantitative estimation of these effluents from the watershed is based on the time course of agricultural and industrial activity, on soil types and processes, and on the processes within the water drainage network. Point and non-point sources have to be distinguished. As the point sources are diversified and easily measurable, most models focus on non-point sources.

5.1. Reviews of Watershed Modeling

Watershed modeling is reviewed in many books, of which the following are suggested as advanced reading: Krenkel and Novotny, 1980; Novotny and Chesters, 1981; Orlob, 1983; Jørgensen, 1986a; Straškraba and Gnauck, 1985; Thomann and Mueller, 1987; Novotny and Olem, 1994. Table 3 lists some major references for models of pollution sources within watersheds.

Tables 4 to 8 and the following text list some models estimating the sources of pollutants reaching lakes, reservoirs and wetlands. The sequence of topics follows Table 2, except for an initial section which lists some general sources for watershed modeling and a section on agricultural pollution which summarizes agricultural sources of different kinds of pollutants.

5.2 Pollution Sources from Agriculture

Major sources of pollution from agriculture can be divided into point sources and non-point sources. Point sources include large farms with animal feedlots, as well as large deposits of fertilizers and organics used for plant protection (pesticides, herbicides etc.). However, often it is not easy to delineate a clear-cut border among the two categories because of a continuous transition between sizes of farms. In this case all agricultural pollution can be considered non-point source pollution.

The following categories of pollutants are delineated:

a) organic matter from animal feedlots (cows and hogs),
b) ammonia (toxic concentrations),
c) fertilizers washed to water-bodies during and after application, and
d) pesticides used for plant protection washed to water bodies.

TABLE 3. References reviewing water quality modeling in general and models estimating sources of different kinds of pollutants within the watersheds in particular. m. = model, models, mg. = modeling, w.q. = water quality.

YEAR AUTHOR	MODEL CHARACTERIZATION
1980 KRENKEL & NOVOTNY	w. q. management m.
1981 LOUCKS *et al.*	monograph on water resource systems
1981 DUNCAN & RZOSKA	land use impact on lake and reservoir ecosystems including use of m. (symposium)
1981 NOVOTNY & CHESTERS	handbook of non-point pollution
1982 CULLEN & O'LOUGHLIN	non-point source pollution in Australia
1983 ORLOB	review of w. q. mg.
1983 JOLÁNKAI	non-point pollution sources
1985 STRAŠKRABA & GNAUCK	review of mg. aquatic ecosystems
1986a JφRGENSEN	fundamentals of ecological modeling
1986 JOLÁNKAI	mg. non-point source pollution
1987 THOMANN & MUELLER	principles of surface w. q. m. (low attention to sources)
1987 HEATHWOLE *et al.*	basin-scale w. q. m.
1988 JØRGENSEN & VOLLENWEIDER	principles of lake management
1989 JØRGENSEN & GROMIEC	w. q. submodels
1990 GALAT	estimates of fluvial mass transport to reservoirs
1990 BENDORICCHIO and MALAGOLI	review of m. for urban pollution
1991 BREBBIA *et al.*	m. for water resource planning and management
1991 CORVIN & WAGGONER	TETRANS - m. of solute transport
1992 FALCONER	w. q. m.
1993 HAESTAD METHODS	m. for control of quantity and q. of land erosion
1994 NOVOTNY & OLEM	management of diffuse pollution

TABLE 4. Models for agricultural sources for various kind of pollution. ag. = agricultural, agr. = agriculture, mg. = modeling, w. q. = water quality, np. = non-point.

YEAR AUTHOR	MODEL CHARACTERIZATION
1981 TUBBS & HAITH and HAITH & TUBBS, 1981	simulation of ag. np. sources
1981 HOLÝ *et al.* and HOLÝ *et al.*, 1982	CREAMS - erosion and w. q. m.
1982 HAITH	ag. np. sources
1982 GOLUBYEV & SHVYTOV	ag. effects on water bodies
1983 MESSNER & BREZONIK	ag. nitrogen m.
1988 RITTER	review of reducing np. source pollution from agr.
1988 DeCOURSEY	symposium on w.q. mg. of ag. np. sources
1989 YOUNG *et al.* AGNPS	ag. np. pollution m.
1990 BENDORICCHIO & MALAGOLI	review of m. for ag. pollution
1994 TIM & JOLY	GIS and hydrologic/w. q. m. of np. pollution
1994 RATSEP *et al.*	ag. impact in the Northern Hemisphere

This section summarizes papers which do not distinguish among these categories; nitrogen and phosphorus as well as toxic organics will be dealt separately in the subsequent section.

Fertilizer losses during application may be considered as consisting of losses due to temporary storage of fertilizers, spreading of fertilizers from air planes, application on bare soil, application on snow and application on crops. The losses during each kind of application depend on various circumstances, i.e. on weather conditions during application (particularly important during airplane application), on soil and groundwater characteristics, on the ability of the vegetation cover in the application period to take up nutrients. In addition all kinds of losses depend on the form of fertilizers (pulverized or pellets) and on their chemical composition. Associated with fertilizer use is also increased pollution by salts that are present in fertilizers in different amounts depending on origin and chemical industry. The majority of investigations are local and only few are sufficient to be used for management models.

5.3 Organic Matter Loading

Modeling sources and fates of organic matter loading from population is the oldest type of environmental modeling and is extensively covered in the summary references in section 5.1. Therefore, only references considered important and/or interesting are listed herein.

TABLE 5. Models for organic matter (OM) non-point source pollution.

YEAR	AUTHOR	MODEL CHARACTERIZATION
1983	JOLÁNKAI	m. of OM non-point source pollution
1983	GROMIEC	sources of organic matter and its fates in streams and rivers
1988	NOVOTNY	urban runoff pollution m. - review
1990	COLE *et al.*	organic loading of New Mexico reservoirs
1993	HAESTAD METHODS	m. for sediment load of reservoirs

5.4 Nitrogen Load

Typically both major nutrients (phosphorus and nitrogen) are covered under the heading of eutrophication. However, this traditional view needs some reformulation because pollution by nitrogen forms, especially nitrites and nitrates, is more important for public health than for its impact on eutrophication. At least in the Northern temperate region with excessive amounts of nitrogen in waters, phosphorus is considered as the nutrient critical for organic matter production by blue-greens (Cyanobacteria), algae and aquatic macrophytes in most inland water bodies. The situation is different in brackish and marine waters which are mostly nitrogen limited. Nitrogen, or more exactly its ratio to phosphorus, seems to play a role mainly for the appearance of blue-green algae and not for the total amount of algae present. Therefore, at present there is no need to control nitrogen in respect to eutrophication. However, it is becoming a pollutant *per se* in contrast to phosphorus. The concentrations of phosphorus will make no primary harm *per se* and are important only in connection with the organic matter production creating pollution of water bodies.

The main sources of dissolved nitrogen in water presently are agriculture, industry and municipal sewage. The ratio of the total load from these diverse sources depends on the local situation: however, for temperate region agriculture is generally recognized as the dominant source. Of the various forms of nitrogen nitrate seem to be dominant in temperate fresh waters and most of the attention has been devoted to it. The situation is different in brackish and marine waters where nitrogen limitation is prevailing. Since Vollenweider's pioneering studies (Vollenweider, 1968) most estimates of nutrient sources are based on simple mass balance models based on unit loads from various sources and multiplication by the corresponding number of units. However, some other approaches are used as well (see Table 6).

TABLE 6. Nitrogen load models. N = nitrogen, different compounds, Nt. = nitrate, m. = models, mg. = modeling.

YEARS	AUTHOR	MODEL CHARACTERIZATION
1980	PROCHÁZKOVÁ	relation of Nt. concentration increase to mineral fertilization
1982	ZWIRNMANN	non-point sources of N from municipal water supply sources
1983	MESNER & BREZONIK	agricultural N m.
1983	HARTIGAN	reservoir mg. for Nt. standards
1983	LEE *et al.*	pollutant prediction in complex terrain
1983	DICKERHOFF ~ DELWICHE & HAITH	predicting nutrient losses from complex watersheds
1984	BOLLA & KUTAS	submodels for nutrient loading estimate
1988	JANSSON & ANDERSSON	Nt. leaching from agriculture
1989	RECKHOW *et al.*	nutrient budget m. for state-level applications
1990	WHITEHEAD	modeling Nt. from agriculture

As mentioned in section 5.1 the nitrogen forms reaching a standing water body might be modified by the processes of transformation (into other compounds), sedimentation, uptake and release by organisms and by sediments.

5.5 Phosphorus Load

Phosphorus is entering surface water bodies predominantly from two sources, (1) sewage (mainly due to the use of phosphate-containing detergents) and (2) from agriculture Most models are focusing on the latter sources (Table 7). In spite of major local differences, the former source is generally dominating in industrialized countries, which is one additional difference to nitrogen. Phosphorus is readily available to the soil-plant system. In many places the concentrations of phosphorus in water seem to be directly correlated to the amount of detergents in use. According to experience, during fertilizer application in agriculture most damaging are losses during improper storage and during airplane application under windy weather conditions. Soil erosion accompanying agricultural practices or due to other reasons is also an important source of phosphorus in water.

TABLE 7. Selected phosphorus load models. m. = model, models, mg. = modeling.

YEAR AUTHOR	MODEL CHARACTERIZATION
1980 LOEHR *et al.*	P management strategies
1980 LOGAN	role of soil in mg. non-point P-sources
1980 RECKHOW & SIMPSON	error analysis of P-loads
1983 CHOW~FRASER & DUTHIE	P-loading in dystrophic lakes
1983 JOLÁNKAI & PESTI	rainfall versus P-export rate m.
1983 LEE *et al.*	pollutant prediction in complex terrain
1983 DICKERHOFF~DELWICHE & HAITH	predicting nutrient losses from complex watersheds
1984 BOLLA & KUTAS	submodels for nutrient loading estimation
1984 NURNBERG	internal P-load in lakes with anoxic hypolimnia
1985 GROBLER & SILBERBAUER	P-export in South Africa
1986 GIORGINI & ZINGALES	m. of agricultural non-point source pollution
1986 SALAS & LIMON	empirical m. for the estimates of P-load for tropical conditions
1986 PRAIRIE & KALFF	catchment size and P-export
1987 CULLEN *et al.*	m. of non-point P-sources in Australia
1989 RECKHOW *et al.*	nutrient budget m. for state-level applications
1990 COLE *et al.*	atmospheric P-transport
1990 TOETZ	loading m. of TP in a reservoir
1990 LEVINE & JONES	GIS-based P-load m.
1992 KALLIO	m. of P transport from agricultural watersheds

Management strategies for reduction of P-load to water consist therefore primarily of:

1) A restriction on use of P - containing detergent or use of low-P detergents, substitution of P in detergents for other, less environmentally damaging compounds (attempts with silica and zeolite are not fully satisfactory, as both create other water quality difficulties).
2) Implementation of fertilizer application strategies leading to minimization of phosphorus losses.
3) Control of soil erosion and soli conservation.

5.6 Acidification

The sources of acidification were recently summarized in several good books, of which the following are recommended: Cowgill (1990), Bresser and Salomons (1990), and Jørgensen (1992). An earlier technical and political overview was by Anonymous (1982).

Modeling of acidification is relatively recent, and since the beginning there has been a concentrated effort by a team centered at the International Institute of Applied System Analysis (IIASA) in Laxenburg (Austria) that has been systematically developing a model for an integrated analysis of acidification in Europe. This is now the most comprehensive analytical tool for most aspects of acidification, starting with sources in industry through distribution by air mass movement to acid rain and its primary and secondary effects on aquatic ecosystems. Kämäri *et al.* (1984) developed the modeling methodology of surface water acidification which was later included into RAINS, Alcamo *et al.* (1985) were the first to overview the entire problem. The systematic work of IIASA in this direction (summary in the book by Alcamo *et al.*, 1990) may be considered exemplary as a teamwork development of managements tools with wide application.

Subsequent papers are less comprehensive and usually stress just one aspect. This chapter also includes the list of models analyzing the energy systems for sustainability. Such analyzes can lead to dramatic reductions of pollution (e.g., the model SESAM) if done with a sufficiently broad perspective. An example of an integral approach to emissions is Thomas *et al.* (1990). One theoretical question to be resolved for further decisions concerning resource allocation for acidification abatement was posed by Hauhs (1994) who recently questioned how far reduction of SO_2 and NO_x emissions leads to recovery, also whether focusing on reduction of SO_2 emissions over NO_x is optimal.

TABLE 8. Models on sources of acidification. a. = acidification, m. = model, mg. = modeling, mgmt. = management.

YEAR AUTHOR	MODEL CHARACTERIZATION
1981 PARK *et al.*	predicting fate of coal-derived pollutants
1983 ELIASSEN & SALTBONES, see also EMEP/CCC, 1984	m. EMEP for long-range S-transport
1985 HORDIJK, KAUPPI *et al.* 1985, 1986, ALCAMO *et al.*, 1990, HORDIJK, 1991, ALCAMO, 1992, AMANN & KLAASSEN, 1992	m. RAINS
1986 HORNBERGER *et al.*	sulphate dynamics of a region
1986 TREMBACK *et al.*	regional atmospheric m.
1988 LASCH *et al.*,	PEMU - air pollutant transport
see also SYDOW *et al.*, 1988	extension of PEMU into DSS
1988 McBEAN & OKADA	acid rain conflict resolution
1988 CHRISTIDIS	atmospheric pollutant transport
1989 BARTH	effect of land use on acidity
1990 HOLNICKI	evaluation of air quality in regional scale. Short-term forecasts for SO_2.
1990 KÄMÄRI *et al.*, also KÄMÄRI '90	regional a. m.
1990 BERGER *et al.* MARKAL-QUEBEC	m. of energy production effects
1990 ILLUM SESAM - DENMARK	m. for sustainable energy systems - economic analysis
1990 THOMAS *et al.*	integral solution of emissions
1990 DONSLUD & ERIKSEN	sources and distribution of a.
1991 SCHÖPP	critical loads for acid depositions
1992 FENGER	sources, pollutants and abatement possibilities
1992 JφRGENSEN	mgmt. and mg. of a.
1993 SOGN	chemical equilibrium equations tested
1994 FEGER	influence of soil and mgmt. practices

5.7 Turbidity

The sources of turbidity are mainly erosion processes which are highly dependent on weather, soil types and vegetation cover and therefore are strongly geographically variable. Maxima of sediment load to waters are found in semi-arid regions with fine soil, scarce vegetation and highly fluctuating precipitation. Moreover, different landscape management practices and human activities result in highly varying sediment yields. The importance of suspended sediments for river and reservoir limnology was recently recognized particularly in South Africa and Australia (Davies and Walmsley, 1984; Grobbelaar, 1991). Sediment particles are not only important as such, but also as carriers of various pollutants such as phosphorus, heavy metals and toxic organics (Oliver and Charlton, 1984 and Charlton and Oliver, 1986 - PCB's; Carignan and Nriagu, 1985 - trace metals; Baker, 1991 - organic contaminants). Mathematical models of sediment yield are listed in Somlyódy and van Straten (1986). Models for natural erosion, highway as well as stream bank erosion, channel and shoreline erosion are summarized in Novotny and Chesters (1981) and Novotny and Olem (1994).

5.8 Salinization

Due to intensive irrigation water use, particularly in more arid regions, salinization is becoming a major problem. A dramatic example mentioned previously is the Aral Sea that was turned into a bowl of dust soil after the tributary rivers had been nearly dried out for irrigation purposes. For management of salinization several models exist of which the model SWAGMAN (Salt, Water And Groundwater Management) by the CSIRO Division of Water Resources, Australia seems to be the most advanced model; it is continuously being enlarged and updated while finding additional applications (Anonymous, 1992). An earlier model of salinization management practices by Skogerboe *et al.* (1983) has not been used as widely .

6. MODELS FOR POLLUTION CONSEQUENCES FOR LAKES AND RESERVOIRS

The models focusing on estimating the fate of pollutants it the lake or reservoir are more extensive than models for pollution sources. These will be dealt with in the following sections. Before proceeding to different pollutants, models of lake/reservoir hydrodynamics will be reviewed which will help to describe the behavior of a polluting substance in the water body as if it behaved as a conservative substance, i.e., without any changes due to chemical and biological interactions. These non-conservative changes can be superimposed on the movement of water masses and their ingredients in a lake/reservoir.

6.1. Models of Lake/Reservoir Hydrodynamics

Models representing in some way the lake/reservoir hydrodynamics (Table 9) can be divided into one-, two-, three-box and multibox ones, the last category being either one-, two- or three-dimensional. The simplest single - box models consider the whole water body as a completely mixed reactor, which is the simplest possible approximation for a shallow lake or the mixing zone of a stratified reservoir/lake. For a more adequate approximation of a stratified lake two-box models distinguish the mixing zone and the hypolimnion, while three-box ones add explicitly either the transition zone, metalimnion, or the sediments. The usefulness of this distinction according to the dimensions modeled depends on the goal and objective of modeling. Only vertical, depth differences are considered in the one-dimensional representation, depth and longitudinal in two-dimensional representation and the full three-dimensional representation considers depth, length and cross-section. Three-dimensional solutions for whole lake or reservoir management do not yet exist but partial models might be useful, e.g., for spreading of heated or contaminated outfall discharges. These solutions require large computer memory size and speed. More common is the one- or two- dimensional solution, which usually in some ways takes the next missing dimension implicitly into account.

6.2 Consequences of Organic Matter Pollution

Organic matter load of lakes, reservoirs and wetlands presently includes two major sources - external load and internal load. The external load is represented by organic matter brought in by the inflows and its modeling was discussed in Section 5.3 of this chapter. The internal load is due to production of organic matter by primary producers in the water body itself from nutrients utilized mainly by algae (see Section 6.3 on eutrophication modeling). Dying or dead algae consume oxygen similarly to the dissolved and/or particulate organic matter brought in by the inflows. The BOD_5 equivalent of a unit of algal biomass was recognized to represent on an average 0.025 mg/l O_2 per ug/l chlorophyll-a (Straškrabová *et al.*, 1983). The decomposition processes of organic matter are the major driving forces for oxygen conditions in the water body, relevant for both chemical and biological processes. Certain chemical processes, among others include the binding and release of several ions important from the water quality point of view, depend on whether water is oxic, i.e., contains some dissolved oxygen, or anoxic, oxygen free. Oxygen concentration is also directly relevant for organisms; organisms not adapted to anoxia can survive only for a short time (e.g., few hours up to a few days) in water with oxygen below some tolerable limit. For example, Central European fish species are unable to live below 2 mg/l O_2 and need at least 4 mg/l O_2 for normal life, but fishes from black waters of Amazona are tolerant to anoxic conditions.

TABLE 9. List of important models of lake/reservoir hydrodynamics used in water quality management. res. = reservoir, m. = model, models, mg. = modeling. For additional models see, e.g., Straškraba (1994).

YEAR AUTHOR	MODEL CHARACTERIZATION
1978 IMBERGER *et al.*	m. DYRESM for the specified res. with much improved theory over the MARKOFSKY & HARLEMAN (1973) m.
1980 FORD & STEFAN	thermal prediction based on an energy budget
1981 SENGUPTA *et al.*	An early, oversimplified m.
1981 IMBERGER & PATTERSON	DYRESM5, 1-D m. with improved inclusion of mixing processes
1982 HARLEMAN	review of res. hydrothermics
1982 IMBERGER	DYRESM6, improved version of DYRESM5
1983 GALEGOS *et al.*	mg. hydrodynamics in agricultural res.
1983 EDINGER & BUCHAK	LARM2 a longitudinal-vertical, time-varying hydrodynamic res. m.
1983 ORLOB	review of numerical simulation m. of res. thermics
1984 BUCHAK & EDINGER	generalized longitudinal-vertical m.
1984 PATTERSON *et al.*	m. of the vertical density structure
1985 IMBERGER	review of processes of res. thermics
1986 GRAY	review of physics based res. mg.
1986 HENDERSON-SELLERS	surface energy budgets for res. mg.
1986 SPIGEL *et al.*	mg. daily changes of the mixed layer
1986 SHANAHAN & HARLEMAN	coupled hydrophysical- ecological m. for shallow lakes
1987 HENDERSON-SELLERS	review of one-dimensional stratification m.
1987 PATTERSON	m. for convective motions in res. sidearm
1988 HENDERSON-SELLERS	sensitivity of thermal stratification m.
1988 PATTERSON & HAMBLIN	mg. thermal stratification during winter ice cover
1988 HOCKING *et al.*	algorithm for selective withdrawal
1988 AMBROSE *et al.*	WASP4, a hydrodynamic and water quality simulation m.
1989 HENDERSON-SELLERS & DAVIES	review of thermal stratification m.
1993 LIVINGSTONE	one box m. with ice cover for lake oxygenation

Decomposition processes are not the sole determinants of oxygen conditions in a lake or reservoir. In the upper layer (epilimnion) reaeration through the surface and by oxygen produced during photo-synthesis are additional sources. However, in the deeper layer (hypolimnion) the situation is different. Therein the phenomenon called hypolimnetic oxygen deficit (HOD) driven mostly by decomposition processes is a decisive factor. Models of HOD listed in Table 10 are empirical, based on statistical analysis of local data. The same phenomenon, i.e., oxygen consumption by organic matter is important also during "reservoir aging" - see Section 8.1.

TABLE 10. Empirical models of the hypolimnetic oxygen deficit (HOD) and the oxygen conditions. m. = model, mg. = modeling.

YEAR AUTHOR	MODEL CHARACTERIZATION
1980 CORNETT & RIGLER	m. of HOD
1980 CHARLTON	m. of HOD
1982 KROGERUS	mg. benthic oxygen deficit
1982 VOLLENWEIDER & JANUS	M. of HOD
1984 CORNETT & RIGLER	HOD
1986 WALKER & SNODGRASS	m. for sediment oxygen demand
1987 CROSS & SUMMERFELT	oxygen demand m.
1987 STAUFER	effect of water transport on HOD
1991 CHAPRA & CANALE	m. of oxygen and P in stratified lakes
1992 MOLOT *et al.*	m. of end-of-summer oxygen profiles

Oxygen conditions in a water body can be modeled in a more complex way by several simulation lake/reservoir water quality (eutrophication) models listed in Table 11, specifically by Baldasano *et al.* (1981), in the model MORDOR by Martin *et al.* (1985) and Martin and Wlosinski (1985), Scott *et al.* (1985), Livingstone (1993), Stefan and Fang (1994a, 1994b) and Hamilton and Schladow (in print).

6.3 Consequences of Contamination by Nitrates

Nitrate is the dominant nitrogen compound in most waters in industrialized countries, the other nitrogen compounds being organic nitrogen, ammonia and nitrites. Depending on conditions, the proportion between these compounds varies, as they undergo physical, chemical and biological changes in water. Sedimentation of N containing particles of both biotic and abiotic origin, chemical transformations between compounds mostly due to the activity of bacteria and biological uptake and release mechanisms, are processes usually included into models. An early investigation of the necessary degree of complexity of the nitrogen model able to follow experimental data is in Harleman (1978). A careful review of experimental nitrogen models was performed by Leonov (1980) and Leonov and Toth (1981). The experiences from experiments were used in the construction of mathematical models of the behavior of nitrogen in lakes and reservoirs.

Nitrites can arise from nitrate reduction in anoxic environments, including, e.g., contaminated ground water, lake/reservoir anoxic hypolimnia, human intestines with unstabilized bacterial flora (e.g., in intestines of babies), or as an intermediate product of nitrification. They are toxic, especially to organisms with haemoglobin. In the presence of nitrogenous organic compounds, nitrites can become precursors of carcinogenic nitrosamines. Ammonia concentrations higher than approximately 250 mg/m^3 are chronically toxic to fish and invertebrates at pH $\geq$ 9. Consequently, ammonia is considered as a controlled toxic substance. Ammonia is effectively retained by soil and in water. It is preferably used by phytoplankton.

6.4 Models of Lake and Reservoir Water Quality and Eutrophication

As pointed out previously two major sources of organic matter cause water quality deterioration: organic matter from external inputs and organic matter produced within the water body as a consequence of eutrophication (primary productivity). In developed countries the former source of organic matter is usually, to a certain degree, under control and is technically easily remedied. Eutrophication remains a more complex and difficult problem to remedy.

Organic matter production in water bodies is affected by many natural variables, some of which may be manageable by technical means (like the mixing depth of a waterbody, its transparency and fish populations), as well as by variables that are at present dominantly connected with human activities. The loads of a water body by the critical nutrient belong mainly in the last category . The earlier controversy about which nutrients are critical for aquatic production initiated mainly by the soap industries has been resolved. It is now generally recognized that micronutrients play a role solely in some geologically unbalanced regions with little human effects. From macronutrients phosphorus is generally considered as critical (due to nitrogen loads created by human activities which mostly exceed saturation levels). Carbon can become limiting only in situations where saturation of the other two macronutrients mentioned and the atmospheric supply or supply from decomposition processes are low (Boers and van der Molen, 1993). However, the ratio of nitrogen to phosphorus is also significant (Smith, 1982) and nitrogen becomes limiting during some periods and for some lakes. In the tropics, both phosphorus and nitrogen have been reported as limiting nutrient (Ryder and Rast, 1989; Melack and McIntyre, 1991).

In respect to management three possible groups of water quality models can be distinguished:

a) *Vollenweider type empirical relations* that are elaborated mainly for two purposes:

i) relation of the phosphorus concentration in the lake/reservoir to the phosphorus load. Additional variables considered include hydraulic load and average or mixing depth of the water body.

ii) relation of chlorophyll-a (as a measure of the phytoplankton biomass) to phosphorus concentration. These models are widely used in management and have been extensively made stochastic by different authors.

Vollenweider type models have been summarized in several treatises: OECD (1982), Janus and Vollenweider (1984), Chapra and Reckhow (1983), Walker (1986 and his extensive reports reported therein). There is also an excellent recent summary by Rast and Ryding (1989).

There is one major inadequacy of the latter models, which, however, plays a role only at high concentrations of phosphorus: it is both theoretically and practically clear, that the concentration of phytoplankton biomass cannot grow indefinitely with the increasing concentration of phosphorus; there is necessarily a saturation effect which should be included.

Empirical models of the phosphorus - chlorophyll relationships taking into account the saturation effect and its important consequences for management have been advocated by Straškraba (1976b, 1978, 1985) and have been broadly supported by recent papers (Prairie *et al.*, 1989; McCauley *et al.*, 1989). Rather than being adequately approximated by a hyperbola the relationship between the algal biomass (measured as CHA) and phosphorus concentration is in the total range of phosphorus concentrations found best described by a sigmoid. The management consequence is that above some critical concentration (around 50 $mg.m^{-3}$ PO_4-P and 100 $mg.m^{-3}$ TP) a decrease of the phosphorus concentration does not produce a proportional response of algae. When P concentrations are fairly high as observed in many rivers in highly populated regions, a reduction of P-concentration to one half or even more may have no effect, and either a very high reduction is necessary or other means for eutrophication abatement have to be sought.

Different local sets of data result in different parameters of the empirical relationships, which is an expression of the fact that additional variables not included into the model that are different among the data sets affect the relationship. Among these important variables are the mixing depth of the given water body as well as its transparency due to non-living (=non-algal) matter. Another important factor is the effect of the composition of grazers on algal biomass, which is now known to create up to two-fold differences (see Section 8.2).

b) *Simplified generalized dynamic models* - e.g., the AQUAMOD series included in Table 11.

c) *Detailed eutrophication (lake ecology) models*, usually coupled with hydrodynamic models: QUAL (+ versions), WASP4, DYRESM_WQ (most models in Table 11). Two basic groups of lakes have to be distinguished: 1) shallow lakes, which are not stratified, and stratified, deep lakes. In the text the first group is mentioned wherever appropriate.

Dynamic eutrophication modeling is extensive and has been reviewed many times: Jørgensen (1983a, 1983b, 1986a), Straškraba and Gnauck (1985), Rossi (1991), Straškraba (1994) for reservoirs.

An important methodical issue in the class b) and c) is the model validation. Two approaches are used:

a) validation of forecasts, and

b) validation of trends.

TABLE 11. Review of dynamic models of lake/reservoir water quality. e. = eutrophication, m. = model, models, mg. = modeling, res. = reservoir, w. q. = water quality, mgmt. = management.

YEAR AUTHOR	MODEL CHARACTERIZATION
1979 PARK *et al.*	MS CLEANER dynamic simulation m. of the res. ecosystem
1980 DVOŘÁKOVÁ & KOZERSKI	AQUAMOD3 three-layer m., extending AQUAMOD2 of STRAŠKRABA (1979) by the sediment-water exchange
1980 RADTKE & STRAŠKRABA	SELFOPT, m. with automatic selection of algae of different size
1980 JøRGENSEN	review of lake and res. mgmt. and the use of m.
1980 LEONOV & LEONOV 1982 BALSECT	mg. P transformations in the shallow L. Balaton, model
1980 LOS, see also LOS *et al.* 1982	first attempt to simulate w. q. mgmt. measures in lakes
1982 LIJKLEMA & HIELTJES	dynamic phosphate budget model
1982 STEFAN & CARDONI	RESQUAL II - shallow res. w. q. m.
1982 BENNDORF & RECKNAGEL	SALMO - application to res.
1982 SCHINDLER & STRAŠKRABA	GIRL OLGA, dynamic optimization m. for e. abatement
1982 KRENKEL & FRENCH	state-of-the-art of res. w. q. mg.
1983 FEDRA	Monte Carlo uncertainty estimate of the model by IMBODEN & GÄCHTER, 1978
1983a JøRGENSEN	review of e. m.
1983 SULLIVAN *et al.*	notebook of processes for aquatic ecosystem simulations
1983 FERRARA & GRIFFIN (1984)	trophic state simulation in res. also GRIFFIN & FERRARA
1983 FEDRA	Monte-Carlo simulation with the m. by Imboden & Gächter
1983 COLLINS & WLOSINSKI, see manual)	CE-QUAL-R1, res. w. q. m. also ANONYMUS, 1986 (user
1985 RILEY & STEFAN, see also HANSON *et al.* ,1986 and RILEY & STEFAN, 1988	MINLAKE, an extension of the lake res. e. m. RESQUAL II
1985 MARTIN & WLOSINSKI	comparison of 1-D and 2-D m.
1985 HOLME *et al.*	simulating run-of-river res.
1985 STRAŠKRABA & GNAUCK	AQUAMOD1, AQUAMOD2, AQUAMOD3 and a systematic treatment of aquatic ecosystem m. and w. q. mgmt. m.
1985 MARTIN *et al.*, see also MARTIN & WLOSINSKI, 1985	MORDOR, m. of oxygen conditions
1985 SCOTT *et al.*	res. oxygen m.
1986b VAN STRATEN	identification, uncertainty assessment and prediction in lake e.
1986a JøRGENSEN *et al.*	generality of e. m.
1986 WALKER	m. FLUX, PROFILE and BATHTUB for e. assessment
1986 VIRTANEN *et al.*	3-D w. q. m.
1986 SOMLYÓDY & VAN STRATEN	shallow lake e. mg. and mgmt.
1986a VAN STRATEN SIMBAL	the most simple Balaton m.

TABLE 11.cont.

YEAR AUTHOR	MODEL CHARACTERIZATION
1986 KUTAS & HERODEK	BEM, a Balaton m. more complex than SIMBAL
1987 LOEHR	m. of hydrodynamics and primary production of an impounded river
1987 PARSONS *et al.*	3-D water mgmt. m.
1988 RILEY & STEFAN	further development of MINLAKE
1988 MARTIN	2-D m. application
1988 AMBROSE *et al.*	WASP4 hydrodynamic and w. q. simulation m.
1988 SOMLYÓDY & WETS	stochastic optimization e. m.
1988 LOS & BRINKMAN	BLOOM, shallow lake e. m.
1988 POSTMA	DELWAG, m. of physical transport processes
1988 VARIS	e. m. with N-fixing cyanobacteria
1989 KMET & STRAŠKRABA	GIRL, a general model of an aquatic ecosystem
1989 LEONOV	simulation m. of w. q. in the Ivankovo Reservoir
1990a and b JANSE & ALDENBERG, see also VAN LIERE & JANSE, 1992	PCLOOS, e. m. for shallow Loosdrecht lakes
1990 GNAUCK *et al.*	use of modelling language SONCHES for aquatic ecosystem mg.
1990 LAVRIK *et al.*	multichamber m. of res. w. q.
1990 POURCHER & SALENCON	plankton mg.
1991 LOS	BLOOMII, extension of BLOOM
1991 JANSEN & ALDENBERG see also JANSE *et al.* 1992	PCLOOS, e. m. for shallow Loosdrecht lakes
1991 HOSOMI *et al.*	e. control using a m.
1992 HAMILTON	DYRESM-WQ, extension of the 1-D DYRESM (see Table 9) for w. q.
1992 SOMLYÓDY & VARIS	thorough review of lake w. q. m.
1993 JANSE & VAN DER VLUGT	m. of e. and its control measures
1993 JANSE *et al.*	PCLAKE, extension of PCLOOS for e. control of shallow lakes
1993 LIVINGSTONE	oxygen m. for lakes with ice cover
1994 VAN DER MOLEN *et al.*	DELWAG-BLOOM-SWITCH for mg. of e. contol of shallow lakes
1994a,b REICHERT	AQUASIM, programing system for w. q. simulations.
1994 STEFAN & FANG, also STEFAN & FANG, 1994a	oxygen m.
in print HAMILTON & SCHLADOW	mg. oxygen in an Australian res.

The first approach elaborated by Jφrgensen (1983b) is based on making a forecast based on available data or a part of it and comparing the output of the resulting model with independent observations after the forecasted changes were realized. The successful forecast made for Lake Glumso by Jφrgensen *et al.* (1986b) has demonstrated that eutrophication models can be used to set up management plans.

The second approach advocated by Scavia and Chapra (1977) and Straškraba (1979) is based on running the model under various conditions and comparing the resulting behavior of model output based on the changes of input variables with the behavior (or its trend) derived from direct observations of different water bodies in nature. Already Thomann (1982) called for comparing the dynamic and empirical model results.

6.5 Models for Responses of Streams and Lakes to Acidification

The response of lakes and reservoirs to acidification is complex and includes physico-chemical changes in the composition of lake water and reaction of organisms to the low pH. Chemical changes are due to the pH dependence of a number of chemical reactions, which mostly take place already in groundwater and in streams, before entering the lake. Important are changes of solubility of various metals (Fe, Cr) and especially of aluminum (Al). The compounds which are harmless to organisms under normal (near neutral) pH become toxic at lower values of pH. As a consequence, organisms in the lake, in addition to being subject to intolerably low values of pH are also toxified by aluminum compounds. Many algae, zooplankton crustaceans and most fish species disappear from acidified lakes, only few which are tolerant to low pH remain. The diversity of acidified lakes and the numbers of organisms are low and water is clear because it is deprived of algae and other particles.

The selection of models listed in Table 12. covers the topic, ranging from chemical effects through algal growth up to fish distribution. Most models are exploratory, only a few are applicable for solution of broader questions important for management. Such models are also mentioned in Table 8.

Management options are mostly confined to one single procedure - liming of the respective water bodies and/or watersheds (Olem, 1991). First it is to be noted that this is typically an unsustainable curative procedure - the amounts of lime used on a continuous basis are very high and cannot be supplied indefinitely.

Recently an idea was advocated that addition of nutrients or sludge (Organic Matter) to acidified lakes may have a fairly positive effect. Creating artificial eutrophication increases pH

TABLE 12. List of some important models for lake and stream acidification, its consequences and management. a. = acidification, ad. = acidified, m. = model, models.

YEAR AUTHOR	MODEL CHARACTERIZATION
1981 KERCHER *et al.*	m. for estimating effects of air pollution on vegetation
1981 VESELKA *et al.*	m. vegetation yield reduction by a.
1981 BROWN & SADLER	chemistry and fishery of a. lakes
1983 HELLIWELL *et al.*	speciation and toxicity of aluminum
1983 HARWELL & WEINSTEIN	m. of the effect of air pollution on forest ecosystems
1984 CHESTER	ecological effects of deposited sulphur and nitrogen
1984 CHEN *et al.*	integrated lake - watershed a. m.
1984 LUNG see also LUNG, 1987	simplified lake a. m.
1985 COSBY *et al.*	MAGIC, m. for watershed a.
1985 GHERINI *et al.*	ILWAS, Integrated Lake-Watershed a. study model
1985 SCHNOOR	M. of acid precipitation impact
1985 SVERDRUP & WARFVINGE	lake reacidification m.
1986 SVERDRUP *et al.*	prediction of recovery of reacidified lakes
1986 HORNBERGER *et al.*	effects of acid deposition
1986 NEAL *et al.*	effect of acidic deposition and conifer afforestation on streams
1986 LIU & SCHNOOR	a. of seepage lakes
1987 REUSS *et al.*	comparison of m. REUS-JOHNSON, BIRKENES, MAGIC and ILWAS
1987 McBean & ELLIS	screening m. for acid rain mgmt. strategies
1987 KELLY *et al.*	acid neutralization in lakes
1987 RECKHOW *et al.*	fish response to a.
1987 RHEA & YOUNG	proton binding in lake sediments
1988 BAKER & BREZONIK	in-lake alkalinity generation
1988 LUNG *et al.*	sulphate and alkalinity in ad. lakes
1988 LAM *et al.*	regional analysis of watershed a.
1988 HOOPER *et al.*	BIRKENES - m. of stream a.
1988 BOBBA & LAM	hydrological m. for ad. watersheds
1988 JENSEN	effect of acidity on mercury uptake
1988 NIKOLAIDIS *et al.*	soft water a.
1989 NIKOLAIDIS *et al.*	lake alkalinity responses to acid depositions
1989 ANONYMOUS	ground water and surface water a.
1989 SIEGFRIED & SUTHERLAND	zooplankton biomass in ad. lakes
1990 ROSE *et al.*	comparison of ILWAS, MAGIC and ETD a. m.
1990 KÄMÄRI	impact m. to assess regional a.
1990 KÄMÄRI *et al.*	regional a. m.
1991 WYLIE & JONES	sensitivity of reservoirs to a.
1993 POSCH *et al.*	m. for critical loads of sulfur and nitrogen
1994 COSBY *et al.*	recovery of freshwater ecosystems

and reverts some of the unfavorable chemical processes which has a positive feedback for support of aquatic life. In regions with more eutrophic conditions, acid rains with the same acidity as in the acidified regions resulted in no observable acidification. This is mainly due to the buffering capacity of waters in more calcareous and/or agriculturally intensively cultivated regions. Fertilizers not only contain the necessary nutrient minerals, but are bound to salts which increase the buffer capacities of the respective waters. The same is true for groundwater and surface streams. However, this methodology may not be applicable in low buffer regions, where only the lake will be supplied with some nutrients, while the entering streams, groundwater and rain water will have a low buffer content and will be low pH waters. Therefore, to use this idea as a management strategy will need a more comprehensive analysis.

6.6 Turbidity Problems Caused By Siltation

Particularly serious problems are caused by turbidity and siltation in reservoirs, due to their rapid filling. The life time of some reservoirs built on turbid rivers is anticipated to be only about 50 years or less, and many earlier reservoirs have already ceased to function. Turbid water is not suitable for drinking, and the amount of sediment (sludge) created during drinking water treatment increases significantly the cost of treatment.

In eutrophic lakes and reservoirs, turbidity has also a positive effect by decreasing the light available for phytoplankton and in this way decreasing algal production and biomass. Mineral or mixed mineral - soil particles interfere with zooplankton feeding and provide adsorption sites for phosphorus.

To simulate the effects of turbidity, models include processes of sedimentation, the effect of suspended solids on light extinction for the algal population, relations between zooplankton feeding and particles, as well as the physical and chemical processes of phosphorus binding and availability of bound phosphorus to algal uptake. Also, some soil particles have a high oxygen consumption capability, so that the lake and/or reservoir can rapidly switch to anoxia, with negative consequences for water treatment (smells, odors, increased iron and manganese). Sedimentation is a complex process which depends not only on the size and specific weight of particles but also to a significant degree, on the water body hydrodynamics. There are no specific models for management of turbidity. However, lake/reservoir hydrodynamic models listed in Table 9 can help find solution of related specific questions. Several models of particle sedimentation also exist (e.g., Galvez and Niell, 1993) but only one, derived from DYRESM, is combining a hydrodynamic model with processes of sedimentation (Schladow and Fischer, in print).

Management of turbid inflows is accomplished in some impoundments by manipulation of the outlets as to create short-cut currents (undercurrents - Straškraba, 1986). The result is a rapid passage of turbid water through the reservoir, without allowing much sedimentation and additional mixing of turbid water with the surface, productive layers. Turbid inflows are usually also rich in nutrients. Associated with in-lake phosphorus precipitation is enhanced sedimentation connected with flocculation. The Wahnbach-type plant located on the inflow (section 7.4) reduces the inflow turbidity, which counteracts to some degree the positive effect of a reduced phosphorus load on decreasing algal biomass by creating improved light conditions for algal growth.

6.7 Water Quality Consequences of Salinization

Salinization is not a specific problem of lakes and reservoirs, but rather of ground waters and streams. It is due mainly to soil salinization in connection with irrigation in arid and semi-arid regions. However, excessive fertilizer application in developed countries seems to also lead to a continuously increasing salt content. As an example, in the Czech Republic the trend of increasing nitrate concentrations that seem to be mainly of agricultural origin is accompanied by a similar steady rise of total salts. Salty inflows change water density and therefore the flow and mixing conditions in lakes and/or reservoirs. Because the effects of salinity on water density are well described, it is not difficult to include this effect into any density driven hydrodynamic model. Explicit inclusion of salinity exists in the model DYRESM, which is due to its origin in semi-arid and arid conditions where salinity effects play a major role.

7. WATERSHED MANAGEMENT ALTERNATIVES

This section contains discussions on various management options to be performed in the watersheds (Table 13), some of which have been enumerated in Section 6 in connection with various types of pollutants.

7.1 Cleaner Production Processes

The most important management alternative that is, however, mostly out of reach of water management agencies, can be summarized as "*Cleaner Production.*" It consists of changing the processes inside the production plant in order to reduce pollution. The major benefit is for the producer: in addition to saving on fees for creating pollution there is considerable saving of energy, water and various materials used in the production process. As an example, the introduction of the cleaner production approach in fifty galvanizing plants in The Netherlands resulted during the first year in a drop of pollution to 55% and in the second year, after gaining more experience, to 37% of the original pollution. In two large enterprises in the Czech Republic the primary savings by introducing cleaner production techniques amounted to 50 million Czech Crowns (about 2 million $), an amount worth consideration by the producers. Therefore, by focusing on this technology the activity shifts to the production area. The interest is created primarily by considerable money savings; perhaps some good feeling that the environment was also improved may also play a role. No mathematical models in this area are known, but surely many models of the production processes in various plants do exist and can help to discover the possibilities. However, the best "model" which water agencies can use is to teach industry to use this kind of approach and take over the initiative for pollution abatement.

TABLE 13. Management options in the watershed.

OPTION	TYPE OF POLLUTION
CLEANER PRODUCTION	All types of pollution
SEWAGE PURIFICATION PLANTS	Organic matter (OM)
TERTIARY TREATMENT	P - elimination
PREIMPOUNDMENTS	P - and turbidity reduction
WAHNBACH PLANT	P, OM and turbidity
AGRICULTURAL PRACTICES	prevention of losses of N and P

A still wider approach analyses the whole production - consumption cycle of a product as to minimize waste. This usually exceeds the possibilities of one producer and demands synchronization of efforts by several enterprises. The environment is also benefiting from such efforts by the progressive and environmentally conscious representatives of industry, who in this way also increase their competitive capabilities.

7.2 Sewage and Wastewater Treatment

See Chapters by Henze and Grau.

7.3 Pre-impoundments (Headwater - Inlet Reservoirs)

Lakes and reservoirs are effective traps for phosphorus and this ability is used successfully to decrease the P-load by constructing small pre-impoundments (including ponds) at the reservoir inflows. A number of empirical models based on local data exist for relating the P-retention of a lake to its hydraulic and P load, mean depth and oxygen conditions at the bottom (Table 14). However, most data are for deep lakes; for small shallow reservoirs the conditions may be different.

There are few such empirical retention models for reservoirs (Turner *et al.*, 1983), however, it is not clear whether reservoirs behave differently in this respect from lakes. A model for organic matter retention (and production) in reservoirs was elaborated by Straškrabová (1976), see also the study by Groeger and Kimmel (1984).

More advanced prediction models for the dimensioning of preimpoundments to arrive at optimum P retention were developed by the group Uhlmann - Benndorf at Technical University Dresden in Germany (Uhlmann *et al.*, 1971; Benndorf, 1973; Benndorf *et al.*, 1975; Uhlmann *et al.*, 1977) which were also successfully applied in South Africa (Twinch and Grobler, 1986).

7.4 Wahnbach Plant

Bernhardt (1967) calculated the contribution of point and non-point sources in the watershed of the Wahnbach Reservoir, a drinking water supply reservoir for the industrial region surrounding Bonn (Germany), and concluded that elimination of all point sources will not decrease the phosphorus concentration at the inflow to the reservoir to such a degree, that eutrophication is prevented. Moreover, it would be very costly to construct tertiary treatment plants at all point sources. He arrived at the conclusion that the best way to remedy the problem was to instal the P-elimination plant directly at the inflow to the reservoir. This is now called the Wahnbach procedure - an advanced plant reducing concentrations of orthophosphate (reactive soluble) P by about 92%, total phosphorus by more than 96%. Simultaneously, turbidity and organic matter is reduced. The resulting decrease of the chlorophyll-a content is 95% (data given for 7 year averages).

TABLE 14. Phosphorus retention of lakes and reservoirs (from Straškraba and Gnauck, 1985, modified). π - theoretical hydraulic retention time [yr], z_{av} - mean depth [m], RP_{TP} - retention of total phosphorus [fraction], RP_{OP} - same for inorganic (orthophosphate) phosphorus, RT - theoretical retention time [days].

1975 KIRCHNER & DILLON ($r^2 = 0.884$)

$RP_{TP} = 0.426*\exp(-0.271\ z_{av} / \pi) + 0.574*\exp(-0.00949\ z_{av} / \pi)$

1975 CHAPRA ($r^2 = 0.884$)

$RP_{TP} = 16 / (16 + z_{av} / \pi)$

1976 LARSEN & MERCIER

$RP_{TP} = 1 / (1 + \sqrt{(1 / \pi)})$ ($r^2 = 0.884$)

$RP_{TP} = 11.73 / (11.73 + z_{av} / \pi)$ ($r^2 = 0.865$)

1978 OSTROFSKY

$RP_{TP} = 0.201*\exp(-0.0425 + z_{av} / \pi) + 0.547*\exp(-0.00949 + z_{av} / \pi)$

1979 RECKHOW, all lakes ($r^2 = 0.885$)

$RP_{TP} = 1 - 1 / (\exp(0.0025 * z_{av} / \pi) + (0.35*\pi + 0.11\ z_{av}^2) * \exp(-0.05\ z_{av}^2 / \pi) + \pi / z_{av} * \exp(-100\ L\ \pi / z_{av})))$

lakes with $z_{av} / \pi < 50$ m/yr and oxic hypolimnion ($r^2 = 0.876$)

$RP_{TP} = 1 - (1 / (18\ z_{av} / (10 + z_{av}) + 1.05\ z_{av} / \pi * \exp(0.012 * z_{av} / \pi))$

lakes with $z_{av} / \pi > 50$ m/yr and oxic hypolimnion ($r^2 = 0.949$)

$RP_{TP} = 1 - 1 / (2.77 + 1.05\ z_{av} / \pi * \exp(0.0011\ z_{av} / \pi))$

lakes with anoxic hypolimnion ($r^2 = 0.948$)

$RP_{TP} = 1 - (1 / (0.17 * \pi + 1.13))$

1993 STRAŠKRABA *et al.*, preimpoundments in Germany, data by Wilhelmus *et al.* (1978)

$RP_{OP} = 85 (1 - \exp(-0.0807\ RT))$

1993 DILLON & EVANS, comparison of models

1995 STRAŠKRABA *et al.*, reservoirs in Czech Republic and world as given in particular in TURNER *et al.* 1983

$RP_{TP} = 76.1 (1 - \exp(-0.0282\ RT)$

as significantly differentiated from lakes

$RP_{TP} = 66 (1 - \exp(-0.00419)$

Note the coefficients of determination, indicating that there is some spreading of phosphorus retention for the same conditions.

No specific model was constructed for the whole plant, but individual processes were quantified in the papers by Bernhardt and Schell (1979, 1993).

7.5 Agricultural Practices

The following agricultural practices are useful for reducing washout of fertilizers from fields:

a) avoidance of fertilization and erosive agricultural practices in a buffer strip around the water body,

b) restriction of the use of nitrogen fertilizers to application rates not exceeding 100 $kg.ha^{-1}$ farmland. $year^{-1}$,

c) using distributed dosage of fertilizers primarily during the period of most rapid growth,

d) using slowly dissolving forms of fertilizers, e.g., in pellets,

e) leaving natural organic matter residues plowed in the fields, to slow nitrate elution

f) keeping the time during which fields are left without vegetation cover after harvest as short as possible, and to use close grown crops, and

g) restricting application of fertilizers onto frozen soil or to fields where nothing has been planted.

Protective buffer strips of grassland and forest-brush prevent surface soil particles from washing down to streams and lakes/ reservoirs. However, they are unable to reduce leaching of nutrients from distant, higher elevation places.

An earlier review on the effect of agricultural practices on water quality is included in Schaller and Bailey (1983).

8. MODELS FOR IN-LAKE AND IN-RESERVOIR MANAGEMENT ALTERNATIVES

Before proceeding to models for evaluation of individual in-lake alternatives a specific problem of reservoirs called reservoir aging or trophic upsurge will be defined. Subsequently, models dealing with this process will be discussed.

8.1 Reservoir Aging

According to Purcell (1939) reservoir aging is a process of rapid changes during the first 10 years after first filling of the reservoir. Aging, based on the full meaning of the term, is a continuous process not restricted to the first years of existence: the reservoir is filling with sediments, is getting shallower, more macrophytic vegetation appears, etc. The changes during the reservoir aging in the strict sense are characterized by increased organic production; for these reasons the process was termed "trophic upsurge" by Ostrofsky (1978). Water quality deterioration exhibited by increased manganese and iron concentrations is characteristic for this period, particularly if low oxygen conditions occur in the hypolimnion, . The period of "aging" is followed by less dramatic changes, sometimes representing a quasi-equilibrium situation. Although detailed data are not available, the actual length of this period seems to depend on the retention time of the reservoir - it appears shorter in flow-through reservoirs. The standard explanations are based on the processes of decay of organic matter from the submerged soils and vegetation, and on the increased nutrient input from submerged soils which are usually disturbed (see the summary by Kimmel and Groeger, 1986). Straškraba *et al.* (1990, 1993) introduced a hypothesis of the role of biotic interrelations during the differential development of short living plankton and long living fish populations.

In spite of the importance of this phenomenon for management of newly build reservoirs, there are just a few models of this phenomenon and they are of partial importance only. Most comprehensive are the models by Ostrofsky (1978), Ostrofsky and Duthie (1978), Grimard and Jones (1982) and Duthie and Ostrofsky (1982) which focus on phosphorus budget as the major cause of the upsurge. There are many laboratory investigations on the decay of organic matter from vegetation and soils, expressed quantitatively as the oxygen consumption rates (Davis *et al.*, 1973; Gunnison *et al.*, 1980, 1983; Krogerus, 1988; Crawford and Rosenberg, 1984; Mouchet, 1984; James *et al.*, 1988). Attempts to use this approach for reservoir water quality prediction are by Kozerski (1975), Küchler and Kozerski (1975), Thérien and Spiller (1981), Thérien *et al.* (1982) and Pereira *et al.* (1992). Brandl (1973) expressed quantitatively the dependence of changes in zooplankton biomass during the first years of a new reservoir on the depth of the reservoir (sampling point) and retention time.

8.2 Models of Lake and Reservoirs Dynamics

Mathematical models for different water quality management techniques within the body of the - reservoir are listed in Table 15.

TABLE 15. Models for in-lake water quality management.

MEASURE	MEANS	REFERENCE
HYDRAULIC REGULATION	Selective withdrawal	STRAŠKRABA, 1986
BIOMANIPULATION (FISH MANAGEMENT)	Zooplankton control - - phytoplankton reduction	GULATI *et al.*, 1990 JANSE *et al.* 1995
ARTIFICIAL MIXING	1. Destratification	SYMONS *et al.*, 1987, HENDERSON-SELLERS, 1981, ZIC & STEFAN, 1988 WÜEST *et al.*, 1992, SCHLADOW, 1992, 1993
	2. Hypolimnetic aeration	BERNHARDT, 1967
	3. Epilimnetic mixing	STRAŠKRABA, 1986
	4. Layer or metalimnetic mixing	KORTMAN *et al.*, 1994
PHOSPHORUS	1. Alum precipitation	COOKE & KENNEDY, 1988, COOKE *et al.*, 1993, WELCH & SCHRIEVE, 1994
INACTIVATION	2. Iron precipitation	BOERS *et al.*, 1992
	3. Sediment covering	PETERSON, 1980
SEDIMENT AERATION	Sediment injection	RIPL, 1976, 1980
SEDIMENT REMOVAL		HANSON & STEFAN, 1985
LIGHT REDUCTION	Shading, covering, suspensions, colors	JφRGENSEN, 1980

9. MODELS FOR QUALITY MANAGEMENT OF RESERVOIR OUTFLOW WATER

Water stored in reservoirs may be withdrawn from the downstream river after release from the reservoir or it may be taken directly from the reservoir. It is desirable to have multiple outlets at different depths for withdrawals for drinking water. The quality of the outflowing water is directly related to the horizontal and vertical distribution of water of good quality within the reservoir and the flexibility of operation of the multiple outlets. During large water releases, the outflowing water changes its quality because the outflow extracts water from other layers or sometimes a shortcut current comes directly from the inflow. Major downstream damages are due to anoxic waters, waters high in methane, ammonium and others, which produce corrosion of outlets, turbines and other steel and concrete structures, prevent the use of water for drinking and other purposes or even cause organism kills. A review of improved methods of reservoir releases is given in Bohac *et al.* (1985). Most of them propose to oxygenate the outflowing water. An approach to optimizing reservoir releases to minimize downstream damages is presented by Sale *et al.* (1982).

Table 16. lists some eco-technological methods of managing reservoir outflow quality.

TABLE 16. Ecotechnological techniques for management of reservoir outflows.

TECHNIQUE	REFERENCE
SELECTIVE WITHDRAWAL	FONTANE *et al.*, 1981, GAILLARD, 1984, PAŘÍZEK, 1984, CASSIDY, 1989, FILHO *et al.*, 1990
AERATION/OXYGENATION AT HYDROPOWER OUTLET WORKS	CASSIDY, 1989
SPILL-WATER REAERATION	CASSIDY, 1989
CZECH METHOD OF OXYGENATION	HAINDL, 1973,
EPILIMNETIC PUMPS	QUINTERO & GARTON, 1973, GARTON *et al.*, 1978, MOBLEY & HARSHBARGER, 1987

With reservoir water being released directly to the river, problems may arise if the water is released from the hypolimnion of an eutrophic reservoir because such water may be deoxygenated; contain large concentrations of organic compounds, iron and manganese; and have a high phosphorus content. The consequence may be large scale fish kills, eutrophication of the river accompanied by odors and taste problems of such magnitude that the water is not treatable for drinking supply. Several management options of re-oxygenation are available such as spill water oxygenation and the use of hydraulic means such as spillways with supercritical flow and hydraulic jump.

10. MODELS FOR PREDICTION OF THE EFFECT OF AN IMPOUNDMENT ON WATER QUALITY OF ITS OUTFLOW

Several management options of re-oxygenation are available such as spill water oxygenation and the use of hydraulic spillways. Mathematical models to predict the downstream effects are listed in Table 17.

TABLE 17. Mathematical models for quantitative prediction ofreservoir effects upon outflow water quality. m. = model, mg. = modeling.

MODEL	REFERENCE	FEATURES OF MODEL
RESTEMP	STRAŠKRABA & GNAUCK, 1985	empirical m. to predict outflow temperature from temperate reservoirs based on their elevation, retention and outlet depth
DYRESM	IMBERGER *et al.*, 1978	m. to predict outflow temperature by means of a hydrodynamic reservoir m. with retention greater than 10 days.
RESERVOIR SEDIMENTATION	GALVEZ & NIELL,'93	Empirical m. to predict sedimentation rate in temperate reservoirs
PHOSPHORUS RETENTION IN RESERVOIRS	NEWBOLD, 1987 BENNDORF *et al.*, 1975 STRAŠKRABA *et al.* in press	The m. by Newbold on changes in phosphorus concentrations due to reservoir effects, with consequences for primary production. The m. by Benndorf for P capture in preimpoundments. The m. by Straškraba for P retention in reservoirs.
RESERVOIR OUTFLOW ORGANIC MATTER	STRAŠKRABOVÁ, 1976	The m. on the effect of a temperate reservoir on the BOD of the outflowing river with different conditions of retention times depth and phytoplankton production level.
RESERVOIR RELEASES	HORSTMAN *et al.*, 1983	mg. hypolimnetic discharge scenarios
	FONTANE *et al.*, 1981	optimal control by selective withdrawal

11. WETLAND MANAGEMENT MODELS

Wetland modeling is a difficult subject because a wetland often includes diversified terrestrial and aquatic components, ranging from meadows and cultivated fields to forests and swamps, from shallow open waters with important benthic components to regions overgrown by emergent, submerged and floating vegetation, all tightly interwoven. Earlier reviews of research concerning natural wetlands are summarized in Logofet and Luckyanov (1982) and Gopal *et al.*, 1982. Several recent books are summarizing the scientific backgrounds of wetlands and also some modeling activities: Mitsch and Gosselink (1986 and a recent updated version by Mitsch, 1994), Mitsch *et al.* (1988), Patten *et al.* (1990, 1994). After the positive functions of natural wetlands on water quality were identified, attempts were initiated to optimize these functions by creating constructed (highly technical) wetlands for water purification. These two types, i.e.; natural and constructed wetlands, have to be strictly distinguished. The recovery of a previously existing wetland or creation of a (near-)natural wetland should be driven by rules of natural wetlands.

Earlier modeling started with individual components of wetlands, like modeling aquatic plant growth and decomposition, modeling tree growth in wetland conditions, oxygen conditions in overgrown water bodies, wetland hydrology and other. There are too many such component models to be enumerated herein and they would be of limited utility in integrated watershed modeling. Table 18. lists models for natural freshwater wetlands directed more to their integral representation and in particular to their water quality functions. No attention will be given to artificial wetlands, simply because the author is not familiar enough with their functioning and modeling. A detailed discussion and description of these systems is to be found in Mitsch and Jϕrgensen (1989) and in the journal *Ecological Engineering* (Elsevier since 1992). A useful summary of the use of constructed wetlands for water quality improvement is Moshiri (1993).

12. INTEGRATED (REAL TIME) MANAGEMENT OF A LAKE, RESERVOIR OR WETLAND

We have been stressing in this paper the interrelations between water quality within the watershed, lake, reservoir or lacustrine wetland, and the outflow water quality. The latter are most important for reservoirs being newly constructed, causing downstream water quality changes, which might

TABLE 18. A selection of models for natural freshwater wetlands. m. = models, mg. = modeling, wtl. = wetlands

YEAR	AUTHOR	MODEL CHARACTERIZATION
1982	AUBLE *et* al.	Hierarchical m. for wtl.research
1982	DAY *et al.*	Research and management m.
1982	MITSCH *et al.*	mg. of N. American wtl.
1983	MITSCH	ecological mg. for wtl. management
1983	MITSCH *et al.*	wtl. classification, mg. and management
1983	TAYLOR & CROWDER	accumulation of atmospherically deposited metals
1987	RABE & CLEMENS	phosphorus removal capability
1988	MITSCH *et al.*	state-of-the-art of wtl. m.
1988	STRAŠKRABA *et al.*	overview of wtl. m.
1988	JφRGENSEN	detailed review of wtl. m.
1988	JφRGENSEN *et al.*	modeling nutrient retention
1988a	MITSCH	ecological engineering and ecotechnology with wtl.
1988b	MITSCH	productivity - hydrology - nutrient m. of forested wtl.
1990	FINN	impact assessment m.
1990	KUENZLER	wtl. as sediment and nutrient traps
1990	MITSCH & JØRGENSEN	wtl. mg. and management
1990	ONDOK	mg. ecological processes in wtl.
1990	SKLAR *et al.*	conceptualization of complex wtl. m.
1991	VIAROLI *et al.*	accumulation of nutrients and heavy metals in wtl.
1992	DORGE	nitrogen transformations in wtl.
1993	VIAROLI	regeneration and loss of nitrogen

be negative for the users of the downstream river. However, water quality changes in the lake also create potential dangers to downstream water uses. An outflow from a wetland will be very important if the wetland is used for decreasing water contamination downstream. Therefore, integrated management has to focus simultaneously on all three components named herein: watershed, water body, and outflow. Mathematical models for water resource systems, ecotechnological and optimization models are listed in Straškraba (1994). A review of modeling needs for managing freshwater resources is given in SASR (1992). Table 19 summarizes the approaches and tools for integrated watershed management.

TABLE 19. Different modeling approaches and tools for integrated watershed management. w. q. = water quality, DSS = decision support, mgmt. = management, m. = model, models, w. = water.

YEAR	AUTHOR	APPROACH
1981	MATSUMURA & YOSHIUKI	dynamic formulation of w.q. optimization
1983	THORNTON *et al.*	Monte-Carlo approaches
1983a	BECK & VAN STRATEN	uncertainty and forecasting
1983b	BECK & VAN STRATEN	m. for water mgmt.
1984	JENQ	linear programing m. for lake e. mgmt. strategies
1985	RECKHOW	decision theory applied to lake mgmt.
1986	TERUGGI *et al.*	WQLEDGER - integrative software for stream pollution
1986	TERUGGI & VENDEGNA	FOREL - mitigation measures of environmental impact
1986	JOHNSON	DDS for w. resource mgmt.
1986	DeSIMONI *et al.*	software for planning of a multi-purpose river basin
1987	ANTUNES *et al.*, also CAMARA *et al.*, 1987	qualitative simulation of w. resource systems
1987	VINCENT	modeling and mgmt. in the presence of uncertainties
1987	ELLIS	stochastic water quality optimization
1987	SEIP *et al.*	multiattribute analysis
1987	BECK	analysis of uncertainty for w. q. m.
1987	WENGER & RONG	fuzzy set m. in watershed mgmt.
1987	RATHKE *et al.*	DDS REH for multiobjective optimization in w. q. mgmt.
1987	GROBLER *et al.*	DDS to assist e. control strategy of lakes
1988	SCHWABL	DSS for environmental planning
1988	VARIS *et al.*	DAVID, system for the method of influence diagrams
1988	BRAUN *et al.*	computer aided decisions in w. q. mgmt
1989	KADEN *et al.*	DSS for water mgmt.
1989	RECKNAGEL	scenario analysis with the m. SALMO (see Tab. 11)
1989	BAKULE & STRAŠKRABA	new formulations for modeling internal and external w. q. control
1989	LAIKARI	river basin mgmt. - symposium
1989	VIRTANEN	overview of m. for river basin

TABLE 19. cont.

YEAR	AUTHOR	APPROACH
1989	GNAUCK *et al.*, see also MODEL *et al.*, 1995	REH - multicriterial decision system applied to w. q. mgmt.
1990a	FEDRA	DSS combined with GIS
1990b	FEDRA	DSS for environmental decisions
1990	PAGE	analysis of environmental expert systems
1991	SEIP	decisions with multiple environmental constraints
1991	DAVIS *et al.*	DSS for analyzing impact of catchment policies
1991	DAVIS & CLARK	expert systems in natural resource mgmt.
1991	DUCKSTEIN & SHRESTHA	multicriterion risk and reliability analysis
1991	SEIXAS *et al.*	mg. structural changes in environmental systems
1991	LOUCK & daCOSTA	review of DDS for w. resources planning
1991	ULRIK *et al.*	MASAS, DDS for micropolutant evaluation
1991	RECKNAGEL *et al.*	DELAQUA for optimal lake w. q. control
1991	SCHREINER *et al.*	IMES, DDS for selection, validation and uncertainty assessment of a number of w. q. m.
1991	GANOULIS	w. resources engineering risk assessment
1991	OKADA *et al.*	m. for basin-wide w. q. mgmt.
1991	ANDREU *et al.*	AQUATOOL, DSS for w. resources mgmt.
1991	BRAAT *et al.*	EXPECT, integrated m. for environmental scenario analysis and impact assessment
1991	FEDRA *et al.*, see also FEDRA 1991	MEXSES, rule based expert system for environmental impact assessment
1992	HENDERSON-SELLERS	Decision Support System prototypes for environmental problems
1992	JANSSEN *et al.*	UNCSAM 1.1. - software for sensitivity and uncertainty analysis
1992	SOMLYÓDY & VARIS	review of DDS for w. q. mg.
1993	GEORGAKAKOS & YAO, see also YAO & GEORGAKAKOS, 1993	new control concepts for uncertain systems
1993	BOUZAHER *et al.*	metamodels for non-point pollution policy
1993	SUMMERS *et al.*	quantifying uncertainties of w. q. m.
1993	JOKIEL *et. al.*	w. q. mg. using GIS
1993	DeNIJS *et al.*	WATNAT, extension of PCLOOS/PCLAKE for waterways and for toxic component
1993	DAVIS	expert systems in environmental modeling
1993	HENDERSON-SELLERS *et al.*	review of modern tools for w. q. mgmt.
1993	HAESTAD METHODS	HEC-3 m. for multipurpose operation of reservoir systems
1994	MUNDE	fuzzy information in multicriteria m.
1994	DePINTO *et al.*	GEO-WAMS, GIS-based watershed analysis and mg. system, includes WASP4

TABLE 19. cont.

YEAR	AUTHOR	APPROACH
1994	KULARATHNA & SOMLYÓDY	review of river basin w. q. m.
1994	SOMLYÓDY *et al.*	evaluation of different w. q. mgmt. strategies for a watershed
1994	VARIS	Review of approaches to w. q. mg.

13. RESEARCH NEEDS FOR ECOTECHNOLOGICAL WATER QUALITY MODELS

Mathematical models are now extensively used in water quality management. Some ideas about the extent of their use can be obtained from the survey by Alasaarela *et al.* (1993) who assembled questionnaires from 100 institutions. They reported that 105 models (both produced in the surveyed institutes or elsewhere) were applied to 800 cases with about 500 person work years being used. Modeling teams consisted on the average from 4 persons.

We can distinguish several directions of further development of ecotechnology as concerns water quality and of the corresponding ecotechnological models:

a) ecological (limnological)
b) economic and technological
c) methodological

The ecological knowledge of aquatic ecosystems is still restricted on one hand to local empirical knowledge of the basic interrelations between the environmental effects and some physical, chemical and biological variables and on the other hand to fairly crude knowledge of processes involved. Often the empirical interrelations are very specific for the set of data analyzed and they mostly are univariate. Differences between various water bodies in respect to other variables are not considered. On the other hand the character of water quality in aquatic ecosystems is multivariate: several variables determine water quality simultaneously and they operate in different water bodies in different proportions.

Therefore, simple empirical relations provide for any situation just a broad range of possibilities and only crude estimates are possible. For reservoirs (man-made impoundments) the additional difficulty is that many limnological investigations do not distinguish between natural and man made impoundments, which neglects the systematic differences reservoirs have in certain respects. The consequences of differences in shape, the surface outflows in lakes and deep outflows in most reservoirs, shorter retention times etc. are often neglected.

The knowledge of processes decisive for water quality changes in water bodies is mainly derived from experiments on components taken out of the ecosystem context. The consequence is that within the ecosystem the process may behave differently due to variables not considered in the experiment and different behavior of the cultivated and wild populations. The extremely rapid capabilities of organisms for acclimation and adaptation are neglected and their consequences are unknown. More knowledge is needed on the ecosystem self organization capability and on self structuring of aquatic ecosystems under changing conditions. Moreover, the multivariate character of the processes and the synergetic effects of variables are difficult to study and therefore inadequately known.

The economic considerations are inadequate to cope with the present situation of diminishing resources. It is now recognized that the present pricing system needs profound changes to reflect the full price of resources and of the environment covering the needs of future generations. Nevertheless, the development of a corresponding economic value system is slow and the acceptance of its need by the world society still slower. Technology is driven by the current value system, environmental costs are not respected and the notion of unlimited possibilities still dominates the engineering world. "Ecological economics" is under development (see the special journal "*Ecological Economics*" that appeared recently). However, apparently the economic valuation itself is not the full remedy, as it is impossible to give an objective estimate of the future values of, for example, some minerals as these highly depend on the (unknown) future technology. Besides the "cleaner production" approach mentioned herein previously, we need to develop other similar philosophies. New market economy drives as those of transferable pollution discharge permits (e.g., Novotny, 1988) have to be developed. Also, we have to focus more on what is now emphasized as the "participation principle", i.e., including discussions on various decisions relevant for the status of the environment all responsible parties, i.e., representatives of the dischargers, of the responsible environmental agencies, of interested public (see the summary document of the Rio summit on the Environment - Agenda 21).

From the methodological point of view there are many inadequacies not only in management model formulations but also in model solutions. This is particularly valid for optimization problems, where the numerical approaches are cumbersome and biased. As to the form of models best suited for management, we need to combine different approaches. We also need effective means for model combination, without a need for reformulation of the original models. The transition to automatized real time operational management is still difficult due to inadequacy of both the specific models and mathematical means.

Abbott *et al.* (cited according to SAST, 1992) distinguish from hydraulics modeling point of view (including water quality aspects) five generations of models briefly characterized as follows:

1st generation: computerized formulae of the early 1960s;

2nd generation: one-off numerical models for one specific site, not generally applicable (since 1960's);

3rd generation: generalized numerical modeling system for computer and numerical method experienced specialists, usually mainframe based but also running on PC's,

4th generation: user friendly software products, PC or work station based, widely applicable by professionals,

5th generation: "intelligent" modeling systems for technically skilled but non-expert users.

Most water quality models listed in this chapter are in the third generation category, only some being of the fourth generation. The water quality model systems of the fifth generation are under intensive development, but are still at an experimental stage as seen from Table 19.

14. CONCLUSIONS

1) The first principle to be stressed herein is that far more attention has to be given to preventive, rather than to curative methods that were until now in the center of interest of water management agencies. Many models can provide advice as to where the preventive actions can be most effective.

2) Sustainability and the ability of today's models to cope with the rules of ecotechnology suggested in this paper need to be incorporated into management models to direct managers in the right direction.

3) One major bias exists with all existing models: they have been created largely in the temperate regions and sufficient information about their functioning in tropical conditions is not available. An example of such uncertainties is the assumption often expressed in literature that tropical soils have a different ratio between the adsorption capacity for phosphorus and nitrogen and that algal production in tropical waters should be limited by nitrogen rather than by phosphorus. However, several recent investigations did not support this conclusion. The estimation of the effect of temperature on various aquatic processes is also uncertain. For the temperate region more or less well verified empirical models have been developed. However, organisms are adapting to prevailing environmental conditions, including temperature. Certain observations show that the temperature reaction of the organism is geographically dependent. Therefore, models developed in the temperate region are most likely not applicable to other regions.

4) Lakes, reservoirs and wetlands as other natural (or near-natural) ecosystems are adaptive systems, undergoing self-organization, i.e.; both quantitative and qualitative changes of structure and function. Neglecting this capability in our management practices and models leads to their failures, sometimes with enormous negative consequences for the environment, economy and humanity. As an example, the introduction of some aquatic plants intended for water quality improvement resulted in their unwarranted spreading, making them pests in the new environment. Management options directed to combat eutrophication by decreasing some noxious algae sometimes lead to changes in the structure of the algal community such that other algae began to cause water quality difficulties.

5) The existence of models useful for complex management of water quality is uneven. Most developed are the models for pollution sources and empirical and dynamic models of water quality in lakes subject to eutrophication. The reactions to pollution loads of reservoirs as opposed of lakes are less understood. Recent evidence indicates significant differences between lakes and reservoirs, particularly reservoirs with shorter retention times. However, refinement of these models is needed as the confidence intervals of our predictions are quite broad. This is connected with the multiaxial character of the ecosystems: many variables are determining their

reactions simultaneously. The present empirical models concentrate just on one or two variables, neglecting the rest. Dynamic models cover a broader range of influences but the response of the ecosystem is multilateral.

Models of acidification are also well-developed although that its effects on lake processes are not yet sufficiently generalized.

Least developed are the models for wetlands. This is due to their high complexity, with varying proportions and interactions between water, sediments, aquatic plants and terrestrial components. Heterogeneity of wetlands is much higher than that of lakes and reservoirs. The reactions of individual aquatic plant species are more diversified than those of lake algae, and geographic differences in aquatic flora are high.

The speed of model development does not cope with the rate of humankind producing new types of pollutants. Models for the effect of heavy metals and toxic organics are an example.

As to modeling water quality management options, we have relevant models for different mixing techniques and for phosphorus precipitation, but modeling of biomanipulation and of the sediment - related techniques needs further applicable development.

6) To make integrated watershed management modeling more effective, we need to change our approaches. We are now in a period of need for more concentrated team efforts devoted to modeling of particular management problems. As the listing of models in this paper proves, environmental modeling is, with exceptions, still mostly in the science domain, is individualistic, focused on new ideas and new independent developments and belonging mostly to the second and third generation of models for water quality management. A weak point of the modeling methodology is the lack of developed validation techniques. In these respects, development of new approaches and theories is urgently needed. The examples of the few third/fourth generation models successfully used for integrated watershed management (models like RAINS, WASP4, QUAL) clearly document that such wide applicability can only be achieved by a concentrated effort of a team of outstanding and highly specialized researchers-consultants. The teams have to be selected for a predetermined topical area, capable to cover different aspects of the problem, starting from problem identification through elaboration of holistic submodels and organization

of necessary databases to assembly of user friendly Decision Support Systems. A team of researchers-consultants hired by an agency for some purpose does not yet guarantee success. The conditions for a successful effort are: i) selection of team members has to be made from an array of creative researchers-consultants with extensive experiences in their respective fields, and ii) an unifying, centripetal force (be it internal conviction, a good team manager or strong economic motivation) exists.

7) The best available modeling approach for integrated management is represented by the fifth water quality model generation realized mostly as advanced Decision Support Systems. Teams like those assembled at the International Institute for Applied System Analysis (Laxenburg, Austria) or teams similar to those assembled in the past by the U.S. Environmental Protection Agency should take upon the task of developing such comprehensive models.

ACKNOWLEDGEMENT

Preparation of this paper was supported by a grant from the Grant Agency of the Czech Republic NO 204/04/1672. The views presented in this chapters are solely those of the author.

REFERENCES

Ahlgren, I., Frisk T., Kamp-Nielsen L. (1988) "Empirical and theoretical models of phosphorus loading, retention and concentrations lake trophic state," Hydrobiologia 170: 285-303

Alasaarela, E., Virtanen M., Koponen J. (1993) "The Bothnian Bay project - past, present and future," Aqua Fennica 23: 117-124

Alcamo, J. (1992) Transboundary air pollution in Central Europe and Eastern Europe. In: Alcamo J. (ed.) Coping with Crisis in Eastern Europe's Environment. Cromwell Press Ltd. Broughton Gifford Wiltshire UK, pp.87-101

Alcamo, J., Hordijk L., Kämäri J., Kauppi P., Posch M., Runcas E. (1985) "Integrated analysis of acidification in Europe." J. Environ. Manage. 21: 47-61

Alcamo, J., Shaw R., Hordijk L. (1990) The RAINS Model of Acidification. Kluwer and International Institute for Applied Systems Analysis Dordrecht

Aldenberg , T., Janse J.H., Kramer P.R.G. (1994 in press) "Fitting the dynamic model PCLake to a multi-Lake survey through bayesian statistics," Ecol. Modelling:

Amann, M., Klaassen G. (1992) Optimized Reductions of Nitrogen Emissions in Europe. IIASA SR-92-7, International Institute for Applied Systems Analysis Laxenburg Austria

Ambrose, R.B.Jr., Wool P.E.T.A., Connolly J.P., Schanz R.W. (1988) WASP4, A Hydrodynamic and Water Quality Model - Model Theory, User's Manual, and Programmer's Guide. Environmental Research Laboratory Office of Research and Development U.S. Environmental Protection Agency Athens Georgia

Andreu, J., Capilla J., Sanchis E. (1991) AQUATOOL, a computer assisted support system for water resources research management including conjuctive use. In: Loucks D.P., Da Costa J.R. (ed.) Decision Support Systems: Water Resources Planning. NATO ASI Series Vol. G26, Springer Berlin

Anonymous (1982) Acidification Today and Tomorrow. Swedish Ministry of Agriculture Uddevalla

Anonymous (1986) CE-QUAL-R1: A Numerical One-dimensional Model of Reservoir Water quality. User's Manual/Instruction. US Army Corps of Engineers Environmental Laboratory Waterways Experiment Station, Report E-82-1

Anonymous (1989) Models for Analysis of Groundwater and Surface Water Acidification. Publications from the Swedish Environmental Protection Agency

Anonymous (1992) SWAGMAN gets smart. Waterlink 11: 6

Anonymous (1993) The Rio Declaration on environment and development. The Global Partnership for Environment and Development. A Guide to Agenda 21. Post Rio Edition United Nations New York, pp.3-9

Antunes, P., Seixas J., Camara A., Pinhero M. (1987) "A new method for qualitative simulation of water resources systems. 2. Applications," Water Resour. Res. 23: 2019-2022

Auble, G.T., Patten B.C., Bosserman R.W., Hamilton D.B. (1982) A hierarchical model to organize integrated research on the Okefenokee Swamp. In: Logofet D.O., Luckyanov N.K. (ed.) Ecosystem Dynamics in Freshwater Wetlands and Shallow Water Bodies, Vol. 2. Centre of International Projects GKNT Moscow pp.203-217

Baker, R.A. (1991) Organic Substances and Sediments in Water. Volume II. Processes and Analytical. Lewis Publishers Inc., Chelsea, MI

Baker, L.A., Brezonik P.L. (1988) "Dynamic model of in-lake alkalinity generation," Water Resour. Res. 24: 65-74

Bakule, L., Straškraba M. (1989) "Modelling internal and external control in lake and reservoir ecosystems," Arch. Hydrobiol., Beih. Ergebn. Limnol. 33: 213-214

Baldasano, J.M., de Broissia M., Coupal B. (1981) Simulation of dissolved oxygen concentration in a reservoir. In: Thérien N. (ed.) Simulating the Environmental Impact of a Large Hydroelectr-ic Project. The society for computer simulation. La Jolla California, pp.45-53

Barth, H. (1989) Effects of Land Use in Catchments on the Acidity and Ecology of Natural Surface Waters. Air Pollution Research Report 13, CEC Brussels

Beck, M.B. (1981) Operational Water Quality Management: Beyond Planning and Design. International Institute for Applied Systems Analysis Laxenburg Austria

Beck, M.B. (1987) "Water quality modelling: a review of the analysis of uncertainty," Water Resour. Res. 23: 1393-1442

Beck, M.B., van Straten G. (1983a) Uncertainty and Forecasting of Water Quality. Springer Berlin

Beck, M.B., van Straten G. (1983b) Water Management Models. Elsevier Amsterdam The Netherlands

Bendoricchio, G., Malagoli M. (1990) "Overview on non point pollution models: experiences and applications," In: Fenhann J., Larsen H., Mackenzie G.A., Rasmussen B. (ed.) Environemntal Models: Emissions and Consequences/Riso International Conference 22-25 May 1989. Elsevier Amsterdam The Netherlands, pp.345-357

Benndorf, J. (1973) "Prognose des Stoffhaushaltes von Staugewässern mit Hilfe kontinuierlicher und semikontinuierlicher biologischer Modelle," Int. Revue ges. Hydrobiol. 58: 1-18

Benndorf, J., Recknagel F. (1982) "Problems of application of the ecological model SALMO to lakes and reservoirs having various trophic status," Ecol. Modelling 17: 129-145

Benndorf, J., Zesch M., Wiesner E.M. (1975) "Prognose der Phytoplanktonentwicklung in geplanten Talsperren durch Kombination von wachstumskinerischen Modellvorstellungen und Analogiebetrachtungen zu bestehenden Talsperren." Int. Revue ges. Hydrobiol. 60: 737-758

Berger, C., Lessard E., Loulou R., Waaub J.-P. (1990) "Implementation of the MARKAL-Quebec energy model with emission factors, abatement technologies, and SO2 constraints." In: Fenhann J., Larsen H., Mackenzie G.A., Rasmussen B. (ed.) Environemntal Models: Emissions and Consequences/Riso International Conference 22-25 May 1989. Elsevier Amsterdam The Netherlands, pp.215-227

Bernhardt, H. (1967) "Aeration of Wahnbach Reservoir without changing the temperature profile," J. Amer. Water Works Assoc. 59: 943-964

Bernhardt, H., Schell H. (1979) "The technical concept of phosphorus-elimination at the Wahnbach estuary using floc-filtration (The Wahnbach system)," Z. f. Wasser- und Abwasser- Forschung 12: 78-88

Bernhardt, H., Schell H. (1993) "Effects of energy input during orthokinetic aggregation on the filterability of generated flocs," Wat. Sci. Tech. 27: 35-65

Bobba, A.G., Lam D.C.L. (1988) "Application of a hydrological model to the acidified Turkey Lakes watershed," Can. J. Fish. Aquat. Sci. 45 (Suppl. 1): 81-87

Boers, P.C.M., van der Does J., Quaak M., van der Vlugt J., Walker P. (1992) "Fixation of phosphorus in lake sediments using iron(III) chloride: experiences, expectations," Hydrobiologia 233: 211-212

Boers, P.C.M., van der Molen D.T. (1993) "Control of eutrophication in lakes: the state of the art in Europe," Eur. Wat. Pollut. Control 3: 19-25

Bohac, Ch.E., Shane R.M., Harshbarger D., Goranflo H.M. (1985) "Recent progress on improving reservoir releases," Lake Reserv. Manage. 2: 187-190

Bolla, M., Kutas T. (1984) "Submodels for the nutrient loading estimation on River Zala," Ecol. Modelling 26: 115-143

Bouzaher, A., Lakshminarayan P.G., Cabe R., Carriquiry A., Gassman P.W., Shogren J.F. (1993) "Metamodels and nonpoint pollution policy in agriculture," Water Resour. Res. 29: 1579-1587

Braat, L.C., Bakema A.H., de Boer K.F., Kok R.M., Meijers R., van Minnen J.G. (1991) EXPECT. Outline of an Integrated Model for Environmental Scenario Analysis and Impact Assessment. National Institute of Public Health and Environmental Protection Bilthoven, Report Nr. 259102001

Brandl, Z. (1973) "Relation between the amount of net zooplankton and the depth of station in the shallow Lipno Reservoir," In: Hrbáček J., Straškraba M. (ed.) Hydrobiological Studies 3. Academia Praha, pp.7-51

Braun, P., Rudolph P., Albrecht K.-F. (1988) "Computer aided decision for water quality management," In: Sydow A., Tzafestas S.G., Vichnevetsky R. (ed.) Mathematical Research. Systems analysis and simulation. Bd.46 Vol.1. Akademie Verlag Berlin, pp.75-78

Brebbia, C.A., Ouazar D., Ben Sari D. (1991) Computer Methods in Water Resources II. Vol. 3. Computer Aided Engineering of Water Resources. Computational Mechanics Publications Southampton

Bresser, T., Salomons W. (1990) Acidic Precipitation. Vol. 5. International Overview and Assessment. Vol. 5. Springer Berlin

Brown, D.J.A., Salder K. (1981) "The chemistry and fishery status of acid lakes in Norway and their relationship to European sulfur emission," J.Appl. Ecol. 18: 434-441

Buchak, E.M., Edinger J.E. (1984) Generalized, longitudinal - vertical hydrodynamics and transport: Development, programming and applications. Doc. No 84-18-R. Waterways Experiment Station Vicksburg Mississippi

Carignan, R., Nriagu J.O. (1985) "Trace metal deposition and mobility in the sediments of two lakes near Sudbury, Ontario," Geochim Cosmochim. Acta 49: 1753-1764

Cassidy, R.A. (1989) "Water temperature, dissolved oxygen, and turbidity control in reservoir releases,". In: J.A. Gore, G.E. Petts (ed.) Alternatives in Regulated River Management. CRC Press Boca Raton Florida, pp.27-62

Camara, A., Pinhero M., Antunes P., Seixas J. (1987) "A new method for qualitative simulation of water resources systems," Vol. 1. Theory. Water Resour. Res. 23: 2015-2018

Chapra, S.C. (1975) Comment on "An empirical metod of estimating the retentiv of phosporus in lakes" by W.B.Kirchner and D.J. Dillon. Water Resour. Res. 2: 1033-1034

Chapra, S.C., Canale R.P. (1991) "Long-term phenological model of phosphorus and oxygen for stratified lakes," Water Res. 25: 707-715

Chapra, S.C., Reckhow K.H. (1983) Engineering Approaches to Lake Management. Mechanistic Modeling. Butterworths Boston

Charlton, M.N. (1980) "Hypolimnion oxygen consumption in lakes: discussion of productivity and morphometric effects," Can. J. Fish. Aquat. Sci. 37: 1531-1539

Charlton , M.N., Oliver B.G. (1986) "Chlorinated organic contaminants on suspended sediment in Lake St. Clair," Water Poll. Res. J. Canada 21: 380-389

Chen, C.W., Gherini S.A., Dean J.D., Hudson R.J.M., Goldstein R.A. (1984) Development and calibration of the integrated lake-watershed acidification study model. Modeling of Total Acid Precipitation Impacts, pp.175-203

Chester, P.F. (1984) "Ecological effects of deposited sulphur and nitrogen compounds - general discussion," Phil. Trans. Roy. Soc. London B 305: 564-566

Chow-Fraser, P., Duthie H.C. (1983) "An interpretation of phosphorus loadings in dystrophic lakes," Arch. Hydrobiol. 97: 109-121

Christidis, Z.D. (1988) "Modeling of atmospheric pollutant transport and deposition in lakeshore enviroments," In: Zannetti P. (ed.) Computer Techniques in Environmental Studies. Springer Great Britain, pp.373-388

Cole, J.J., Caraco N.F., Likens G.E. (1990) "Short-range atmospheric transport - a significant source of phosphorus to an oligotrophic lake," Limnol. Oceanogr. 35: 1231-1237

Cole, R.A., Ward T.J., Bolton S.A. (1990) "Estimating intermittent runoff concentrations of organic matter and the allochthonous organic loading of New Mexico Reservoirs," Lake Reserv. Manage. 6: 187-196

Collins, C.D., Wlosinski J.H. (1983) Verification of the reservoir quality model, CE-QUAL-R1, using daily flux rates. Lake and Reservoir Management. U.S. EPA Washington, pp.206-212

Committee on Restoration of Aquatic Ecosystems (1994) Restoration of Aquatic Ecosystems. National Academy Press, Washington, DC

Cooke, G.D., Kennedy R.H. (1988) Water quality management for reservoirs and tailwaters. Report 2. In-reservoir water quality management techniques. Technical Report E-88-X. U.S. Army Engineer Waterways Experiment Station Vicksburg Mississippi

Cooke, G.D., Welch E.B., Martin A.B., Fulmer D.G., Hyde J.B., Schrieve G.D. (1993) "Effectiveness of Al, Ca, and Fe salts for control of internal phosphorus loading in shallow and deep lakes," Hydrobiologia 253: 323-335

Cornett ,R.J., Rigler F. (1980) "The areal hypolimnetic oxygen deficit: an empirical test of the model," Limnol. Oceanogr. 25: 672-679

Cornett, R.J., Rigler F.H. (1984) "Dependence of hypolimnetic oxygen consumption on ambient oxygen concentration: fact or artifact?," Water Resour. Res. 20: 823-830

Corwin ,D.L., Waggoner B.L. (1991) TETrans: A user-friendly, functional model of solute transport. In: Barnwell T.O., Ossenbruggen P.J., Beck M.B. (ed.) Watermatex '91. Pergamon Press Oxford, pp.57-66

Cosby, B.J. et al. (1985) "Modeling the effects of acid deposition: Estimation of long-term water quality responses in a small forested catchment," Water Resour. Res. 21: 1591-1595

Cosby, B.J., Bulger A.J., Wright R.F. (1994) "Predicting recovery of freshwater ecosystems: trout in Norwegian lakes." In: Steinberg C.E.W., Wright R.F. (ed.) Acidification of Freshwater Ecosystems: Implications for the Future. Wiley J. and Sons Ltd. Chichester, pp.355-373

Cowgill, U.M. (1990) "Acid precipitation: A review." In: Veziroglu T.N. (ed.) Environmental Problems and Solutions. Greenhouse Effect, Acid Rain, Pollution. Hemisphere Publishing Corporation USA, pp.111-138

Crawford, P.J., Rozenberg D.M. (1984) "Breakdown of conifer needle debris in a new northern reservoir, Southern Indiana Lake, Manitoba," Can. J. Fish. Aquat. Sci. 41: 649-658

Cross, T.K., Summerfelt R.C. (1987) "Oxygen demand of lakes: Sediment and water column BOD," Lake Reserv. Manage. 3: 109-116

Cullen, P., Farmer N., O'Loughlin E. (1987) "Estimating nonpoint sources of phosphorus to lakes," Verh. int. Verein. Limnol. 23: 588-593

Cullen, P., O'Loughlin E.M. (1982) "Non-point sources of pollution." In: O'Loughlin E.M., Cullen P. (ed.) Prediction in Water Quality. Australian Academy of Science Canberra, pp.437-453

Davies, B.R., Walmsley R.D. (1985) "Perspectives in Southern Hemisphere Limnology. Developments in Hydrobiology" Vol. 28 (Reprinted from Hydrobiologia, Vol. 125). Junk Publishers The Hague The Netherlands

Davis, E., Smith R.B., Goos G. (1973) Predicting reservoir quality changes - A laboratory investigation. In: Reinelt E.R., A.H. Laycock, W.M. Schultz (ed.) Proceedings of the Symposium on the Lakes of Western Canada, 1973. The University of Alberta Edmonton Alberta, pp.40-55

Davis, J.R. (1993) "Expert systems and environmental modelling." In: Jakeman A.J., Beck M.B., McAlleer M.J. (1993) Modelling Change in Environmental Systems. Wiley Chichester, pp.505-517

Davis, J.R., Clark J.L. (1991)" A selective bibliography of expert systems in natural resource management," AI Appl. Nat. Resour. Manage. 3: 1-19

Davis, J.R., Nanninga P.M., Biggins J., Laut P. (1991) "Prototype decision support system for analyzing impact of catchment policies," J. Water Resourc. Plng Mgmt. ASCE 117: 399-414

Day, J.W., Sklar F.H., Hopkinson Ch.S., Kemp G.P., Conner W.H. (1982) "Modelling approaches to understanding and management of freshwater swamp forests in Louisiana (U.S.A.)." In: Logofet D.O., Luckyanov N.K. (ed.) Freshwater Ecosystem Dynamics in Wetlands and Shallow Water Bodies, vol.2. UNEP/SCOPE Moscow, pp.73-105

De Coursey, D.G. (Ed.) (1988) Proceedings of the International Symposium on Water Quality Modelling of Agricultural Non-Point Sources. Utah, June 19-23, 1988. U.S. Department of Agriculture Utah

de Nijs, A.C.M., Janse J.H., Wortelboer F.G., Kramer P.R.G., Aldenberg T. (1993) WATNAT Model Documentation, Version 1.0. Rijksinstituut voor Volksgezondheid en Milieuhygiene, Laboratory for Water & Drinking Water Research Bilthoven The Netherlands

De Simoni, G., Fanello R., Guariso G. (1986) "Computer aided planning of a multipurpose river basin." In: Zannetti P. (ed.) Envirosoft 86. Computational Mechanics Publications Great Britain, pp.421-433

DePinto, J.V., Calkins H.W., Densham P.J., Atkinson J.F., Guan W., Lin H., Rodgers P.W. (1994) "An approach for integrating GIS and watershed analysis models,". Microcomputers in Civil Engineering 9: 251-262

Dickerhoff~Delwiche, L.L., Haith D.A. (1983) "Loading functions for predicting nutrient losses from complex watersheds," Water Res. Bull 19: 951-959

Dillon, P.J., Evans H.E. (1993) "A comparison of phosphorus retention in lakes determined from mass balance and sediment core calculations," Wat. Res. 27: 659-668

Donslud, B., Eriksen P.B. (1990) "Disposal of residues from electricity production. Practical application of a mathematical model to predict the impact on ground water quality." In: Fenhann J., Larsen H., Mackenzie G.A., Rasmussen B. (ed.) Environemntal Models: Emissions and Consequences/Riso International Conference 22-25 May 1989. Elsevier Amsterdam The Netherlands, pp.321-331

Dorge, J. (1992) Modelling nitrogen transformations in freshwater wetlands. Estimating nitrogen retention and removal in natural wetlands in relation to their hydrology and nutrient loadings. ISEM's 8th International Conference on the State-of-the Art in Ecological Modelling, September 28 - October 2, 1992, Kiel Germany, pp.74

Duckstein, L., Shrestha B.P. (1991) "Multicriterion risk and reliability analysis in hydrologic system design and operation." In: Ganoulis J. (ed.) Water Resources Engineering Risk Assessment. Springer Berlin, pp.363-392

Duncan, A., Rzoska J. (1981) Land Use Impact on Lake and Reservoir Ecosystems. Proc. MAB Project 5. Facultas-Verlag Wien

Duthie, H.C., Ostrofsky M.L. (1982) "Use of phosphorus budget models in reservoir management," Can. Water Res. J. 7: 337-347

Dvořáková, M., Kozerski H.-P. (1980) "Three-layer model of an aquatic ecosystem," Journal International Society for Ecological Modeling 2: 63-70

Edinger, J.E., Buchak E.M. (1983) Developments in LARM2: a longitudinal-vertical, time-varying hydrodynamic reservoir model. Technical Report E-83-1. U.S. Army Engineer Waterways Experiment Station Vicksburg Mississippi

Eliassen, A., Saltbones J. (1983) "Modeling of long-range transport of sulfur over Europe: A two-year model run and some model experiments," Atmospheric Environment 17: 1457-1473

Ellis, J.H. (1987) "Stochastic water quality optimization using imbedded chance constraints," Water Resour. Res. 23: 2227-2238

EMEP/CCC (1984) Summary Report 2/84. Norwegian Institute for Air Research Lilestrom Norway

Falconer, R.A.(Ed.) (1992) Water Quality Modelling. Ashgate and The Institution of Water and Environmental Management UK

Fedra, K. (1983) Environmental Modeling under Uncertainty: Monte Carlo Simulation. IIASA RR-83-28. International Institute for Applied Systems Analysis Laxenburg Austria

Fedra, K. (1990a) Decision support and geographic information systems. In: Scholten H.J., Stilwell J.C.H. (ed.) Geographic Information Systems for Urban and Regional Planning

Fedra, K. (1990b) Interactive environmental software: Integration, simulation, and visualization. In: Pillman W., Jaeschke A. (ed.) Informatik für den Umweltschutz Vienna 19-21 September 1990. Proceedings.

Fedra, K. (1991) A Computer-Based Approach to Environmental Impact Assessment. IIASA RR-91-13, International Institute for Applied Systems Analysis Laxenburg Austria

Fedra, K., Winkelbauer L., Pantulu V.R. (1991) Expert Systems for Environmental Screening. IIASA RR-91-19, International Institute for Applied Systems Analysis Laxenburg Austria

Feger, K.H. (1994) "Influence of soil development and management practices on freshwater acidification in Central European forest ecosystems." In: Steinberg C.E.W., Wright R.F. (ed.) Acidification of Freshwater Ecosystems: Implications for the Future. Wiley Chichester, pp.67-82

Fenger, J. (1992) Sources, pollutants and abatement possibilities. In: Jϕrgensen S.E. (ed.) Management of Lake Acidification. Guidelines of Lake Management. Vol 5. International Lake Environment Committee Shiga Japan, pp.19-46

Ferrara, R.A., Griffin T. (1983) "Model complexity for trophic state simulation in reservoirs," J. International Society of Ecological Modelling 5: 101-102

Filho, M.C.A., de Jesus J.A.O., Branski J.M., Hernandez J.A.M. (1990) "Mathematical modelling for reservoir water-quality management through hydraulic structures: a case study," Ecol. Modelling 52: 73-85

Finn, J.T. (1990) "Impact assessment models." In: Patten B.C., Jϕrgensen S.E., Dumont H., Gopal B., Květ J., Svirezhev Y.M., Koryavov P.P., Löffler H., Tundisi J.G. (eds.) Wetlands and Shallow Continental Water Bodies. Vol.1 Natural and Human Relationships. SPB Academic Publishing The Hague The Netherlands, pp.703-725

Fontane, D.G., Labadie J.W., Loftis B. (1981) "Optimal control fo reservoir discharge quality through selective withdrawal," Water Resour. Res. 17: 1594-1604

Ford, D.E., Stefan H. (1980) "Stratification variability in three morphometrically different lakes under identical meteorolo-gical forcing," Wat. Res. Bull. 16: 243-247

Gaillard, J. (1984) "Multilevel withdrawal and water quality," J. Envir. Engng Div., ASCE 110: 123

Galat, D.L. (1990) "Estimating fluvial mass transport to lakes and reservoirs: avoiding spurious self-correlations," Lake Reserv. Manage. 6: 153-163

Gallegos, C.L., Schiebe F.R., Stefan H.G. (1983) Modelling water movement and thermal structure in agricultural impoundme- nts. St. Anthony Falls Hydraulic Laboratory, Technical Paper, Series A. 155

Galvez, J.A., Niell F.X. (1993) "Sedimentation and minerali-zation of seston in a eutrophic reservoir, with a tentative sedimentation model." In: Straškraba M., Tundisi J.G. and Duncan A. (ed.) Comparative Reservoir Limnology and Water Quality Management. Kluwer Academic Publishers Netherlands, pp.119-126

Garton, J.E., Strecker R.G., Summerfelt C.R. (1978) "Performance of an axial flow pump for lake destratification,' Proc. SE. Ass. Game Fish Commn. 30: 336-347

Georgakakos, A.P., Yao H.M. (1993) "New control concepts for uncertain water resources systems.1.Theory," Water Resour. Res. 29: 1505-1516

Gherini, S.A. et al. (1985) "The ILWAS Model: Formulation and Application," Water, Air, and Soil Pollution 26: 425-431

Giorgini, A., Zingales F. (1986) Agricultural nonpoint source pollution: model selection and application. Contributions to a workshop held in June 1984 in Venice, Italy. Elsevier Amsterdam The Netherlands

Gnauck, A., Mathäus E., Straškraba M., Affa I. (1990) "The use of SONCHES for aquatic ecosystem modelling," Syst. Anal. Model. Simul. 7: 439-458

Gnauck, A., Straubel R., Wittmüss A. (1989) "Mehrkriterialer Steuerungsentwurf zur Wassergütebewirtschaftung von Flussgebieten," Messen-Steuern-Regeln Berlin 32: 294-299

Golubyev, G., Shvytov I. (1982) Modeling agricultural environmental processes in crop production. International Institute of Applied Systems Analysis CP-82-S5. International Institute of Applied Systems Analysis Laxenburg Austria

Gopal, B., Turner R.E., Wetzel R.G., Whigham D.F. (1982) Wetlands: Ecology and Management. Proceedings of the First International Wetland Conference, New Delhi, India, 10-17 Sept. 1980. National Institute of Ecology Jaipur and International Scientific Publications Jaipur India

Gray, W.G. (1986) Physics - Based Modelling of Lakes, Reservoirs, and Impoundments. Americal Society of Civil Engineers, New York

Griffin ,T.T., Ferrara R.A. (1984) "A multicomponent model of phosphorus dynamics in reservoirs," Water Res. Bull. 20: 777-788

Grimard, Y., Jones H.G. (1982) "Trophic upsurge in new reservoirs: a model for total phosphorus concentrations," Can. J. Fish. Aquat. Sci. 38: 1473-1483

Grobbelaar, J.U. (1991) The trophic status of freshwaters with high suspended sediment loads and the availability of adsorbed nutrients in the semi-arid areas of South Africa. In: Tiessen H., Frossard E., (ed.) Phosphorus Cycles in Terrestrial and Aquatic Ecosystems. Regional Workshop 4: Africa. Proceedings of a workshop arranged by the Scientific Committee on Problems of the Environment (SCOPE) and the United Nations Environmental Programme (UNEP) at UNEP (Nairobi, Kenya, March 18-22, 1991). Saskatchewan Institute of Pedology Saskatoon Canada, pp.19-29

Grobler, D.C., Rossouw J.N., van Eeden P., Oliveira M. (1987) "Decision support system for selecting eutrophication control strategies," In: Beck M.B. (ed.) Systems Analysis in Water Quality Management. Pergamon Oxford, pp.219-230

Grobler, D.C., Silberbauer M.J. (1985) "The combined effect of geology, phosphate sources and runoff on phosphate export from drainage basins," Water Res. 19: 975-981

Groeger, A.Q., Kimmel B.L. (1984) Organic matter supply and processing in lakes and reservoirs. EPA 440/5/84-001. Lake and Reservoir Management. U.S. EPA Washington D.C., pp.282-285

Gromiec, M.J. (1983) "Biochemical oxygen demand - dissolved oxygen: river models." In: Jørgensen S.E. (ed.) Application of Ecological Modelling in Environmental Management Part A. Elsevier Amsterdam The Netherlands, pp.131-225

Gulati, R.D., Lammers E.H.R.R., Meijer M.-L., van Donk, E. (1990) Biomanipulation. Tool for Water Management. Kluwer Academic Publishers Dordrecht The Netherlands

Gunnison, D., Brannon J.M., Smith I. Jr., Burton G.A (1980) "Changes in respiration and anaerobic nutrient regeneration during the transition phase of reservoir development." In: Barica J., L. Mur (ed.) Hypertrophic Ecosystems. Junk The Hague, pp.151-158

Gunnison, D., Chen R.L., Brannon J.M. (1983) "Relationship of materials in flooded soils and sediments to the water quality of reservoirs. I. Oxygen consumption rates." Water Res. 17: 1609-1617

HAESTAD METHODS (1993) HEC-3, SEDIMOT-II and STORM. Haestad Methods, Software Reference Handbook. Haestad Waterbury Connecticut

Haindl, K. (1973) Suitable solution of bottom outlets of dams and oxidation outlets for the improvement of water quality in rivers. Proceedings IAHR Istanbul, Turkey

Haith, D.A. (1982) Models for Analysing Agricultural Nonpoint-Source Pollution. IIASA RR-82-17, International Institute of Applied Systems Analysis Laxenburg Austria

Haith, D.A., Tubbs L.J. (1981) "Watershed loading functions for nonpoint sources," J. Envir. Engng Div., ASCE 107: 121-137

Hamilton, D. (1992) Management options for Prospect Reservoir - Final report. Appendix III - User's guide and reference manual for the water quality model DYRESM-WQ. Centre for Water Research University of Western Australia Nedlands Western Australia

Hamilton, D.P., Schladow S.G. (in print) "Modelling the sources of oxygen in an Australian reservoir," Verh. int. Verein. Limnol. 25:

Hanson, M.J., Stefan H.G. (1984) "Side effects of 58 years of copper sulphate treatment of the Fairmont Lakes, Minnesota," Water Res. Bull. 20: 889-900

Hanson, M.J., Stefan H.G. (1985) Shallow lake water quality improvement by dredging. In: Lake and Reservoir Management: Practical Applications. Proceedings of the Fourth Annual Conference and International Symposium, October 16-19, 1984 McAfee, New Jersey, pp. 168-171

Hanson, M.J., Stefan H.G., Riley M. (1986) Dynamic (Mathematical) Modelling of Lake Processes for Management Decisions. Lake and Reservoir Management Volume II. Proceedings of the Fifth Annual Conference and International Symposium an Applied Lake & Watershed Management. November 13-16, 1985 Lake Geneva Wisconsin, pp.225-228

Harleman, D.R.F. (1978) A comparison of water quality models of the aerobic nitrogen cycle. IIASA RM-78-34, International Institute of Applied Systems Analysis Laxenburg Austria

Harleman, D.R.F. (1982) "Hydrothermal analysis of lakes and reservoirs," J. Hydraulic Engng Div.. ASCE 108: 303

Hartigan, J.P. (1983) 'Reservoir modelling to develop out nitrate standard," J. Envir. Engng Div., ASCE 109: 1243-1258

Harwell, M.A., Weinstein D.A. (1983) "Modelling the effect of pollutants on forested ecosystems.' In: Lauenroth W.K., Skogerboe G.V., Flug M.(Eds) (ed.) Analysis of Ecological Systems: State- of- the-Art in Ecological Modelling. Elsevier Amsterdam The Netherlands, pp.497-502

Hauhs, M. (1994) "Does reduction in SO2 and NOx emissions lead to recovery?" In: Steinberg C.E.W., Wright R.F. (ed.) Acidification of Freshwater Ecosystems: Implications for the Future. Wiley J. and Sons Chichester, pp.313-323

Heatwole, C.D., Bottcher A.B., Campbell K.L. (1987) " Basin scale water quality model for coastal plain flatwoods," Transac. ASAE 30: 1023-1030

Helliwell, S., Batley G.E., Florence T.M., Lumsden G.C. (1983) "Speciation and toxicity of aluminum in a model fresh water," Eng. Tech. Letters 4: 141-144

Henderson-Sellers, B. (1981) "Destratification and rearation as tools for in-lake management," Water S.A. 7: 185-189

Henderson-Sellers, B. (1986) "Calculating the surface energy balance for lake and reservoir modelling: a review," Rev. Geophys. 24: 625-649

Henderson-Sellers, B. (1987) "One-dimensional modelling of thermal stratification in oceans and lakes," Environmental Software 2: 78-84

Henderson-Sellers, B. (1988) "Sensitivity of thermal stratification models to changing boundary conditions," Appl. Math. Model. 12: 31-43

Henderson-Sellers, B. (1992) Decision Support Techniques for Lakes and Reservoirs. Water Quality Modelling, Volume IV. CRC Press Boca Raton Florida

Henderson-Sellers, B., Davies A.M. (1989) "Thermal stratif-ication modelling for oceans and lakes," Annual Review Numerical Fluid Dynamics and Heat Transfer 2: 86-156

Henderson-Sellers, B., Davis J.R., Webster I.T., Edwards J.M. (1993) "Modern tools for environmental management: water quality." In: Jakeman A.J., Beck M.B., McAlleer M.J. (1993) Modelling Change in Environmental Systems. Wiley Chichester, pp.519-542

Hocking, G.C., Sherman B.S., Patterson J.C. (1988) "Algorithm for selective withdrawal from stratified reservoir,' J. Hydraulic Engng Div., ASCE 114: 707-719

Holme, D.D., Bakke J.W., Wlosinski J.H. (1985) "Coupling a one-dimensional model for simulating run-of-river reservoirs," Lake Reserv. Manage. 2: 87-92

Holnicki, P. (1990) "A simulation computer model for evaluating air quality." In: Fenhann J., Larsen H., Mackenzie G.A., Rasmussen B. (ed.) Environemntal Models: Emissions and Consequences/Riso International Conference 22-25 May 1989. Elsevier Amsterdam The Netherlands, pp.117-125

Holý, M., Handová Z., Kos Z., Vaška J., Vrána K. (1981) Erosion and Water Quality as Modeled by CREAMS: a Case Study of the Sedlický Catchment. IIASA CP-81-35. International Institute of Applied Systems Analysis Laxenburg Austria

Holý, M., Svetlosanov V., Handová Z., Kos Z., Vaška J., Vrána K. (1982) Procedures, Numerical Parameters and Coefficients of the CREAMS Model: Application and verification in Czechoslovakia. IIASA CP-82-23. International Institute of Applied Systems Analysis Laxenburg Austria

Hooper, R.P., Stone A., Christophersen N., Degrosbois E., Seip H.M. (1988) "Assessing the BIRKENES model of stream acidification using a multisignal calibration," Water Resour. Res. 24: 1308-1316

Hordijk, L. (1985) "A model for evaluation of acid deposition in Europe," In: Sydow A., Thoma M., Vichnevetsky R. (ed.) Systems Analysis and Simulation II., pp.30-39

Hordijk, L. (1991) "Use of the RAINS model in acid rain negotiations in Europe," Environmental Science and Technology 25:

Hornberger, G.M., Cosby B.J., Galloway J.N. (1986) "Modeling the effects of acid deposition: Uncertainty and spatial variability in estimation of long-term sulphate dynamics in a region," Water Resour. Res. 22: 1293-1302

Horstman, H.K., Copp R.S., Browne F.X. (1983) Use of predictive phosphorus model to evaluate hypolimnetic discharge scenarios of Lake Wallenpupack. Lake and Reserv. Manage., EPA Washington DC: 165-170

Hosomi, M., Okada M., Sudo R. (1991) Assessment of eutrophication control programs using an ecological model for a dimictic lake. In: Barnwell T.O., Ossenbruggen P.J., Beck M.B. (ed.) Watermatex '91. Pergamon Press Oxford, pp.339-348

Illum, K. (1990) "SESAM - DENMARK sustainable energy systems analysis model." In: Fenhann J., Larsen H., Mackenzie G.A., Rasmussen B. (ed.) Environemntal Models: Emissions and Consequences/Riso International Conference 22-25 May 1989. Elsevier Amsterdam, pp.281-289

Imberger, J. (1982) Reservoir dynamics modelling. In: O'Loughlin E.M., P. Cullen (ed.) Prediction in Water Quality. Australian Academy of Science Canberra, pp.223-248

Imberger, J. (1985) "Thermal characteristics of standing waters: an illustration of dynamic processes,". Hydrobiologia 125: 7-29

Imberger, J., Patterson J. (1981)" A dynamic reservoir simulation model - DYRESM5." In: Fischer H.B. (ed.) Transport Models for Inland and Coastal Waters. Academic Press New York, pp.310-361

Imberger, J., Patterson J., Herbert B., Loh J. (1978) "Dynamics of a reservoir of medium size." J. Hydraulic Engng Div., ASCE 104: 725-743

James, W.F., Kennedy R.H., Shain W.E., Myers R.K. (1988) "Leaf litter breakdown in a recently impounded reservoir," Water Res. Bull. 24: 831-837

Imboden ,D.M., Gächter R. (1978) "A dynamic lake model for trophic state prediction," Ecol. Modelling 4: 77-98

Janse, J.H., Aldenberg T. (1990a) PCLoos, a eutrophication model of the Loosdrecht Lakes. WOL report no. 1990-1. Report no. 714502001. National Institute of Public Health and Environmental Protection Bilthoven The Netherland

Janse, J.H., Aldenberg T. (1990b) "Modelling phosphorus fluxes in the hypertrophic Loosdrecht Lakes," Hydrobiol. Bull. 24: 69-89

Janse, J.H., Aldenberg I.T. (1991) "Modelling the eutrophication of the shallow Loosdrecht Lakes," Verh. int. Verein. Limnol. 24: 751-757

Janse, J.H., Aldenberg T., Kramer P.R.G. (1992)" A mathematical model of the phosphorus cycle in Lake Roosdrecht and simulation of additional mesures," Hydrobiologia 233: 119-136

Janse, J.H., Van der Does J., Van der Vlugt J.C. (1993) "PCLAKE; Modelling eutrophication and its control measures in Reeuwijk lakes," In: Giussani G., Callieri C. (ed.) Strategies for Lake Ecosystems beyond 2000. Proceedings 5th Int. Conf. on the Conservation and Management of Lakes, Stresa Italy, pp.117-120

Janse, J.H., Van der Vlugt J.C. (1993) "PC-lake; modelling eutrophication and its control measures in Reeuwijk lakes," In: Giussani G., Callieri C. (ed.) Strategies for Lake Ecosystems beyond 2000. Proceedings 5th Int. Conf. on the Conservation and Management of Lakes, 17-21 May, 1993, Stresa Italy, pp.117-120

Janse, J.H., van Donk E., Gulati R.D. (in press) "Modelling nutrient cycles in relation to food web structure in a biomanipulated shallow lake," Neth. J. Aquat. Ecol. 29

Janssen, P.H.M., Heuberger P.S.C., Sanders R. (1992) UNCSAM 1.1.: A software package for sensitivity and uncertainty analysis. RIVM Rept. No. 959101004. Natl. Inst. Environ. Prot. (RIVM) Bilthoven Netherlands

Jansson, P.E., Andersson R. (1988) "Simulation of runoff and nitrate leaching from an agricultural district in Sweden," J. Hydrol. 99: 33-48

Janus, L.L., Vollenweider R.A. (1984) "Phosphorus residence time in relation to trophic conditions in lakes," Verh. int. Ver. Limnol. 22: 179-184

Jenq, T. et al. (1984) "A phosphorus management LP model case study," Water Res. Bull. 20: 511-522

Jensen, A.L. (1988) "Modelling the effect of acidity on mercury uptake by walleye in acidic and circumneutral lakes," Environmental Pollution 50: 285-294

Jirka, G.H., Doneker R.L., Barnwell T.O. (1991) "CORMIX: An expert system for mixing zone analysis," Water Sci. Tech. 24: 267-274

Johnson, L.E. (1986) "Water resource management decision support systems," J. Water Resourc. Plng Mgmt., ASCE 112: 308-325

Jokiel, C., Wagner A., Köngeter J. (1993) Water Quality Modelling Using GIS. Lehrstuhl und Institut für Wasserbau und Wasserwirtschaft, Rheinisch-Westfälische Technische Hochschule Aachen, Aaachen Germany, pp.1-4

Jolánkai, G. (1983a) An approach to solve the "Missing Link" problem of non-point source modelling. In: Jolánkai G., Roberts G. (ed.) Proc. UNESCO-MAB. 5 Workshop on Land Use Impacts on Aquatic System, Oct. 10-14 Budapest Hungary, pp.271-283

Jolánkai, G. (1983b) 'Modelling of non-point source pollution." In: Jφrgensen S.E. (ed.) Application of Ecological Modelling in Environmental Management, Part A. Elsevier Amsterdam The Netherlands, pp.283-385

Jolánkai, G., Pesti G. (1983) A rainfall vs. phosphorus export rate model for Tetves Creek Lake Balaton. In: Jolánkai G., Roberts G. (ed.) Proc. UNESCO-MAB. 5 Workshop on Land Use Impacts on Aquatic System, Oct. 10-14 Budapest Hungary, pp.357-368

Jφrgensen, S.E. (1980) Lake Management. Pergamon Press

Jφrgensen, S.E. (1983a) "Eutrophication models of lakes." In: Jφrgensen S.E. (ed.) Application of Ecological Modelling in Endironmental Management, Part A., Elsevier Amsterdam The Netherlands, pp.227-282

Jφrgensen, S.E.(ed.) (1983b) Application of Ecological Modelling in Environmental Management, Part A. Elsevier Amsterdam The Netherlands

Jφrgensen, S.E. (1986a) Fundamentals of Ecological Modelling. Elsevier Amsterdam The Netherlands

Jφrgensen, S.E. (1986b) "Structural dynamic model." Ecol. Modelling 31: 1-9

Jφrgensen, S.E. (1988) "Modelling of wetlands." In: Marani A. (ed.) Advances in Environmental Modelling, Developments in Environmental Modelling 13. Elsevier Amsterdam The Netherlands, pp.535-543

Jφrgensen, S.E. (1992) Management of Lake Acidification. Guidelines of Lake Management, Vol.5. International Lake Environment Committee Kusatsu Japan

Jφrgensen, S.E., Friis M.B., Henriksen J., Jorgensen L.A., Mejer H.F. (1979) Handbook of Environmental Data and Ecological Parameters. International Society for Ecological Modelling Copenhagen

Jφrgensen, S.E., Gromiec M.J. (1989) Mathematical Submodels in Water Quality Systems. Development in Environmental Modelling. Elsevier Amsterdam The Netherlands

Jφrgensen ,S.E., Hoffmann C.C., Mitsch W.J. (1988) "Modelling nutrient retention by a reedswamp and wet meadow in Denmark." In: Mitsch W.J., Straškraba M., Jφrgensen S.E. (ed.) Wetland Modelling. Elsevier Amsterdam The Netherlands, pp.133-151

Jφrgensen, S.E., Kamp-Nielsen L., Jφrgensen L.A. (1986a) "Examination of the generality of eutrophication models." Ecol. Modelling 32: 251-266

Jφrgensen, S.E., Kamp-Nielsen L., Christensen T., Windolf -Nielsen J., Westergaard B. (1986b) "Validation of a prognosis based upon a eutrophication model," Ecol. Modelling 32: 165-182

Jφrgensen, S.E., Koryavov P.P. (1990) "Modeling ecosystem dynamics." In: Patten B.C., Jorgensen S.E., Dumont H. (ed.) Wetlands and Shallow Continental Water Bodies, Vol. 1. SPB Academic Publishing The Netherlands, pp.691-702

Jφrgensen, S.E., Mitsch W.J. (eds) (1983) Application of Ecological Modelling in Environmental Management, Part B. Elsevier Amsterdam The Netherlands

Jφrgensen, S.E., Nielsen S.N., Jφrgensen L.A. (1991) Handbook of Ecological Parameters and Ecotoxicology. Elsevier Amsterdam The Netherlands

Jφrgensen, S.E., Patten B.C., Straškraba M. (1992) "Ecosystems emerging: toward an ecology of complex systems in a complex future," Ecol. Modelling 62: 1-27

Jφrgensen, S.E., Vollenweider R.A. (Eds) (1988) Principles of Lake Management. Guidelines of Lake Management, Vol. 1. International Lake Environment Comittee Kusatsu Japan

Kaden, S., Becker A., Gnauck A. (1989) "Decision-support systems for water management," IAHS Publ. 180: 11-21

Kalčeva, R., Outrata J., Schindler Z., Straškraba M. (1982) "An optimization model for the economic control of reservoir eutrophication," Ecol. Modelling 17: 121-128

Kallio, K. (1992) Modeling of phosphorus transport from an agricultural basin. Modeling of Phosphorus Transport from an Agricultural Basin. Nordisk Hydrologisk Program Rapport 29. Nordisk Hydrological Program, pp.150-158

Kämäri, J. (Ed.) (1990) Impact Models to Assess Regional Acidification. Kluwer Academic Publishers Dordrecht The Netherlands

Kämäri, J., Brakke D.F., Jenkins A., Norton S.A., Wright R.F. (Eds) (1990) Regional Acidification Models: Geographic Extent and Time Development. Springer Berlin

Kämäri, J., Posch M., Kauppi L. (1984) Development of a model analysing surface water acidification on a regional scale: Application to individual basins in Southern Finland. In: Johansson I. (ed.) IHP Workshop, Uppsala 15-16 September 1984, NHP Report 10 Sweden, pp.151-171

Kauppi, P., Kämäri J., Posch M., Kauppi L., Matzner E. (1985) Acidification of forest soils: A model for analysing impacts of acidic deposition in Europe. Version II. IIASA CP-85-27. International Institute for Applied Systems Analysis Laxenburg Austria, pp.1-28

Kauppi, P., Kämäri J., Posch M., Kauppi L., Matzner E. (1986) "Model development and application for analysing impacts of acidic deposition in Europe," Ecol. Modelling 33: 231-253

Kelly, C.A., Rudd J.W.M., Hesslin R.H., Schindler D.W., Dillon P.J., Driscoll C.T., Gherini S.A., Heskey R.H. (1987) "Prediction of biological acid neutralization in acid sensitive lakes," Biogeochem. 3: 129-140

Kercher, J.R., Axelrod M.C., Bingham G.E. (1981) "Hierarchical set of models for estimating the effects of air pollution on vegetation." In: Mitsch W.J., Bosserman R.W., Klopatek J.M. (ed.) Energy and Ecological Modelling. Proceedings of a Symposium held from 20 to 23 April 1981 at Louisville Kentucky. Elsevier Amsterdam, pp.401-409

Kimmel, B.L., Groeger A.W. (1986) "Limnological and ecological changes associated with reservoir aging." In: Hall G.E., Van Den Avyle M.J. (ed.) Reservoir Fisheries Management: Strategies for the 80's. Southern Division American Fisheries Society Bethesda Maryland, pp.103-109

Kirchner, W.B., Dillon D.J. (1975) "An empirical method of estimating the retention of phosphorus in lakes," Water Resour. Res. 2: 182-183

Kmet, T., Straškraba M. (1989) "Global behavior of a generalized aquatic ecosystem model," Ecol. Modelling 45: 95-110

Kortmann, R.W., Knoecklein G.W., Rich P.H. (1994) "Aeration of stratified lakes: theory and practice," Lake Reserv. Manage. 8: 99-120

Kozerski, H.-P. (1975) "Ein mathematisches Modell des Temperaturregimes und des Sauerstoffhaushaltes im Hypolimnion von Talsperren," Acta Hydrophysica Berlin 19: 201-250

Krenkel, P.A., French R.H. (1982) "State of the art of modeling surface water impoundments," Wat. Sci. Tech. 14: 241-248

Krenkel, P.A., Novotny V. (1980) Water Quality Management. Academic Press New York

Krogerus, K. (1982) Measuring and modelling of benthic oxygen demand. Proc. 10th Nordic Symp. on Sediment, pp.55-71

Küchler, L., Kozerski M.P. (1975) "Vorhersage des Sauerstoff-gehaltes im Tiefenwasser einer neuerrichteten Talsperre für die ersten Jahre des Anstauens," Acta hydrochim. hydrobiol. 3: 473-478

Kuenzler, E. J. (1990) "Wetlands as sediment and nutrient traps for lakes." Proceedings of a National Conference on Enhancing the States' Lake and Wetland Management Programs, May 18-19, 1989, Chicago Illinois, pp.105-112

Kularathna, M., Somlyódy L. (1994) River Basin Water Quality Management Models: A State-of-the-Art Review. International Institute Applied Systems Analysis WP-94-3. International Institute Applied Systems Analysis Laxenburg Austria

Kutas, T., Herodek S. (1986) "A complex model for simulating the Lake Balaton ecosystem." In: Somlyody L., van Straten G. (ed.) Modeling and Managing Shallow Lake Eutrophication. Springer New York, pp.309-322

Laikari, H. (Ed.) (1989) River Basin Management. Proceedings of the fifth RBM Conference, Rovaniemi, Finland, 31 July - 4 August. National Board of Waters and the Environment Finland Water Pollution Research and Control. Advances in Water Pollution Control Series. Pergamon Press Oxford

Lam, D.C.L., Fraser A.S., Swayne D.A., Storey J., Wong I. (1988) "Regional analysis of watershed acidification using the expert systems approach." In: Zannetti P. (ed.) Computer Techniques in Environmental Studies. Springer Great Britain, pp.67-81

Larsen, D.P., Mercier H.T. (1976) "Phosphorus retention capacity of lakes." J. Fish. Res. Board Can. 33: 1742-1750

Lasch, P., Model N., Bellman K. (1988) "The PEMU/Air pollutant transport model based on emittent - receptor - point - transmission calculation." In: Sydow A., Tzafestas S.G., Vichnevetsky R. (ed.) Mathematical Research. Systems analysis and simulation. Bd 46 vol.2. Akademie Verlag Berlin, pp.42-44

Lavrik, V.I., Bilyk A.N., Nikiforovich N.A. (1990) "Multichamber simulation modelling of water quality in reservoirs," Syst. Anal. Model. Simul. 7: 625-635

Lee, H.N., Kau W.S., Kao S.K. (1983) "A model for pollutant concentration prediction in complex terrain," Air Pollution Modelling and its Applications 2, pp.429-436

Leonov, A.V. (1980) Mathematical modeling of phosphorus transformation in relation to eutrophication of Lake Balaton. Proc. Second joint MTA/IIASA Task Force Meeting on Lake Balaton Modelling Vol.1, pp.111-175

Leonov, A.V. (1980) The Chemical-Ecological Modeling of Aquatic Nitrogen Compound Transformation Processes. IIASA WP 80-1-121. International Institute for Applied Systems Analysis Laxenburg Austria

Leonov, A.V. (1982) Transformations and turnover of phosphorus compounds in the Lake Balaton ecosystem 1976-1978. IIASA WP 82-82-27. International Institute for Applied Systems Analysis Laxenburg Austria

Leonov, A.V. (1989) "Phosphorus transformation and water quality in the Ivankovo Reservoir: Study by means of a simulation model," Arch. Hydrobiol., Beih. Ergebn. Limnol. 33: 157-168

Leonov, A.V., Toth D. (1981) The Study of Nitrogen Transformation in Fresh Water: Experiments and Mathematical Modeling. IIASA CP-81-24. International Institute for Applied Systems Analysis Laxenburg Austria

Levine, D.A., Jones W.W. (1990) "Modeling phosphorus loading to three Indiana reservoirs: A geographic information system approach," Lake Reserv. Manage. 6: 81-92

Lijklema, L., Hieltjes H.M. (1982) "A dynamic phosphate budget model for a eutrophic lake," Hydrobiologia 91: 227-233

Liu, J.C., Schnoor J.L. (1986) "Acid precipitation model for seepage lakes," J. Envir. Engng Div., ASCE 112: 677-694

Livingstone, D.M. (1993) "Lake oxygenation: Application of a one-box model with ice cover," Int. Revue ges. Hydrobiol. 78: 465-480

Loehr, J. (1987) "Impact of the hydrodynamic conditions on the primary production on an impounded river," Ecol. Modelling 39: 227-245

Loehr, R.C., Martin C.S., Rast W. (1980) Phosphorus Management Strategies for Lakes. Ann Arbor Sci. Publ. Ann Arbor Michigan

Logan, T.J. (1980) "The role of soil and sediment chemistry in modelling nonpoint sources of phosphorus." In: Overcash M.R., Davidson J.M. (ed.) Environmental Impact of Nonpoint Source of Pollution. Ann Arbor Sci Publ.Inc. Ann Arbor Michigan, pp.189-208

Logofet, D.O., Luckyanov N.K. (1982) Ecosystem Dynamics in Freshwater Wetlands and Shallow Water Bodies. Vols 1 and 2. Centre of International Projects GKNT Moscow

Los, F.J. (1991) Mathematical Simulation of Algae Blooms by the Model BLOOM II, Version 2. Documentation report Delft Hydraulics The Netherlands

Los, F.J., Brinkman J.J. (1988) "Phytoplankton modelling by means of optimization: a 10-year experience with BLOOM II," Verh. int. Verein. Limnol. 23: 790-795

Los, F.J., De Rooij N.M., Smits J.G.C., Bigelow J.H. (1982) Policy Analysis of Water Management for the Netherlands, Vol. IV. Design of Eutrophication Control Strategies. Rand Note, 1500/6-Neth. Rand Corporation Santa Monica California

Loucks, D.P., da Costa J.R. (Eds) (1991) Decision Support System: Water Resources Planning. Nato ASI Series Vol. G26 Springer Berlin

Loucks, D.P., Stedinger J.R., Haith D.A. (1981) Water Resource Systems. Prentice Hall Englewood Cliffs

Lung, W-S. (1984) "Understanding simplified lake acidification models," J. Envir. Engng Div., ASCE 110: 997-1002

Lung, W-S. (1987) "Lake acidification model: practical tool," J. Envir. Engng Div., ASCE 113: 900-915

Lung, W-S., Mills A.L., Movall L.A., Herlihy A.T. (1988) "Modeling fate and transport of sulphate and alkalinity in an acidified lake," Water, Air, and Soil Pollution 37: 157-170

Markofsky, M., Harleman D.R.F. (1973) "Prediction of water quality in stratified reservoirs," J. Hydraulic Engng Div., ASCE 99: 729-745

Martin, J.L. (1988) "Application of two-dimensional water quality model," J. Envir. Engng Div. ASCE 114: 317-336

Martin, J.L., Wlosinski J.H. (1985) "A comparison of reservoir oxygen predictions from one- and two-dimensional models," Lake Reserv. Manage. 2: 98-104

Martin, S.C., Effler S.W., DePinto J.V., Trama F.B., Rodgers P.W., Dobi J.S., Wodka M.C. (1985) "Dissolved oxygen model for a dynamic reservoir," J. Envir. Engng Div., ASCE 111: 647-665

Matsumura, T., Yoshiuki S. (1981) "An optimization problem related to the regulation of influent nutrient in aquatic ecosystems," Int. J. Syst. Sci. 12: 565-585

McBean, E.A., Ellis J.H. (1985) "A screening model for development and evaluation of acid rain abatement strategies," J. Environ. Manage. 21: 287-292

McBean ,E.A., Okada N. (1988) "Use of metagame analysis in acid rain conflict resolution," J. Environ. Manage. 26: 153-162

McCauley, E., Downing J.A., Watson S. (1989) "Sigmoid relationships between nutrients and chlorophyll among lakes," Can. J. Fish. Aquat. Sci. 46: 1171-1175

Melack, J.M., MacIntyre S. (1991) "Phosphorus concentrations, supply and limitation in tropical african lakes and rivers." In: Tiessen H., Frossard E. (ed.) Phosphorus Cycles in Terrestrial and Aquatic Ecosystems. Regional Workshop 4: Africa. Proceedings of a workshop arranged by the Scientific Committee on Problems of the Environment (SCOPE) and the United Nations Environmental Programme (UNEP) at UNEP (Nairobi, Kenya, March 18-22, 1991). Saskatchewan Institute of Pedology Saskatoon Canada, pp.1-18

Messer, J., Brezonik P.L. (1983) "Agricultural nitrogen model: A tool for regional environmental management," Environ. Manage. 7: 177-187

Mitsch, W.J. (1983) "Ecological models for management of freshwater wetlands." In: Jφrgensen S.E., Mitsch W.J. (ed.) Application of Ecological Modelling in Environmental Management, Part B. Elsevier Amsterdam The Netherlands, pp.283-310

Mitsch, W.J. (1988a) "Ecological engineering and ecotechnology with wetlands. Applications of systems approaches." In: Marani A. (ed.) Advances in Environmental Modelling. Developments in Environmental Modelling 13. Elsevier Amsterdam The Netherlands, pp.565-580

Mitsch, W.J. (1988b) "Productivity-hydrology-nutrient models of forested wetlands," In: Mitsch W.J., Straškraba M., Jφrgensen S.E. (ed.) Wetland Modelling. Elsevier Amsterdam The Netherlands, pp.115-132

Mitsch, W.J. (Ed.) (1994) Global Wetlands: Old World and New. Elsevier Amsterdam The Netherlands

Mitsch, W.J., Day J.W., Taylor J., Madden Ch. (1982) "Models of north american freshwater wetlands," Int. J. Ecol. Environ. Sci. 8: 109-140

Mitsch, W.J., Gosselink J.G. (1986) Wetlands. Van Nostrand Reinhold New York

Mitsch, W.J., Jφrgensen S.E. (Ed.) (1989) Ecological Engineering. Wiley, New York

Mitsch, W.J., Jφrgensen S.E. (1990) "Modeling and management." In: Patten B.C., Jφrgensen S.E., Dumont H., Gopal B., Květ J., Svirezhev Y.M., Koryavov P.P., Löffler H., Tundisi J.G. (ed.) Wetlands and Shallow Continental Water Bodies. Vol.1 Natural and Human Relationships. SPB Academic Publishing The Hague The Netherlands, pp.727-744

Mitsch, W.J., Straškraba M., Jφrgensen S.E. (Eds) (1988) Wetland Modelling. Elsevier Amsterdam The Netherlands

Mitsch, W.J., Taylor J.R., Benson K.B. (1983) "Classification, modelling and management of wetlands - a case study in Western Kentucky." In: Lauenroth W.K., Skogerboe G.V., Flug M. (ed.) Analysis of Ecological Systems: State- of- the- Art in Ecological Modelling. Elsevier Amsterdam The Netherlands, pp.761-770

Mobley, M.H., Harshbarger E.D. (1987) Epilimnetic pumps to improve reservoir releases. Miscellaneous Paper E-87-3: 133-135 Proceedings CE Workshop on Reservoir Releases. U.S. Army Engineer Waterways Experiment Station Vicksburg Mississippi

Model, N., Wittmüss A., Gnauck A. (1995) Berechnung von Sanie-rungsstrategien fur Fliessgewässer. Ökosysteme: Modellierung und Simulation. Blottner Verlag Taunusstein

Molot, L.A., Dillon P.J., Clark B.J., Neary B.P. (1992) "Predicting end-of-summer oxygen profiles in stratified lakes," Can. J. Fish. Aquat. Sci. 49: 2363-2372

Moshiri, G.A. (Ed.) (1993) Constructed Wetlands for Water Quality Improvement. Lewis Publishers, Boca Raton, FL

Mouchet, P.C. (1984) "Influence of recently drowned terrestrial vegetation on the quality of water stored in impounding reservoirs," Verh. Int. Verein. Limnol. 22: 1068-1619

Munda, G. (1994) Fuzzy Information in Multicriteria Environ-mental Evaluation Models. Institute for Systems Engineering and Informatics, Joint Research Centre Commission of European Communities Ispra Italy

Neal, C., Whitehead P., Neale R., Cosby J. (1986) "Modelling the effects of acidic deposition and conifer afforestation on stream acidity in the British uplands,' J. Hydrol. 86: 15-26

Newbold, J.D. (1987) "Phosphorus spiralling in rivers and river-reservoir systems: implications of a model." In: Craig J.F., J.B. Kemper (ed.) Regulated Streams. Advances in Ecology. Plenum Press London

Nikolaidis, N.P., Rajaram H., Schnoor J.L., Georgakakos, K.P. (1988) "Generalized soft water acidification model," Water Resour. Res. 24: 1983-1996

Nikolaidis, N.P., Schnoor J.L., Georgakakos K.P. (1989) "Modeling of long-term lake alkalinity responses to acid deposition," J. Wat. Pollut. Control. Fed. 61: 188-199

Nirmalakhandan, N., Arulgranendran V., Mohsin M., Sun B., Cadena F. (1994) "Toxicity of mixtures of organic chemicals to microorganisms," Water Res. 28: 543-552

Novotny, V. (1988) "Diffuse (nonpoint) pollution - a political, institutional, and fiscal problem." J. Wat. Pollut. Control. Fed. 60: 1404-1413

Novotny, V., Chesters G. (1981) Handbook of Nonpoint Pollution. Sources and Management. Environmental Engineering Series, Van Nostrand Reinhold New York

Novotny, V., Olem H. (1994) Water Quality - Prevention, Identification, and Management of Diffuse Pollution. Van Nostrand Reinhold New York

Nurnberg, G.K. (1984) "The prediction of internal phosphorus load in lakes with anoxic hypolimnia," Limnol. Oceanogr. 29: 111-124

OECD (1982) Eutrophication of Waters. Monitoring, Assessment and Control. OECD Paris

Okada, M., Chihara K., Kawashima H., Suzuki M. (1991) "Development of a personal computer-based system to support basin - wide management of water quality in lakes and rivers," Water Sci. Tech. 24: 125-132

Olem, H. (1991) Liming Acidic Surface Waters. Lewis Publ., Chelsea, MI

Oliver, B.G., Charlton M.N. (1984) "Chlorinated organic contaminants on settling particulates in the Niagara River vicinity of Lake Ontario," Environ. Sci. Technol. 18: 903-908

Ondok, J.P. (1990) "Modeling ecological processes." In: Patten B.C., Jørgensen S.E., Dumont H. (ed.) Wetlands and Shallow Continental Water Bodies, Vol. 1. SPB Academic Publishing The Netherlands, pp.659-689

Orlob, G.T. (1983) Mathematical Modeling of Water Quality: Streams, Lakes, and Reservoirs. Wiley Chichester

Orlob, G.T., Selna L.G. (1970) "Temperature variations in deep reservoirs," J. Hydraulic Engng Div., ASCE 96: 391-410

Ostrofsky, M.L. (1978) 'Trophic changes in reservoirs; an hypothesis using phosphorus budget models," Int. Revue ges. Hydrobiol 64: 481-499

Ostrofsky, M.L., Duthie H.C. (1978) "n approach to modelling productivity in reservoirs," Verh. Int. Verein. Limnol. 20: 1562-1567

Page, B. (1990) "An analysis of environmental expert systems applications," Environm. Softw. 5: 177-198

Pařízek, J. (1984) "Využití efektu čisté vrstvy (Utilization of the "clean layer" effect - In Czech)." In: Straškraba M., Brandl Z., Porcalová P. (ed.) Hydrobiologie a kvalita vody údolních nádrží. ČSVTS České Budějovice, pp.72-83

Park, R.A., Collins D.C., Leung O.K. (1978) 'The aquatic ecosystem model MS CLEANER. State of the Art in Ecological Modelling." Proc. Ecol. Modelling, Copenhagen 28 Aug.-2 Sept. 1978. Society for Ecological Modelling Copenhagen Denmark, pp.579-602

Park, R.A., Indyke B.H., Heitzman G.W. (1981) "Predicting the fate of coal-derived pollutants in aquatic environments." In: Mitsch W.J., Bosserman R.W., Klopatek J.M. (ed.) Energy and Ecological Modelling. Proceedings of a Symposium held from 20 to 23 April 1981 at Louisville, Kentucky. Elsevier Amsterdam The Netherlands, pp.115-122

Parsons, J.E., Skaggs R.W., Doty C.W. (1987) "Application of a 3-dimensional water management model." Transac. ASAE 30: 960-968

Patten, B.C. (1994) Wetlands and Continental Water Bodies, Vol. 2, Case Studies. SPB Academic Publishing The Hague The Netherlands

Patten, B.C., Jørgensen S.E., Dumont H., Gopal B., Květ J., Svirezhev Y.M., Koryavov P.P., Löffler H., Tundisi J.G. (ed.) (1990) Wetlands and Shallow Continental Water Bodies, Vols.1 Natural and Human Relationships. SPB Academic Publishing The Hague The Netherlands

Patterson, J.C. (1987) "A model for convective motions in reservoir sidearm." In: Graf W.H., Lemmin U. (ed.) Topics in Lake and Reservoir Hydraulics. Proceedings AIRH-Congress-IAHR Lausanne, pp.68-73

Patterson, J.C., Hamblin P.F. (1988) "Thermal simulation of a lake with winter ice cover," Limnol. Oceanogr. 33: 323-338

Patterson, J.C., Hamblin P.F., Imberger J. (1984) "Classification and dynamic simulation of the vertical density structure of lakes," Limnol. Oceanogr. 29: 845-861

Pereira, A., Tassin B., Jørgensen S.E. (1992) Drowned vegetation decomposition model to an Amazonian Reservoir. 8th Conference of the International Society for Ecological Modelling 1992 Kiel Germany

Peterson, (1982) "Lake restoration by sediment removal," Water Res. Bull. 18: 423-435

Posch, M., Forsius M., Kämäri J. (1993) "Critical loads of sulphur and nitrogen for lakes I: Model description and estimation of uncertainty," Water, Air, and Soil Pollution 66: 173-192

Postma, L. (1988) DELWAQ Users Manual, Version 3.0. Delft Hydraulics The Netherlands

Pourcher, A-M., Salencon M-J. (1990) "Modelisation du plancton dans une retenue oligotrophe: Sainte-Croix sur le Verdon," Hydroecol. Appl. 1/2: 91-134

Prairie, Y.T., Duarte C.M., Kalff J. (1989) "Unifying nutrient - chlorophyll relationships in lakes," Can. J. Fish. Aquat. Sci. 46: 1176-1182

Prairie, Y.T., Kalff J. (1986) "Effect of catchment size on phosphorus export," Water Res. Bull. 22:

Procházková, L. (1980) "Agricultural impact on the nitrogen and phosphorus concentration in water." In: Duncan A., J. Rzoska (ed.) Land Use Impact on Lake and Reservoir Ecosystems, Proc. MAB Project 5. Facultas-Verlag Wien, pp.78-100

Purcell, L.T. (1939) "The ageing of reservoir waters," J. Amer. Water Works Assoc. 31: 1775-1806

Quintero, J.E., Garton J.E. (1973) "A low energy lake destratifies," Transact. ASAE 16: 973-978

Rabe, F. W., Clemens C.L. (1987) "Phosphorus Removal Capability Of Freshwater Wetlands.' Proceedings of Northwest Nonpoint Source Pollution Conference March 1987, pp.368-378

Radtke, E., Straškraba M. (1980) "Self-optimization in phytoplankton model," Ecol. Modelling 9: 247-268

Rathke, P., Straubel R., Wittmüss A., Gnauck A. (1987) "Multi-objective optimization in river water quality management by means of the DSS REH." In: Lewandowski A., Stanchev I. (ed.) Methodology and Software for Interactive Decision Support. Proceedings of the International Workshop held in Albena Bulgaria. Springer New York

Ratsep, R., Nihlgard B., Bashkin V.N., Blažka P., Emmet B., Harris J., Kruk M. (1994) "Agricultural impacts in the northern temperate zone." In: Moldan B., Černý J. (ed.) Biochemistry of Small Catchments, Wiley Chichester England, pp.361-382

Reckhow, K.H. (1979) "Empirical lake models for phosphorus development. Applications, Limitations and Uncertainty." In: Scavia D., Robertson A. (ed.) Perspectives on Lake Eutrophication Modelling, Ann Arbor Science Ann Arbor, pp.193-221

Reckhow, K.H. (1985) "Decision theory applied to lake management. Lake and reservoir management: practical applications." Proceedings of the Fourth Annual Conference and International Symposium, October 16-19 1984, McAfee New Jersey, pp.196-200

Reckhow, K.H. (1994) "Water quality simulation modelling and uncertainty analysis for risk assessment and decision making." Ecol. Modelling 72: 1-20

Reckhow, K.H., Black R.W., Stockton T.B. Jr., Wogt J.D., Wood J.G. (1987) "Empirical models of fish response to lake acidification," Can. J. Fish. Aquat. Sci. 44: 1432-1442

Reckhow, K.H., Chapra S.C. (1983) Enginering Approaches for Lake Management. Volume 1: Data Analysis and Empirical Modeling. Butterworths Pub. Boston Massachusetts

Reckhow, K.H., Hartigan J.P., Coffey S. (1989) "Lake nutrient budget development for state-level applications." Proceedings of a National Conference on Enhancing States' Lake Management Programs, Chicago Illinois, May 12 - 13 1988, pp.45-52

Reckhow, K.H., Simpson J.T. (1980) "A procedure using modelling and error analysis for the prediction of lake phosphorus concentration from land use information." Can. J. Fish. Aquat. Sci. 37: 1439-1448

Recknagel, F. (1989) Applied Systems Ecology. Akademie - Verlag Berlin

Recknagel, F., Beuschold E., Petersohn U. (1991) "DELAQUA - a prototype expert system for operational control and management of lake water quality," Water Sci. Tech. 24: 283-290

Reichert, P. (1994a) Concepts underlying a Computer Program for the Identification and Simulation of Aquatic Systems. Schriftenreihe der EAWAG Nr. 7, Swis Federal Institute for Environmental Science and Technology (EAWAG) Dübendorf Switzerland

Reichert, P. (1994b) "AQUASIM - a tool for simulation and data analysis of aquatic systems." Water Quality International '94. IAWQ 17th Biennial International Conference, 24-29 July 1994 Budapest Hungary, pp.21-30

Reuss, J.O. et al. (1986) "A critique of models for freshwater and soil acidification," Water, Air, and Soil Pollution 30: 909-1005

Rhea, J.R., Young T.C. (1987) "Application of a continuous distribution model for proton binding by humic acids extracted from acidic lake sediments," Environmental Geology and Water Sciences 10: 169-173

Riley, M.J., Stefan H.G. (1975) "MINLAKE: A dynamic lake water quality management model," Ecol. Modelling: 55

Riley, M.J., Stefan H.G. (1988) "Development of the Minnesota Lake Water Quality Management Model "MINLAKE"," Lake Reserv. Manage. 4: 73-84

Ripl, W. (1976) "Biochemical oxidation of polluted lake sediment with nitrate - a new lake restoration method," AMBIO 5: 132-135

Ripl, W. (1980) "Natural and induced sediment rehabilitations in hypertrophic lakes." In: Barica J., L. Muhr (ed.) Hypertrophic Ecosystems. Junk The Hague

Ritter, W.F. (1988) "Reducing impacts of nonpoint source pollution from agriculture: a review." J. Env. Quality 23: 645-68

Rose, K., Brenkert A., Cook R., Gardner H., Hettelingh J.-P. (1990) Systematic Comparison of ILWAS, MAGIC and ETD Watershed Acidification Models: Mapping among Model Inputs and Determinist-ic Results. Environmental Sciences Division Oak Ridge National Laboratory Oak Ridge Tennessee

Rossi, G. (1991) Modelling Lake Pollution. Commission of the European Communities Brussels - Luxenbourg

Ruberti, A., Mohler R. (1975) Variable Structure Systems with Applications to Economics and Biology. Lecture Notes in Economics and Mathematical Systems, Vol. 111. Springer Berlin

Ryding, S.-O, Rast W. (1989) The Control of Eutrophication of Lakes and Reservoirs. UNESCO and The Parthenon Publishing Group Park Ridge New Jersey

Salas, H.J., Limon G. (1986) Memoria del tercer encuentro del proyecto regional desarrello de metodologias simplificadas para la evaluacion de eutroficacion en lagos calidos tropicales. Proc. Workshop, Centro Panamericano de Ingenieria Sanitaria y Ciencias del Ambiente Organizacion Mundial de la Salud Lima Peru

Sale, M.J., Brill E.D. Jr., Herricks E.E. (1982) "An approach to optimizing reservoir operation for downstream aquatic resources," Water Resour. Res. 18: 705-712

SASR (1992) Project No. 6: Research and Technological Development for the Supply and Use of Freshwater Resources: Report on Monitoring and Modeling by I.Krüger Consult AS and Danish Hydraulic Institute. Commission of the European Communities Brussels-Luxembourg

Scavia, D., Chapra S.C. (1977) "Comparison of an ecological model of Lake Ontario and phosphorus loading models," J. Fish. Res. Board Can. 34: 286-290

Schaller, F.W., Bailey G.W. (1983) Agricultural Management and Water Quality. IOWA State University Press

Schindler, Z., Straškraba M. (1982) "Optimální řízení eutrofizace údolních nádrží. (Optimal management of reservoir eutrophication)," Vodohospodářský časopis SAV 30: 536-548

Schladow, S.G. (1992) "Bubble plume dynamics in a stratified medium and its implications for water quality amelioration in lakes," Water Resour. Res. 28: 313-321

Schladow, S.G. (1993) "Lake destratification by bubble-plume systems: design methodology," J. Hydr. Engng Div., ASCE 119: 350-364

Schladow, S.G., Fischer I.H. (in print) "The response of temperate lakes to artificial destratification: physical limnology," Limnol. Oceanogr.(in print)

Schnoor, J.L. (ed.) (1984) Modeling of Total Acid Precipitation Impacts. Acid Precipitation Series Vol 2. Butterworth Publishers Stoneham Massachussetts

Schöpp, W. (Ed.) (1991) Modeling of Critical Loads for Acid Deposition in Austria. IIASA SR-91-4. International Institute of Applied Systems Analysis Laxenburg Austria

Schreiner, S.P., Gaughan M., Schultz H.L., Walenowicz R. (1991) "The integrated model evaluation system (IMES): A database for evaluation of exposure assessment models," Water Sci. Tech. 24: 315-322

Schwabl, A. (1988) "Environmental planning with the aid of a decision support system." In: Zannetti P. (ed.) Computer Techniques in Environmental Studies. Springer Great Britain, pp.587-596

Scott, C.M., Effler S.W., Depinto J.V., Trama F.B., Rodgers P.W., Dobi J.S., Wodka M.C. (1985) "Dissolved oxygen model for a dynamic reservoir," J. Envir. Engng Div., ASCE 111: 647-665

Seip, K.L. (1991) "Decisions with multiple environmental objectives. The sitting of oil drilling wells in Norway." In: Ganoulis J. (ed.) Water Resources Engineering Risk Assessment. Springer Berlin, pp.503-524

Seip, K.L., Ibrekk H., Wenstmp F. (1987) "Multiattribute analysis of the impact on society of phosphorus abatement measures." Water Resour. Res. 23: 755-764

Seixas, M.J.F., Camara A., Antunes M.P., Pinheiro M. (1991) "Accommodating structural change in environmental systems: The approach of qualitative simulation," J. Forecasting 10: 211-230

Sengupta, S.E., Nwadike E., Lee S.S. (1981) "Long-term simulation of stratification in cooling lakes," Appl. Math. Model. 5: 313-321

Shanahan, P., Harleman D.R.F. (1986) "Lake eutrophication model: A coupled hydrophysical-ecological model," In: Somlyody L., van Straten G. (ed.) Modeling and Managing Shallow Lake Eutrophication. Springer New York, pp.256-284

Siegfried, C. A., Sutherland J. W. (1989) Empirical prediction of zooplankton biomass in Adirondack Lakes. Lake Reserv. Manage. 5: 91-97

Sklar, F.H., Costanza R., Day J.W. (1990) "Model conceptual-ization." In: Patten B.C., Jφrgensen S.E., Dumont H. (ed.) Wetlands and Shallow Continental Water Bodies, Vol. 1. SPB Academic Publishing The Netherlands, pp.625-658

Skogerboe, G.V., Walker W.R., Evans R.G. (1983) "Modelling water quality from irrigated an rainfed agricultural lands." In: Jφrgensen S.E., Mitsch W.J. (ed.) Application of Ecological Modelling in Environmental Management, Part A. Elsevier Amsterdam The Netherlands, pp.635-730

Sogn, T.A. (1993) "A test of chemical equilibrium equations and assumptions commonly used in soil-oriented charge balance models for soil and freshwater acidification," Ecol. Modelling 70: 221-238

Somlyódy, L., Kularathna M., Masliev I. (1993) Water Quality Management of the Nitra River Basin (Slovakia): Evaluation of Various Control Strategies. IIASA WP-1-93. International Institute of Applied Systems Analysis Laxenburg Austria

Somlyódy, L.S., van Straten G. (1986) Modelling and Managing Shallow Lake Eutrophication. Springer New York

Somlyódy, L.S., Varis O. (1992) Water Quality Modeling of Rivers and Lakes. IIASA WP-92-041. International Institute of Applied Systems Analysis Laxenburg Austria

Somlyódy, L.S., Wets R.J.B. (1988) "Stochastic optimization models for lake eutrophication management," Operations Research 36: 660-681

Smith, V.H. (1982) "The nitrogen and phosphorus dependence of algal biomass in lakes: An empirical and theoretical analysis," Limnol. Oceanogr. 27: 1101-1112

Spigel, R.H., Imberger J., Rayner K.N. (1986) "Modeling the diurnal mixed layer," Limnol. Oceanogr. 31: 533-556

Stauffer, R.E. (1987) "Effects of oxygen transport on the areal hypolimnetic oxygen deficit," Water Resour. Res. 23: 1887-1892

Stefan, H.G., Cardoni J.J. (1982) RESQUAL II: A dynamic water quality simulation program for a stratified shallow lake or reservoir: application to Lake Chicot, Arkansas. University of Minnesota

Stefan, H.G., Fang X. (1994a) "Dissolved oxygen model for regional lake analysis," Ecol. Modelling 71: 37-68

Stefan, H.G., Fang X. (1994b) "Model simulations of dissolved oxygen characteristics in Minnesota lakes: Past and future," Environmental Management 18: 73-92

Straškraba, M. (1976a) "Development of an analytical phyto-plankton model with parameters empirically related to dominant controlling variables," Abhandlungen der Akademie der Wissenschaften der DDR 1974: 33-65

Straškraba, M. (1976b) "Empirical and analytical models of eutrophication," Proc. Eutrosym. Karl-Marx-Stadt 3: 352-371

Straškraba, M. (1978) "Theoretical considerations on eutrophication," Verh. Int. Verein. Limnol. 20: 2714-2720

Straškraba, M. (1979) "Mathematische Simulation der Produktionssteuerung in Talsperren," Z. f. Wasser - und Abwasser - Forschung 12: 56-64

Straškraba, M. (1984) "New ways of eutrophication abatement." In: Straškraba M., Brandl Z., Porcalová P. (Eds) Hydrobiology and Water Quality of Reservoirs. Czechoslovak Academy of Sciences České Budějovice Czechoslovakia, pp. 37-45

Straškraba, M. (1985) "Managing of eutrophication by means of ecotechnology and mathematical modelling." Lakes Pollution and Recovery, Proc. Internat. Congress, Rome 15-18th April 1985. European Water Pollution Control Association Rome, pp.17-27

Straškraba, M. (1986) "Ecotechnological measures against eutrophication," Limnologica (Berlin) 17: 239-249

Straškraba, M. (1993) "Ecotechnology as a new means for environmental management," Ecological Engineering 2: 311-331

Straškraba, M. (1994) "Ecotechnological models for reservoir water quality management," Ecol. Modelling 74: 1-38

Straškraba, M., Blažka P., Brandl Z., Desortová B., Komárková J., Kubečka J., Procházková L., Seďa J. (1990) "A hypothesis on reservoir aging. Arch," Hydrobiol., Beih. Ergebn. Limnol. 3: 803

Straškraba, M., Gnauck A. (1985) Freshwater Ecosystems. Modelling and Simulation. Elsevier Amsterdam The Netherlands

Straškraba, M., Mitsch W., Jørgensen S.E. (1988) "Wetland modelling - an introduction and overview." In: Mitsch W.J., Straškraba M., Jorgensen S.E. (ed.) Wetland Modelling. Elsevier Amsterdam The Netherlands, pp.1-8

Straškraba, M., Tundisi J.G., Duncan A. (1993) "State of the art of reservoir limnology and water quality management." In: Straškraba, M., Tundisi J.G. & Duncan A. (ed.) Comparative Reservoir Limnology. Kluwer Academic Publishers Dordrecht The Netherlands, pp.213-289

Straškraba, M., Tundisi J.G., Duncan A. (Eds) (1993) Comparative Reservoir Limnology and Water Quality Management. Kluwer Academic Publishers Dordrecht The Netherlands

Straškraba, M., Dostálková I. Hejzlar J., Vyhnálek V. (in press). "The Effect of Reservoirs on Phosphorus Concentration." Intern. Revue der ges. Hydrobiologie: in press

Straškrabová, V. (1976) "Self-purification of impoundments," Water Res. 9: 1171-1177

Straškrabová, V., Desortová B., Šimek K., Vyhnálek V., Bojanovski B. (1983) "Ovlivnění biochemické spotřeby kyslíku v povrchových vodách přítomností řas." (The influence of algae on biochemical oxygen demand in surface waters) (In Czech). Vodní hospodářství, B 33: 165-168

Sullivan, P., Swartzman G., Bindman A. (1983) Process Notebook for Aquatic Ecosystem Simulations. Centre for Quantitative Sciences Seattle Washington

Sverdrup, H., Warfvinge P. (1985) "A reacidification model for acidified lakes neutralized with calcite," Water Resour. Res. 21: 1374-1380

Sverdrup, H., Warfvinge P., Bjerle I. (1986) "A simple method to predict the time required to reacidify a limed lake," Watten 42: 10-15

Sydow, A., Bellman K., Straubel R., Imming I., Kaschenz H., Damrath U., Hofmann G., Andres S. (1988) "PEMU- An impact model based environment protection decision support system." In: Sydow A., Tzafestas S.G., Vichnevetsky R. (ed.) Mathematical Research. Systems Analysis and Simulation. Bd 46 Vol.1. Akademie-Verlag Berlin, pp.37-41

Symons, J.M., Irwin W.H., Clark R.M., Robeck G.G. (1987) "Management and measurement of DO in impoundments," J. Sanit. Engng Div., ASCE 93: 181-209

Taylor, G. J., Crowder A.A. (1983) "Accumulation of atmospherically deposited metals in wetland soils of Sudbury, Ontario," Water, Air, and Soil Pollution 19: 29-42

Teruggi, S., Vendegna V. (1986) FOREL: A Tool for the choice of the mitigating measures of environmental impact on lakes. In: Zannetti P. (ed.) Envirosoft 86. Computational Mechanics Publications Great Britain, pp.303-322

Teruggi, S., Vendegna V., Vaccarone M. (1988) "WQLEDGER: An interactive software for streams pollution budget." In: Zannetti P. (ed.) Computer Techniques in Environmental Studies. Springer Berlin, pp.3-12

Thérien, N., Spiller G. (1981) "A mathematical model of the decomposition of flooded vegetation in reservoirs". Soc. Computer Simulation California, Simulation Conference Proc. Ser. V9 (2), pp.87-98

Thérien, N., Spiller G., Coupal B. (1982) "Simulation de le decomposition de la matiére végetale et des sols inondés dans les reservoirs de la region de la Baie de James," Can. Water Res. J. 7: 375-396

Thomann, R.V. (1982) "Verification of water quality models," J. Envir. Engng Div., ASCE 108: 923-940

Thomann, R.V., Mueller J.A. (1987) Principles of Surface Water Quality Modeling and Control. Harper Collins Publishers New York

Thomas, R., Maas R.J.M., Swart R.J., Kok R.M. (1990) "Environmental information and planning system 'RIM'." In: Fenhann J., Larsen H., Mackenzie G.A., Rasmussen B. (ed.) Environemntal Models: Emissions and Consequences/Riso International Conference 22-25 May 1989. Elsevier Amsterdam The Netherlands, pp.301-308

Thornton, K.W., Ford D.E., Norton J.L. (1983) "Use of Monte Carlo water quality simulation to evaluate reservoir operation and management alternatives." In: Lauenroth W.K., Skogerboe G.V., Flug M.(Eds) (ed.) Analysis of Ecological Systems: State- of- the- Art in Ecological Modelling. Elsevier Amsterdam The Netherlands, pp.641-648

Tim, U.S., Joly R. (1994) "Evaluating agricultural nonpoint - source pollution using integrated geographic information systems and hydrologic water quality model." J. Environ. Qual. 23: 23-35

Toetz, D.W. (1990) "Loading model predictions of total phosphorus in an offset reservoir," Lake Reserv. Manage. 6: 181-186

Tremback, C., Tripoli G., Arritt R., Cotton W.R., Pielke R.A. (1986) "The regional atmospheric modeling system." In: Zannetti P. (ed.) Envirosoft 86. Computational Mechanics Publications Great Britain, pp.601-607

Tubbs, L.J., Haith D.A. (1981) "Simulation model for agricultural non-point-source pollution," J. Wat. Pollut. Control. Fed. 53: 1425-1433

Turner, R.L., Laws E.A., Harris R.C. (1983) "Nutrient retention and transformation in relation to hydraulic flushing rate in a small impoundment," Freshw. Biol. 13: 113-127

Twinch, A.J., Grobler D.C. (1986) "Pre-impoundment as a eutrophication management option: a simulation study at Hartbeespoon Dam," Water S.A. 12: 19-26

Uhlmann, D. (1983) "Entwicklungstendenzen der Ökotechnologie," Wissenschaftliche Zeitschrift der Technischen Universität Dresden 32: 109-116

Uhlmann, D., Benndorf J., Albert W. (1971) "Prognose des Stoffhaushaltes von Staugewässern mit Hilfe kontinuierlicher oder semikontinuierlicher Modelle. I. Grundlagen," Int. Rev. ges. Hydrobiol. 56: 513-539

Uhlmann, D., Benndorf J., Gnauck A. (1977) "Entwicklung von Modellen zur Vorhersage der Phytoplankton-entwicklung und Wasserbeschaffenheit in eutrophierten Staugewässern," Wissen-schaftliche Zeitschrift der Technischen Universität Dresden 26: 271-278

Ulrich, M., Schwarzenbach R.P., Imboden D.M. (1991) "MASAS - Modelling of anthropogenic substances in aquatic systems on personal computers - application on lakes," Environm. Softw. 6: 34-38

van der Molen, D.T., Los F.J., van Ballegooijen L., van der Vat M.P. (1994) "Mathematical modelling as a tool for managment in eutrophication control of shallow lakes," Hydrobiologia 275/276: 479-492

van Liere, L., Janse J.H. (1992) "Restoration and resilience to recovery of the Lake Loosdrecht ecosystem in relation to its phosphorus flow," Hydrobiologia 233: 95-104

van Straten, G. (1986a) "Lake eutrophication models." In: Somlyody L., van Straten G. (ed.) Modeling and Managing Shallow Lake Eutrophication. Springer New York, pp.35-68

van Straten, G. (1986b) Identification, Uncertainty Assessment and Prediction in Lake Eutrophication. Proefschrift ter verkrijging van de graad van doctor in de technische wetenschappen aan de Universiteit Twente, Twente

Varis, O. (1988) "Temporal sensitivity of Aphanizomenon flos-aquae dominance - A whole-lake simulation study with input perturbations," Ecol. Modelling 43: 137-153

Varis, O. (1994) "Water quality models: tools for the analysis of data, knowledge and decisions." Water Quality International '94. IAWQ 17th Biennial International Conference, 24-29 July 1994 Budapest Hungary, pp.13-19

Varis, O., Kettunen J., Sirviö H. (1988) "Influence diagrams in environmental management - a review of the DAVID system." In: Zannetti P. (ed.) Computer Techniques in Environmental Studies. Springer Great Britain, pp.609-618

Veselka, T., Ballou S.W., Dauzvardis P.A. (1981) "An approach to modelling vegetation yield reduction from point source emissions of sulfur dioxide." In: Mitsch W.J., Bosserman R.W., Klopatek J.M. (ed.) Energy and Ecological Modelling. Proceedings of a Symposium held from 20 to 23 April 1981 at Louisville Kentucky. Elsevier Amsterdam The Netherlands, pp.339-346

Viaroli, P. (1993) "Studies on regeneration and loss of nitrogen in two different macrophyte - dominated wetlands." In: Giussani G., Callieri C. (eds). Strategies for Lake Ecosystems Beyond 2000. 5th International Conference on the Conservation and Management of Lakes Stressa (Italy), pp. 203-206

Viaroli, P., Gorbi G., Campanini G. (1991) "Accumulation of nutrient and heavy metals in an experimental macrophyte covered lagoon system." In: Madoni P. (ed.) Biological Approach to Sewage Treatment Process: Current Status and Perspectives. Perugia, pp.379-384

Vincent, T.L. (1987) "Modeling and management in the presence of uncertain inputs," Can. J. Fish. Aquat. Sci.: 267-274

Virtanen, M. (1989) "Mathematical models within river basin management - an overview," Aqua Fennica 19: 145-152

Virtanen, M., Koponen J., Dahlbo K., Sarkkula J. (1986) "Three-dimensional water-quality-transport model compared with field observations," Ecol. Modelling 31: 185-199

Vollenweider, R.A. (1968) The Scientific Basis of Lake and Stream Eutrophication, with Particular Reference to Phosphorus and Nitrogen as Factors in Eutrophication. Technical Report to O.E.C.D. Paris, DAS/CSI/68.27

Vollenweider, R.A. (1975)" Input-output models. With special reference to the phosphorus loading concept in limnology," Schweiz. Z. Hydrol. 37: 53-84

Vollenweider, R.A. (1976) "Advances in defining critical loading levels for phosphorus in lake eutrophication," Mem. Ist. ital. Idrobiol. 33: 53-83

Vollenweider, R.A., Janus L.L. (1982) "Statistical models for predicting hypolimnetic oxygen depletion rates,' Mem. Ist. Ital. Idrobiol. 40: 1-24

Walker, R.R., Snodgrass W.J. (1986) "Model for sediment oxygen demand in lakes," J. Hydraulic Engng Div. , ASCE 112: 25-43

Walker, W.W. Jr. (1981) Empirical methods for predicting eutrophication in impoundments. Report 1, Phase I: Data base development. Technical Report E-81-9. U.S. Army Engineer Waterways Experiment Station Vicksburg Mississippi

Walker, W.W.Jr. (1986) "Models and software for reservoir eutrophication assessment," Lake Reserv. Manage. 2: 143-148

Welch, E.B., Schrieve G.D. (1994) "Alum treatment effectiveness and longevity in shallow lakes," Hydrobiologia 275/276: 423-431

Wenger, R.B., Rong Y. (1987) "Two fuzzy set models for comprehensive environmental decision-making," J. Environ. Manage. 25: 167-180

Whitehead, P.G. (1993) "Dynamic modelling of nitrate in reservoirs and lakes," Wat. Res. 27: 1377-1384

Wilhelmus, B., H. Bernhardt, D. Neumann (1978) "Vergleichende Untersuchungen über die Phosphoreliminierung von Vorsperren - Verminderung der Algenentwicklung in Speicherbecken und Talsperren," DFGW-Schriftenreihe Wasser 16: 140-176

Wüest, A., Brooks N.H., Imboden D.M. (1992) "Bubble plume modelling for lake restoration," Water Resour. Res. 28: 3235-3250

Wylie, G.D., Jones J.R. (1991) "Assessment of the sensitivity of Missouri reservoirs to acidification," J. Freshwater Ecol. 6: 431-437

Yao, H.M., Georgakakos A.P. (1993) "New Control Concepts for Uncertain Water Resources Systems. 2. Reservoir Management," Water Resour. Res. 29: 1517-1525

Young, R.A., Onstad C.A.,Bosch D.D., Anderson W.P. (1989) "AGNPS: a nonpoint-source pollution model for evaluating agricultural watersheds," J. Soil Water Conservation 44: 168-172

Zwirnmann, K.H. (1982) Nonpoint Nitrate Pollution of Municipal Water Supply Sources: Issues of Analysis and Control. IIASA CP-82-S4. International Institute of Applied Systems Analysis Laxenburg Austria

CHAPTER 4

MODELING TOXIC CONTAMINANTS IN AN AQUATIC ENVIRONMENT

S.E. Jørgensen[1]

1. USE OF TOXIC SUBSTANCE MODELS

Models are increasingly used to overview complex problems, and all problems related to the environment are complex problems due to the enormous number of many interacting components. Consequently, models attempt to capture the most essential components and processes related to a specific problem in a specific environment. Of course the model cannot contain all components and processes involved in a problem, but only the most important ones for the focal problem may be considered. If all processes were to be included the model would not meet the required simplification it would become the system itself.

Models have been used in science and technology in many different contexts, but environmental modeling is a particular challenge due to many interacting components in ecosystems (environmental systems). In ecosystems everything is linked to everything and, therefore, it is not possible in principle to omit any component in the model. In practice we can, however, examine the most essential mass- and/or energy flows for the focal problem, and limit our model to these flows and their related components.

The processes and components of minor importance may be taken care of through calibration of the model. Calibration is understood as the process where the model outputs are tested with measurements from the focal ecosystems and it is allowed to change a number of previously selected parameters within some indicated realistic ranges. Mainly, parameters known with the smallest certainty from literature or previous studies are selected for calibration, and by changing these parameters within realistic ranges, it is possible to a certain extent to account for processes and components, that are not included in the model due to their insignificance.

---

[1] DFH, Environmental Chemistry, Copenhagen Ø, Denmark

NATO ASI Series, Partnership Sub-Series, 2. Environment – Vol. 3
Remediation and Management of Degraded River Basins
Edited by V. Novotny and L. Somlyódy

Development of a model must always contain at least six elements:

1) conceptualization (usually a diagram showing the relationship between components and processes);
2) translation of the conceptual diagram into a set of mathematical equations;
3) verification, where the internal logic of the model is tested;
4) sensitivity analysis, where the sensitivity of the most crucial components to changes in model inputs, parameters and equations is found;
5) calibration, where a number of selected parameters are changed to improve the fit between observations and model outputs; and
6) validation, where the model is tested with an independent set of data without any change of the parameters. The validation gives the modeler opportunity to express the accuracy of the model to cope with the examined case study.

Models of toxic substances in our environment seem to be a logical area of application. Toxic substances participate in many interacting processes and are accumulated in many living and non-living components. Therefore, models are the ultimate tool in a survey of the effect and fate of toxic substances. Since the mid seventies, many models have been developed for fate and effects of toxic substances in the environment.

However, models must not be considered as a mysterious new tool to solve pollution problems, but should be considered only as a synthesizing tool, which facilitates overview of our knowledge about a specific problem. Ecotoxicological models are therefore generally less accurate than other ecological models mainly due to limited knowledge on model parameters. It implies that ecotoxicological models often only can give semi-quantitative results. This is, however, extremely useful in an environmental management context, where the question often is, whether the concentration level of a toxic substance is close to an assessed maximum concentration or whether it is a magnitude/magnitudes lower.

It is preposterous to develop a model in a vacuum of knowledge. Models of toxic substances in our environment can therefore only be developed, if we have good data and a basic knowledge about the most frequent processes for toxic substances in ecosystems. The next section is therefore devoted to an overview of the most important processes for contaminants in our environment and to what extent we can describe these processes mathematically. The third section of the chapter focuses on the characteristic features and the classification of ecotoxicological models, followed by a section, that gives an overview of existing models of toxic substances in aquatic ecosystems. The final section attempts to summarize and conclude the state of the art in modeling toxic substances in an aquatic environment.

2. IMPORTANT PROCESSES FOR CONTAMINATS IN AN AQUATIC ECOSYSTEM

2.1 Overview

Toxic substances may participate in many processes in the environment. It is therefore important in a modeling context to describe these processes by well-tested and theoretically sound mathematical equations.

Distribution of toxic compounds between a solid and an aquatic phase is of utmost importance, because it is most often the dissolved fraction, which might cause toxicity because toxic substances are mainly taken up in a dissolved form. Two processes should be mentioned in this context: adsorption (including ion exchange) and precipitation.

Many toxic substances are also able to perform an acid-base reaction, particularly by reaction with water (hydrolysis). Furthermore, they may (for instance heavy metals) exist in several oxidation states. The redox potential determines therefore the form of many toxic substances, particularly the oxidation state of heavy metals.

Heavy metals in general form many and strong complexes. As our aquatic environments contain many natural and man-made ligands, the possibility for formation of heavy metal complexes is almost infinite. These processes are of particular importance, because many of the complexes are less toxic than the pure ionic forms of heavy metals as shown in Jørgensen (1990).

Toxic substances interact also with living organisms. They are taken up by plants and animals directly from the media, i.e., for aquatic ecosystem water. Furthermore, they may be a micro-component in food and therefore taken up by digestion processes. The efficiency of the uptake through the intestines is in most cases strongly dependent on the partition coefficient, K_{ow}. Excretion processes make it possible to eliminate toxic substances from organisms.

Adsorption

Adsorption may be described by simply using a concentration factor known as a linear partition coefficient, giving the ratio of concentration between the solid and aquatic phase. This approach is a particular good approximation, when the concentration in the aquatic phase is low. In some cases available data may follow the more accurate Freundlich's and Langmuir's adsorption isotherms defined as

Langmuir Isotherm

$$Ss = DCw/(Km + Cw) \tag{1}$$

Freundlich Isotherm

$$Ss = K\,Cw^{1/n} \tag{2}$$

where

Ss is the weight of the considered component sorbed per g of sorbent,
Cw is the weight in solution per liter or ml of solution, and
D, K, Km, and n are constants.

At low substrate concentrations, n is often close to 1, and K becomes a partition coefficient.

Smith et al. (1977) have shown, in a limited number of case studies, that for a given sorbent the logarithm of the partition coefficient (it is here defined as the ratio of the concentration of considered compound in sediment and in water) and the logarithm of water solubility are linearly related. Similar relations are widely applied in ecotoxicology. Although this relationship seems to be generally valid, compounds that interact via an ion exchange process probably would not fit this plot.

Adsorption processes are of importance for modeling pollution from non-point sources, for modeling the distribution of toxic substances in an aquatic ecosystem, for a quantitative description of the exchange of nutrient or toxic substances between sediment and water, and for modeling soil processes.

In most cases the adsorption process is very fast and equilibrium is attained in minutes, or hours at most. This implies that in models using days as time steps, it is sufficient to describe the process by application of the above mentioned equilibrium terms. If it is required to include the rate of the adsorption process, this can be done by using a first order expression:

$$dSs/dt = Ka(Se - Ss) \tag{3}$$

where Se is Ss at equilibrium.

Sorption includes not only the physical process of adsorption, but also a number of chemical processes named chemosorption. Here the mechanisms are more complicated, but in many cases it is possible to use the same basic mathematical description.

Precipitation

Precipitation of all heavy metals involves a low solubility in water due to the formation of hydroxides and sulfides. However, while the precipitation process can be described in most ecological models by means of an equilibrium expression, settling of suspended matter formed by means of the equilibrium expression, can be described as a first-order reaction.

Acid-base reactions and hydrolysis processes

These processes are of great environmental interest, because almost all processes in the environment are dependent on pH. A few illustrative examples of acid-base reactions are included in the following list:

1. Release of heavy metals from sediment and soil increases rapidly with decreasing pH. The influence of H^+ on the solubility process is the major factor. Concentration of sulfide ions decreases with decreasing pH, as sulfide ions react with H^+ and form hydrogen sulfide and dihydrogen sulfide. This increased solubility of heavy metals by lower pH values implies a significant higher heavy metal concentration in acid waters. Organic acids are generally more soluble at higher pH due to hydrolysis. Phenol for instance is almost completely soluble in water above pH 8-9.

2. Most ligands are acid-base systems and have therefore different forms at different pH values.

3. Heavy metal hydroxides have a very small solubility product which implies that most heavy metal ions are precipitated at pH 7.5 or above. Many organic acids are only slightly soluble 1-2 pH units below the pk_a - value.

4. Many metal ions react with water by formation of metal-hydroxides and hydrogen ions:

$$Fe^{3+} + 3H_2O \rightarrow Fe(0H)_3\ (s) + 3H^+ \qquad (4)$$

5. Toxic substances are able to form a number of species as a result of hydrolysis.

Redox processes

Many inorganic ions are dominant participants in environmental *redox processes*. It is obviously important to describe the stability relationships for various inorganic components. This can be done in a simple, graphical representation using pe-pH diagrams (see also Chapter by Salomons). Such diagrams show, in a comprehensive way, how protons and electrons simultaneously shift the equilibrium under various conditions of ph and pH.

pe is defined as the negative logarithm of the relative electron activity

$$pe = -\log[e] \qquad (5)$$

which is a parallel to the pH definition:

$$pH = -\log[H+] \qquad (6)$$

pe is related to E, the electrode potential, by:

$$pe = \frac{F * E}{2.3\,RT} \qquad (7)$$

$$E = E^o + \frac{2.3RT}{nF} \log \frac{[ox]}{[red]} \qquad (8)$$

$$pe^o = \frac{F * E^o}{2.3\,RT} \qquad (9)$$

This implies that pe is a measure of free energy, ΔG:

$$pe = \frac{\Delta G}{-n * 2.3\,RT} \qquad pe^o = \frac{\Delta G^o}{-n * 2.3\,RT} \qquad (10)$$

(ΔG is the standard free energy and ΔG and ΔG^o refer to the half reaction written in the form of reduction), and that pe is related to the equilibrium constant as follows:

$$pe = (\log K)/n \qquad (11)$$

The use of these equations can be illustrated by the following process:

$$Fe^{3+} + e = Fe^{2+} \qquad (12)$$

For this process we have:

$$E^{o} = 0.77; \; pe^{o} = \frac{F * 0.77}{2.3 * R * 298} \quad (t = 25^{o}\,C) \qquad (13)$$

$$log\ K = n * pe^{o} = 1 * \frac{0.77}{0.059} = 13.1 \qquad (14)$$

If in an acid solution $[Fe^{3+}] = 10^{-3}$ and $[Fe^{2+}] = 10^{-2}$ (15)

$$pe = pe^{o} + \frac{1}{n}\ log\ \frac{[Fe^{3+}]}{[Fe^{2+}]} = 13.1 + \frac{1}{1}\ log\ 10^{-1}1 + 12.1 \qquad (16)$$

The importance of inorganic redox processes in the environmental context can be illustrated by examples.

If FeS_2 is exposed to air, e.g., by reduced water level in mines, the following processes will occur:

$$2FeS_2 + 2H_2O + 7O_2 = 2FeSO_4 + 2H_2SO_4 \qquad (17)$$

$$4FeSO_4 + O_2 + 2H_2SO_4 = 2Fe_2(SO_4)_3 + 2H_2O \qquad (18)$$

$$Fe_2(SO_4)_3 + 6H_2O = 2Fe(OH)_3\ (s) + 3H_2SO_4 \qquad (19)$$

As will be seen, formation of considerable amounts of sulfuric acid occurs, resulting in extremely low pH values, which in many cases cause great damage to the aquatic environment.

Edgington and Callender (1970) mention another example. Lake Michigan has ferro-manganese nodules which contain unexpectedly large concentrations of arsenic (up to 345 ppm, but averaging 180 ppm). Under aerobic conditions, the nodules are stable, but under anaerobic conditions arsenic will be released. As arsenic is highly toxic to mammals and also carcinogenic it is obviously of great importance to formulate the redox processes in Lake Michigan sediment to provide predictions for release of arsenic.

Figure 1 attempts to summarize these considerations. If redox processes seem to be of importance for a model, several questions should be answered before the model equations are formulated.

These questions are:

* What is the time step needed for the model?
* What is the rate of the processes compared with the time steps?
* Is there a chain of reactions?
* Which reaction step determines the overall rate?

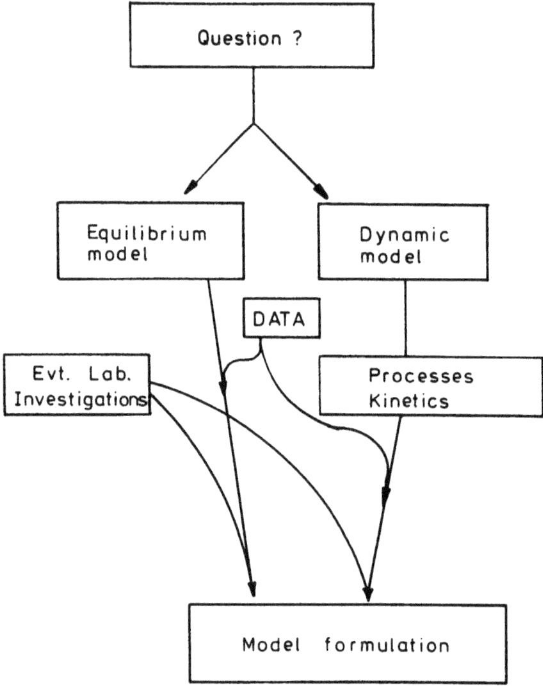

Figure 1. The diagram illustrates how it is feasible to construct step-wise a model of redox processes

Complex formation

These processes are of particular interest for heavy metal pollution. As for other chemical processes, the equilibrium expression is also valid for these reactions. For the process:

$$Me^{n+} + L^{-m} \rightarrow MeL^{(n-m)+} \tag{21}$$

where Me denotes a metal, and L a ligand, the equilibrium expression assumes:

$$\frac{[MeL^{(n-m)+}]}{[Me^{n+}]\ [Al^{-}]} \tag{22}$$

K, the equilibrium constant, is termed the stability constant or formation constant, if the process - as here - is the formation of a complex.

The co-ordination number of a metal is the number of bonds formed from the metal ion to the donor atoms of the ligands. Even coordination numbers (2, 4, 6 and 8) are much more common than odd ones, *coordination numbers 4 and 6 being most frequently encountered.*

Formation *of co-ordination complexes* plays a major role in modeling distribution as well as effects of metal ions in aquatic ecosystems due to the following influences:

1) increase of metal solubility:

$$MeY(S) + L \rightarrow MeL + Y \tag{23}$$

where Me represents a metal which is bound in the sediment to the compound MeY;
L denotes a ligand,

2) alteration of the distribution between oxidized and reduced forms of metals. The equilibrium constant for the process

$$Me^{n+} + me- \rightarrow Me^{(n-m)+} \tag{24}$$

differs from #1 when

$$MeL^{n+} + me- => MeL^{(n-m)+} \tag{25}$$

if the complexes, MeL^{n+} and $MeL^{(n-m)+}$, have different stability constants, which most often is the case.

3) alleviation of toxicity due to alteration of availability to aquatic life,

4) change in adsorption and ion exchange on sediment and suspended matter, and

5) change in stability of metal containing colloids.

The presence of ligands in water will generally increase the transfer of metals from sediment, soil and suspended matter, which increases solubility, but as these complexes show lower toxicity than metal ions, harm to the aquatic life in most cases does not increase proportionally to increased solubility.

Application of *co-ordination chemistry to environmental problems* is very complex, as a great number of ligands are present in aquatic ecosystems with all the ligands competing simultaneously for formation of complexes with metal ions.

Formation of heavy metal complexes is of equal environmental interest, as the toxic effect of heavy metal in most cases is related to free ions. Stability constants of some heavy metal complexes of environmental interest are given in Table 1. As can be seen from these constants, considerable amounts of heavy metals can be bound as complexes by naturally occurring complexes. It implies that the toxic effects are reduced approximately correspondingly.

Table 1. Stability Constants (-pK) of complexes present in water.

Cation	Glycine	ATP	Glutathione	Acetic acid
Ca^{2+}	1.31	3.60		0.39
Mg^{2+}	3.44	4.00		0.28
Zn^{2+}	5.52	4.85	8.30	1.57
Cd^{2+}	4.80		10.50	1.70
Cu^{2+}	8.62	6.13		2.24

Figure 2 shows a model of copper contamination in an aquatic ecosystem which considers the formation of complexes. The main concern is the concentration of copper ions in the water that may reach a toxic level to the phytoplankton. Zooplankton and fish are much less sensitive to copper contamination, so the concentration level of greatest interest is the level that is harmful to phytoplankton. It is therefore necessary to model the partition of copper in ionic form, complex bound form and adsorbed form.

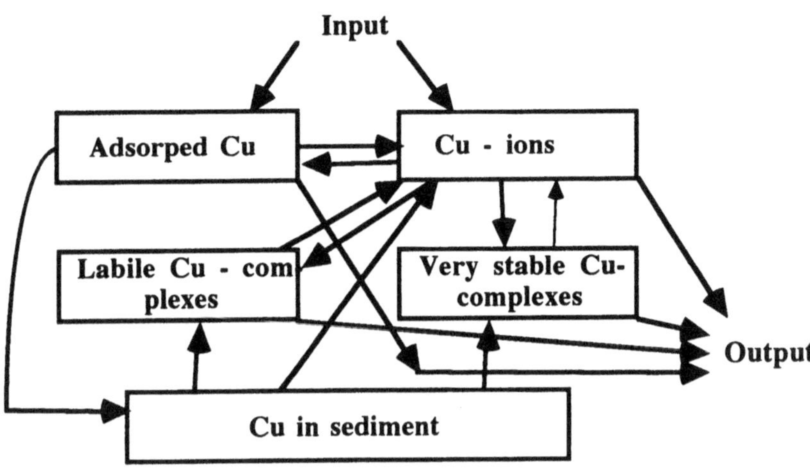

Figure 2. Conceptual Diagram of simple copper model.

Interactions of heavy metals with living organisms

Figure 3 illustrates possible interactions of toxic substances with living organisms. The uptake by living organisms is from the medium (water) and from digested food (= total food - undigested food). Outputs are mortality (transfer to detritus), excretion and predation from the next level in the food chain. Uptake directly from the medium is often described by a simple concentration factor, CF, giving a fixed ratio between the concentration in the organisms and the concentration in the water. If the uptake includes the indirect route through the food chain, the expression biological concentration factor is often used.

Biomagnification describes the process by which pollutants are passed from one trophic level to the next and exhibits most frequently increasing concentrations in organisms related to their tropic status. Magnification, i.e., the increasing concentration through the food chain, is caused by retention of heavy metal. This process may be described by the so-called biological or ecological magnification factor, which indicates magnification from one level in the food chain to the next. Figure 4 illustrates a conceptual diagram of a food chain model of lead.

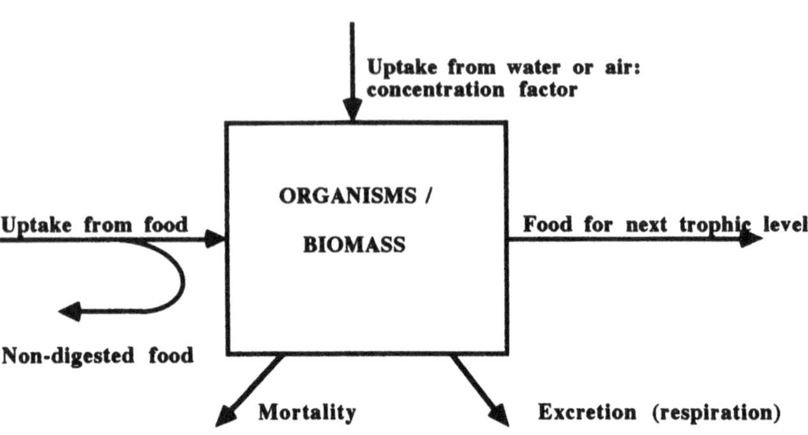

Figure 3. Processes of interest for modeling the concentration of a heavy metal at one trophic level.

Allometric relationships may also be used to find parameters for models of heavy metals interacting with living organisms. The surface determines the extent of contact between an organism and its environment. Therefore uptake and excretion of toxic substances are related to surface. It is illustrated in Figure 5. Rates are increasing by $w^{2/3}$ or l^1. This implies that specific rates (g excreted per unit of time per g body weight) versus length of various organisms gives a slope on approximately - 1 in a log-log diagram. The validity of this relationship is exemplified on Figure 5 for excretion rate. Other examinations have, however, shown that it has a general validity for toxic substances; see Jørgensen (1984), (1990), (1994) and (1995).

Biodegradation

The substrate concentration is low in most cases of ecotoxicological interest, and the rate of biodegradation may be expressed as a second-order or a first-order reaction scheme. The latter is valid when the concentration of microorganisms is constant, for instance under constant environmental conditions.

However, it is rather difficult to estimate the reaction rate of biodegradation in ecology because it is influenced by many ecologica factors. Since many factors considerably affect the reaction therefore also the concentration of microorganisms is affected as well.

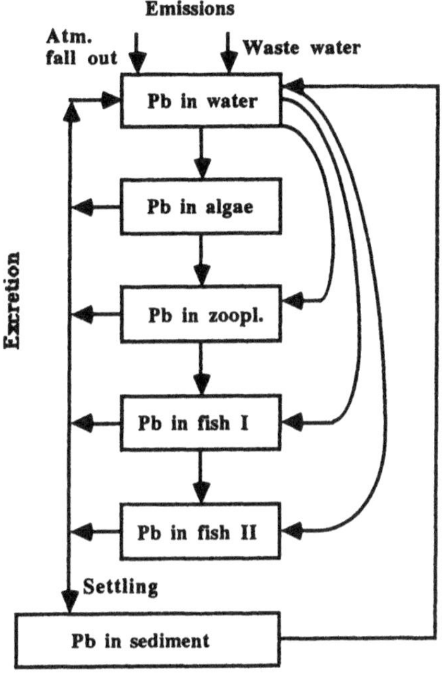

Figure 4.

Conceptual diagram of the biomagnification (bioaccumulation) of lead through a food chain in an aquatic ecosystem.

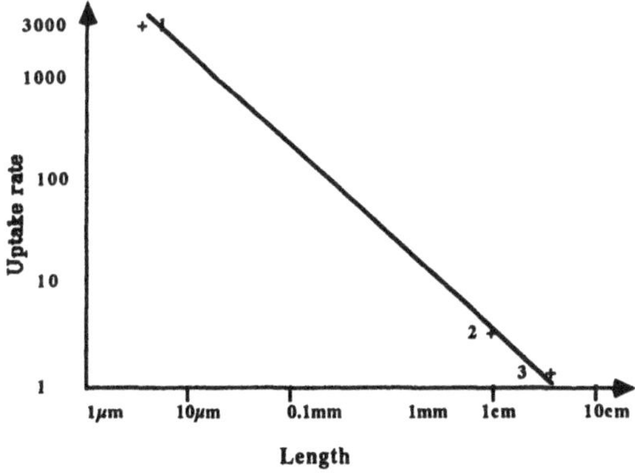

Figure 5.

Uptake rate (μg/g 24h) of cadmium plotted versus the length of various animals: (1) phytoplankton, (2) clams, (3) oysters. Jørgensen (1984).

The most important factors are:

1) *Temperature.* This factor may be taken into consideration by use of the same equations as used in eutrophication or BOD models.
2) *Adaptation.* Previous exposure of natural microbial population to a particular chemical compound often has an influence on the degradation rate. A rather detailed examination is required to model the adaptation period and in most cases it is therefore necessary to presume that the adaptation is either complete or has not yet been initiated.
3) *Sorption.* This process was considered previously and its influence on biodegradation is simply considered by inclusion of sorption isotherms in the model.
4) *Redox potential.* This environmental property determines which terminal electron acceptor is used and, hence, also determines the composition of microorganisms. Significantly increased (see, for example, Larson and Perry 1981) as well as decreased rates of biodegradation have been observed in anaerobic environments.
5) *Ionic strength.* The degradation rate will in general decrease as the ionic strength increases.
6) *Nutrients.* Microbial growth is affected by major nutrients as well as by essential micronutrients, including trace metals. The composition of microorganisms indicates that the biomass contains approximately 0.5 -1.0% phosphorus and 5-9% nitrogen. Oil pollution problems can be solved by microbiological degradation of hydrocarbons, but the presence of nutrients in sufficient concentrations to ensure microbial growth is a necessary prerequisite.
7) *Soil Composition.* Rates of biodegradation vary considerably for various types of soil due to differences in availability of organic nutrients for microorganisms.

Some qualitative rules for the relationship between molecular structure and rate of biodegradation are be definedas folllows:

1) *Aromatic compounds* are in general less biodegradable than aliphatic compounds.
2) The higher *the molecular weight*, the less biodegradable the compound.
3) The *more substituent groups* the molecule contains the less biodegradable the compound.
4) *Double bonds* will in general increase biodegradability
5) *-O- and -N=* will in general decrease biodegradability.

A quantitative relationship between biodegradation kinetics and chemical structure exists, as also pointed out by Jørgensen, 1990, but it is weak, particularly because the environmental factors mentioned previously play a major role.

The degree of biodegradation after 7 days of incubation, BD7, or the biological half life time is often mentioned in the literature. These values may be found in Jørgensen et al. (1991) and Jørgensen (1991).

Water-air exchange

The distribution of a chemical in the spheres is of particular interest in fate models.

Distribution Air-Water:

$$\frac{C^*}{To^*} = FL = \frac{H * 18}{1000R * T} \ (approximation) \tag{25}$$

or at 1 atm. and 20°C

$$FL = 7.49^* \ 10^{-4} \ ^* H \tag{26}$$

where H is Henry's constant (atm.). C^* and To^* are equilibrium concentrations (weight/volume) in air and water respectively.

Rate of Evaporation:

$$Ev = -D^* \ \frac{dTo}{dx} = b * KL * A \ (To - To^*), \tag{27}$$

where

D is the molecular diffusion coefficient ($L^2 \ T^{-1}$),

A is the area,

b is a coefficient, that adjusts the aeration coefficient, KL, to other compounds than oxygen (b = 1.0 for oxygen). b can be found in Jørgensen (1991) for other compounds.

Several empirical equations for calculation of KL can be found in Jørgensen and Gromiec (1989) or Jørgensen and Johnsen (1989).

Note that

$$b = \frac{KL - compound}{KL - oxygen} = \frac{D - compound}{D - oxygen} \tag{28}$$

Distribution water/solid (sediment)

Distribution coefficient, FS, may be used at low concentrations (see also Langmuir's and Freundlich's adsorption isotherms):

$$FS = C^{*} / To^{*} \tag{29}$$

where

C^{*} and To^{*} are the concentrations in the solid and water phase at equilibrium.

Since the unit of C^{*} is M/M (for instance mg/g dry matter) and To^{*} has the unit of ML-3 (for instance mg/l), FS gets the unit L3 M-1 (l/kg for instance).

FS may be found from:

$$FS = Foc^{*} foc \tag{30}$$

where Foc is the distribution coefficient for organic matter and foc is the fraction of organic carbon in the considered solid.

The following relationship between Foc and Kow exists (Brown and Flagg, 1981):

$$\log Foc = -0.006 + 0.937 \log Kow \tag{31}$$

for activated sludge:

$$\log FS = 0.39 + 0.67 \log Kow \text{ (l/kg as unit for FS)} \tag{32}$$

Foc and FS-values for selected chemicals can be found in the literature.

3. CHARACTERISTIC FEATURES AND CLASSIFICATION OF ECOTOXICOLOGICAL MODELS

Ecotoxicological models differ from ecological models in general by:

1) being most often more simple,
2) require more parameters,
3) using parameter estimation methods more widely, and
4) possibly inclusion of an effect component.

Ecotoxicological models can be divided into six classes. Classification presented herein is based on differences in the modeling structure. The decision as to which model class to apply is based upon the ecotoxicological problem that the model aims to solve. Definitions of the model classes are given below. It is also indicated where it is most appropriate to use each of the model types.

3.1. Food Chain or Food Web Dynamic Models

This class of models considers the flow of toxic substances through the food chain or food web. Such models will be relatively complex and contain many parameters, which often have to be estimated. This type of model will typically be used when many organisms are effected by the toxic substance, or the entire structure of the ecosystem is threatened by the presence of a toxic substance. Because of the complexity of these models, they have not been widely used. They are similar to the more complex eutrophication models that consider the flow of nutrients through the food chain or even through the food web. Sometimes they are even constructed as submodels of a eutrophication model, see for instance Thomann et al., 1974. Figure 4 shows a conceptual diagram of an ecotoxicological food chain model for lead. The input of lead is from atmospheric fallout and waste water to an aquatic ecosystem, where it is concentrated through the food chain. This is called 'bioaccumulation.' A simplification is hardly possible for this model, because the aim of the model is to describe and quantify the bioaccumulation through the food chain.

3.2. Static Models of the Mass Flows of Toxic Substances

If seasonal changes are minor, or of minor importance, a static model of the mass flows will often be sufficient to describe the situation and even to indicate expected changes if the input of toxic substances is reduced or enlarged. This type of model is based upon a mass balance. It will often, but not necessarily, contain more trophic levels, and the modeler often is concerned with the flow of the toxic substance through the food chain. If there are some seasonal changes, this type, which is usually simpler than type one, can still be advantageously used, for instance, if the modeler is concerned with the worst case and not the changes.

3.3. A Dynamic Model of a Toxic Substance in a Trophic Level

It is often only the toxic substance concentration in one trophic level which is of concern. This includes the zero trophic level, which is understood as the medium - either soil, water or air. Figure 2 gives an example. It is a model of copper contamination in an aquatic ecosystem. The main concern is the copper concentration in water, as it may reach a toxic level for phytoplankton.

Figure 6 gives another example. Therein the main concern is the DDT concentration in fish, because they may become so high that, according to WHO's standards, the fish is unfit for human consumption. The model can therefore be simplified by not including the entire food chain but only the fish. Some physical-chemical reactions in the water phase are still of importance and they are included as shown on the conceptual diagram in Fig. 6.

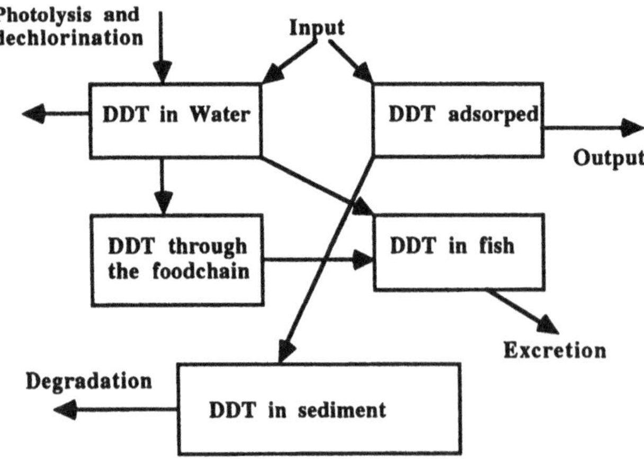

Figure 6. Conceptual diagram of a simple DDT-model.

As seen from these examples, simplifications are often feasible when the problem is well defined, including which component is most sensitive to toxic matter, and which processes are most important for concentration changes.

3.4. Ecotoxicological Models in Population Dynamics

Population models are biodemographic models and have therefore numbers of individuals or species as state variables. The simple population models consider only one population. The growth of the population is a result of the differences between natality and mortality:

$$dN/dt = B^{*}N - M^{*}N = r^{*}N, \tag{33}$$

where

N is the number of individuals,

B is natality i.e., the number of new individuals per unit of time and per unit of population,

M is mortality, i.e., the number of organisms that died per unit of time and per unit of population, and

r is the increase in the number of organisms per unit of time and per unit of population.

The concentration of toxic substance in the environment or in organisms influences the natality and mortality, and if the relation between a toxic substance concentration and these population dynamic parameters is included in the model, it becomes an ecotoxicological model of population dynamics.

Population dynamic models may include two or more trophic levels and ecotoxicological models will include the influence of the toxic substance concentration on natality, mortality and interactions between these populations. In other words, an ecotoxicological model of population dynamics is a general model of population dynamics with the inclusion of the relation between toxic substance concentrations and some of the model parameters.

3.5. Ecotoxicological Models with Effect Components

Though Class 4 models already include relationships between concentrations of toxic substances and their effects, they are limited to population dynamic parameters. In comparison Class 5 models include more comprehensive relationships between toxic substance concentrations and effects. These models may include not only lethal and/or sublethal effects but also effects on biochemical reactions or on the enzyme system. These effects may be considered on various levels of the biological hierarchy from cells to ecosystems.

In many problems it may be necessary to go into more detail on the effect to answer the following relevant questions:

1. Does the toxic substance accumulate in the organism?
2. What will be the long term concentration in the organism when uptake rate, excretion rate and biochemical decomposition rate are considered?
3. What is the chronic effect of this concentration?
4. Does the toxic substance accumulate in one or more organs?
5. What is the transfer between various parts of the organism?
6. Eventually will decomposition products cause additional effects?

Detailed answers to these questions may require a model of the processes that take place in the organism, and a translation of concentrations in various parts of the organism into effects. This implies, of course, that the intake which equals (uptake by the organism) times (efficiency of uptake) is known. Intake may either be from water or air, which also may be expressed by the use of concentration factors, which are the ratios between the concentration in the organism and in the air or water.

However, if all of the above mentioned processes should be taken into consideration for just a few organisms, the model will easily become too complex, contain too many parameters to calibrate, and require more detailed knowledge than is possible to provide. Therefore, most models in this class will not consider too many details of the partition of the toxic substances in organisms and their corresponding effects, but rather be limited to the simple accumulation in the organisms and their effect. In most cases, accumulation is rather easy to model and the following simple equation is often sufficiently accurate:

$$dC/dt = (ef^*Cf^*F + em^*Cm^*V)/W - Ex^*C = (INT)/W - Ex^*C \tag{34}$$

where

C is the concentration of the toxic substance in the organism;

ef and em are the efficiencies for the uptake from food and medium respectively (water or air);

Cf and Cm are the concentrations of the toxic substance in food and medium respectively;

F is the amount of food uptake per day;

V is the volume of water or air taken up per day;

W is the body weight either as dry or wet matter; and

Ex is the excretion coefficient (1/day).

As seen from the equation INT covers the total intake of toxic substance per day. This equation has a numerical solution, and the corresponding plot is shown in Figure 7:

$$C/C(max) = (INT^{*}(1 - exp(Ex^{*}t)))/(W^{*}Ex) \quad (35)$$

where

C(max) is the steady state value of C:

$$C(max) = INT/(W^{*}Ex) \quad (36)$$

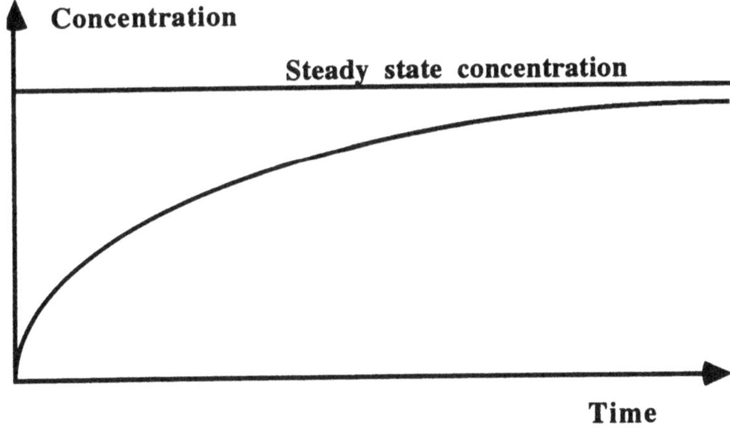

Figure 7. Concentration of a toxic substance in an organism versus time.

3.6. Fate Models with or without a Risk Assessment Component

The last type of ecotoxicological models focuses on the fate of the toxic substances, i.e., where in the ecosystem will the toxic substance be found? In what concentration?

The complete solution of an ecotoxicological problem requires in principle four (sub)models, of which the fate model may be considered the first model in the chain as shown on Figure 8. The 4 components presented on the figure are (see Morgan 1984):

1) A fate or exposure model.
2) An effect model, translating the concentration into an effect; see type 5 above.
3) A model for human perception processes.
4) A model for human evaluation processes.

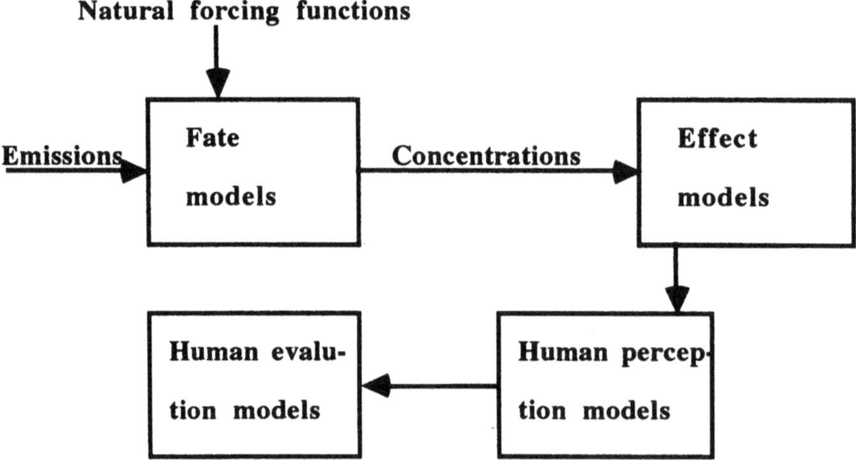

Figure 8. The four submodels of a total ecotoxicological model are shown.

The first two submodels are in principle "objective", predictive models, corresponding to Types 5 and 6, while the latter two are value oriented. Development of submodels (1) and (2) are based upon physical, chemical and biological processes. They are very similar to other environmental models and arev based upon mass transfer, mass balances, physical, chemical and biological processes.

The second submodel requires accurate knowledge to the effects of toxic components. Submodels (3) and (4) are different from the generally applied environmental management models and are presented in some detail below, covering risk assessment.

Fugacity multi-media models are a special type of fate models. They attempt to answer the following questions: In which of these six compartments, corresponding to the spheres, can we expect the greatest problem for chemicals emitted to the environment? Which concentration will be expected in each compartment? What are the implications of this concentration? The fugacity models are to a large extent based on physical-chemical parameters and the use of these parameters to estimate other required parameters. Applicability of the fugacity model is therefore very dependent on the use of the estimation methods. The risk assessment component, associated with the fate model, comprises human perception and evaluation processes (see Figure 8). These submodels are explicitly value laden, however, they must build on objective information of concentrations and effects.

Matthies et al. (1992) have developed a fate model, that exemplifies this type of model very well. Based upon basic information about the chemical, the model can make predictions on concentrations in air, river water and soil. The conceptual diagram of the river model is shown in Fig. 9.

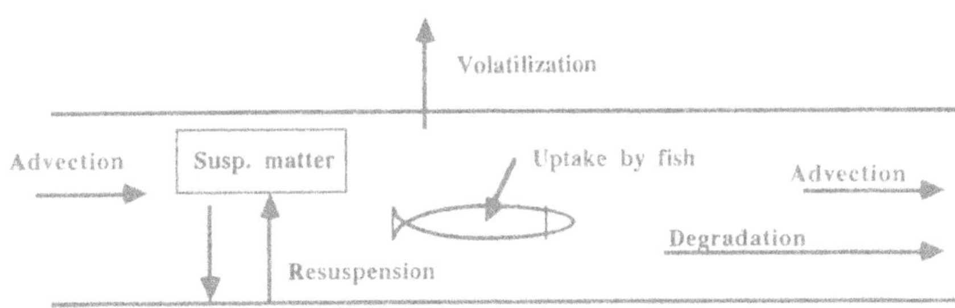

Figure 9. Conceptual diagram of the river model, that is an integrated submodel of the fate model, developed by Matthies et al., 1992.

The conceptual diagram shows that the model includes a description of the following processes:

1. Dilution.
2. Degradation, including hydrolysis, photolysis, photolysis and biodegradation.
3. Volatilization according to equations whicha are more elaborate than those presented above. The description considers partition coefficient between air and water, velocity of the current, wind velocity wind and film resistances.
4. Sedimentation and resuspension are considered by simple first order reaction schemes.
5. Uptake by fish is covered by using BCF.

The model is very versatile and can easily be used for simulations, provided that the characteristic parameters for the chemicals are known, and the estimation methods can assist the user. The model represents a modeling state which can be achieved by a medium complex model with respect to risk assessment and environmental management of toxic substances.

Factors that may be important to consider in this context are:

1. Magnitude and time constant of exposure.
2. Spatial and temporal distribution of concentration.
3. Environmental conditions determining the process rates and effects.
4. Translation of concentrations into magnitude and duration of effects.
5. Spatial and temporal distribution of effects.
6. Reversibility of effects.

Uncertainties related to the information on which the model is based and uncertainties related to the development of the model, are crucial in risk assessment. Uncertainty in risk problems may be classified into one or more of the following 5 categories:

1. Good direct knowledge and statistical evidence on important components (state variables, processes and interrelations of the variables) of the model is available.
2. Good knowledge and statistical evidence on the important submodels are available, but the aggregation of the submodels are less certain.
3. No adequate knowledge of the model components for the considered system is available, accurate data are available for the same processes from a similar system and it is estimated that these data may be applied for use directly or with minor modifications, to the model development.

4. Some, but insufficient, knowledge is available from other systems. Attempts are made to use these data without necessary transferability. Attempts are made to eliminate gaps in knowledge by use of additional experimental data as far as it is possible within the limited resources available for the project.
5. The model is to a large extent based on the subjective judgement of experts.

The acknowledgement of the uncertainty is of great importance and may be taken into consideration, either qualitatively or quantitatively. Another problem is of course where the uncertainty should be taken into account. Should the economy or the environment benefit from the uncertainty? Unfortunately, up to now most decision makers have used the uncertainty to the benefit of the economy. This is of course completely unacceptable. The same decision makers would never consider in a infrastructure project whether uncertainty should be used for the benefit of the economy or the strength of a bridge.

Until 10-15 years ago researchers had developed very little understanding of the processes by which people actually perceive exposures and effects of toxic chemicals, but these processes are just as important for risk assessment as the exposures and effects processes. The characteristics of risks and effects are of importance for the perceptions of people. These characteristics may be summarized in the following lists:

Characteristics of risk:

Voluntary or involuntary?

Are the levels known to the exposed people or to science?

Is it novel, or old and familiar?

Is it common or dreaded (for instance does it involve cancer)?

Does it involve death?

Are mishaps controllable?

Are future generations threatened?

Global, regional or local?

Function of time? How (whether for instance increasing or decreasing)?

Can it easily be reduced?

Characteristics of effects?

Immediate or delayed?

On many or a few people?

Global, regional or local?
Involve death?
Are effects of mishaps controllable?
Observable immediately?
Function of time?

A factor analysis has been performed by Slovic et al. (1982) which shows, among other results, a not surprising correlation between people's perception of dreadful and unknown risks. Broadly speaking there are two methods of selecting risks we will deal with.

The first may be described as the *"rational actor model"*, involving people that look systematically at all risks they face and make choices about which they will live with and at what levels. For decision making this approach would use some single, consistent, objective functions and a set of decision rules.

The second method may be named the *"political/cultural model"*. It involves interactions between culture, social institutions and political processes for the identification of risks and determination of those which people will live with and at what level.

Both methods are unrealistic, as they are both completely impractical in their pure form. Therefore, we must select a strategy for risk abatement founded on a workable alternative based on the philosophy behind both methods.

Several risk management systems are available, but no attempt will be made here to evaluate them. However, some recommendations should be given for the development of risk management systems:

1. Consider as many characteristics listed above as possible and include human perceptions of these characteristics in the model.
2. Do not focus too narrowly on certain types of risks. This may lead to suboptimal solutions. Attempt to approach the problem as broad mindedly as possible.
3. Choose strategies that are pluralistic and adaptive.
4. Benefit-cost analysis is an important element of the risk management model, but it is far from being the only important element and the uncertainty in evaluation of benefit and cost should not

be forgotten. The variant of this analysis applicable to environmental risk management may be formulated as follows:

net social benefit - social benefits of the project - "environmental" costs of the project.

5. Use multi-attribute utility functions, but remember that people in general have troubles in thinking about more than 2-3, at the most 4, attributes in each outcome.

Presentation of the six classes of models above clearly show the advantages and limitations of ecotoxicological models. The simplifications used in classes two, three and six (at least without risk assessment components) often offer great advantages. They are sufficiently accurate to give a very applicable picture (overview) of the concentrations of toxic substances in the environment, due to the application of large safety factors. The application of the various available estimation methods renders it feasible to construct such models even if your knowledge of the parameters is limited. The estimation methods have obviously a considerable uncertainty, but the high safety factor helps in accepting this uncertainty. On the other hand our knowledge about the effects of toxic substances is very limited - particularly at the organism and organ level. It must not be expected, therefore, that models with effect components and give more than a first rough picture of what is known today in this area.

Because of the character of ecotoxicological models it is recommended that a few points are clarified before entering the modeling procedure, as presented in modeling textbooks:

1. **Obtain the best possible knowledge about the processes of toxic substances under consideration.**
2. **Attempt to get parameters of the toxic substance processes in the environment from literature.**
3. **Estimate all parameters**
4. **Compare the results from 2 and 3 and attempt to explain discrepancies, if present.**
5. **Use widely sensitivity analysis to estimate which processes and state variables would be feasible and relevant to include into the model.**

4. AN OVERVIEW OF TOXIC SUBSTANCE MODELS FOR AQUATIC ECOSYSTEMS

A number of toxic substance models for aquatic ecosystems are reviewed in Table 2. Most models reflect the proposition that adequate knowledge of the problem and the ecosystem can be used to make reasonable simplifications. Model characteristics shown in the table are state variables and/or processes considered in the model.

Table 2 Examples of Toxic Substance Models

Heavy metal	Model Characteristics	Reference
Cadmium	Food chain	Thomann et al. 1974
Mercury	6 st. var:water, sediment, susp. matter, invertebrates, plant, fish	Miller, 1979
Methyl mercury	1 trophic level: intake, excretion, metabilism, growth	Fagerström &Aasell 1973
Vinyl chloride	Chemical processes in water	Gillett et al., 1974
Pesticides in fish	Ingestion, CF, adsorption	Leung, 1978
Hydrophobic organics	Exchange air-water, sorption, hydrolysis, photolysis, hydrodynamics	Schwarzenbach & Imboden, 1984
PAH	Transport, degradation, bioaccumulation	Bartell et al., 1984
Heavy metals	CF, excretion, bioaccumulation	Aoyama et al., 1978
Zinc	CF, hydrodynamics	Seip, 1978
Copper	Complex formation, adsorption, sub-lethal effects	Orlob et al., 1980
Metals	Thermodynamic equilibrium	Jϕrgensen, 1991
Lead	Hydrodynamics, precipitation, toxic effects on algae and fish	Lam and Simons, 1976
Cadmium	Settling, sediment, steady state food chain model	Thomann, 1984
Heavy metals	Hydraulic submodel, adsorption	Nyholm et al., 1984
Aluminum	Survival of fish populations	Breck et al., 1988
Chromium	Hydrodynamics and accumulation in mussels	Jϕrgensen et al., 1991
Mercury	Hydrodynamics and accumulatuion in fish	Jϕrgensen et al., 1991
Mirex and lindane	Fate in Lake Ontario	Halfon, 1986

The most difficult part of modeling the effect and distribution of heavy metals is to obtain relevant knowledge about the behavior of toxic substances in the environment and to use this knowledge to make feasible simplifications. It gives the modeler a particular challenge in the selection of the right and balanced complexity, and there are many examples of simple models of toxic substances, that are able to solve the focal problem.

It can be seen from Table 2 that most models have been developed during the last decade. Before around 1975, toxic substances were hardly associated with environmental modeling, as the problem seems straight forward. Many pollution problems associated with toxic substances were deemed to be easily solvable simply by elimination of the source. During the seventies it was acknowledged that the environmental problems associated with toxic substances are very complex problems due to the interaction of many sources and numerous simultaneously, interacting processes and components. Several accidental releases of toxic substances into the environment have reinforced the need for models. The result has been that several ecotoxicological models have been developed in the period from the late seventies until today. The list of models in Table 2 gives an overview of available models, but the list should not be considered complete because the table is not a result of a complete literature review. The aim of the table is to give an idea of the spectrum of available models of toxic substance pollution.

5. A CASE STUDY: A MERCURY MODEL FOR MEX BAY, ALEXANDRIA (EGYPT)

A case study taken from a coastal zone, Mex Bay near Alexandria, has been selected to illustrate the use of ecotoxicological models as a management tool. The case study would only have been slightly different if the ecosystem was a river or a lake. Many lakes are suffering from excessive mercury concentrations and almost exactly the same model could be applied for these cases. Mex Bay is located west of Alexandria and is suffering from serious pollution problems due to the discharge of waste water from many industries, such as cement, tanneries, an oil refinery and a chlorine alkali plant.

The most serious pollution problem of the bay is the mercury contamination of fish. The concentration of mercury in most fish caught in the bay exceed the limit for human food set by WHO (1 ppm). The surface area is 29 km^2 and the mean depth is 10 m.

A comprehensive investigation of the mercury pollution of the bay has been carried out at

Alexandria University. The results on which the development of the model is based are published in Jørgensen (1990, 1994).

5.1 The Model

A static model is used to describe the mercury contamination of the bay. The model is based on a mass balance for the bay. Figure 10 shows the following principal processes:

1) Discharge of municipal and industrial waste water.
2) Atmospheric fallout - dry and wet deposition.
3) Volatilization.
4) Exchange with open sea.
5) Sedimentation.
6) Release from the sediment.
7) Fishery.

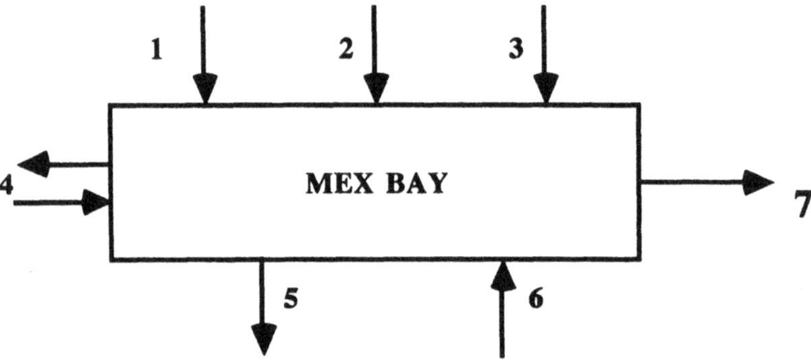

Figure 10. The model is developed on a basis of the mass balance principles applied on the seven processes shown herein.

The model consists of five submodels which are interrelated as shown in the conceptual diagram Figure 11.

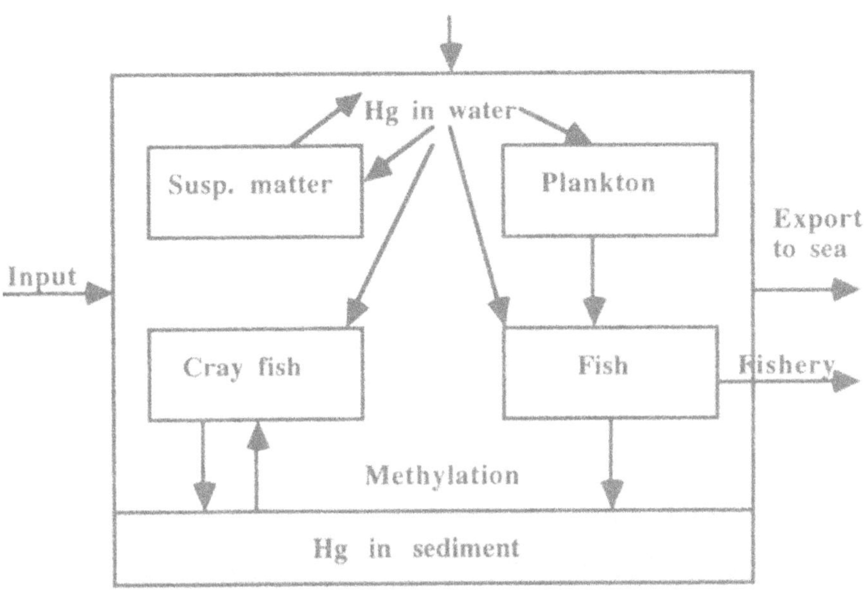

Figure 11. The total model.

Submodel 1 deals with the mercury concentration in water. It describes the mercury concentration as a function of the distance from the outlet by the use of basic hydrodynamics. Details are included in Jørgensen (1990, 1994). The change in mercury concentration with time is a result of: (dispersion - advection - settling + methylation). As the discharge of mercury has been almost constant for several years, we are able to transform the partial differentiation equation to a differential equation, where the independent variable is the distance from the outlet. This simplification is often valid for contamination of aquatic ecosystems, which have received an almost constant input of toxic compounds. We are furthermore not interested in daily fluctuations but in the general pollution picture.

Submodel 2 considers the concentration of suspended matter in water. It describes the concentration of suspended matter as a function of the distance from the outlet, using the same hydrodynamics as for mercury. It is again possible to transform partial differential equations into a differential equations, as the discharge of suspended matter has been constant for a longer period. We are furthermore not interested in the changes on a day-to-day basis, but on the general pollution picture of Mex Bay.

Submodel 3 describes the concentration of mercury in phytoplankton. The model distinguishes between organic mercury and inorganic mercury in phytoplankton. They are both described simply as a concentration factor times the concentration in water.

Submodel 4 deals with mercury in the sediment. The concentration in the sediment is a result of a settling (from submodel 1) and the methylation (also included as a first order reaction in submodel 1). As the mercury concentration in the sediment is a function of these two processes, which are considered constant with the time (see again submodel 1), the concentration in the sediment is considered a constant at a given station - it is only dependent on the distance from the outlet and depth.

Submodel 5 considers mercury in fish and distinguishes between inorganic and organic mercury. The mercury concentration of fish, HgF is determined by:

1) the uptake from water: = CF* Hg in water* dw/dt
2) the uptake from food: $a*w^{b}*$ Hg in food* eff
3) the excretion: excretion coefficient* Hg in fish

where

CF is the concentration factor (fish/water),
w is the weight of the fish, which implies that dw/dt is the growth of the fish,
a and b are characteristic constants describing food uptake by the fish,
eff is the efficiency of the mercury uptake from the food (it is different for inorganic mercury).

The change in mercury concentration of the fish is determined by:

d HgF/ dt = uptake from water + uptake from food - excretion. (37)

The growth of the fish is found by:

$$dw/ dt = a*w^{b} - r*w^{c} \quad (38)$$

where a and b are the constants mentioned above, while r and c are other constants. In accordance with several investigations, b=0.68 and c= 0.8.

Two possible food chains are included in the model. For each station the mercury concentration of sediment and phytoplankton is determined by use of submodels 3 and 4. A probability generator determines in which of the stations the "average" filter feeders (Sardina pilchardus), the "average benthic invertebrates (Penaeus kerat- hurus) and the "average" Pelagic fish (Boops boops) are on a given day.

The station determines the mercury concentration of food for these three species. Their concentrations are currently determined by use of the above mentioned equations and the concentration of carnivorous predators is determined by use of the same set of equations, but now using the mercury concentration of their average food sources. The ratio of the three species that comprise the food, is determined by use of analysis of the stomach contents of the fish compared with general food sources. The ratio of the three species, that comprise the food, is determined by use of analysis of the stomach contents of the fish compared with general knowledge of the species preferred food items. The state variables and forcing functions of the model are listed in Table 3.

At this stage the model has only been calibrated. Submodel 1 and 2 are second-order differential equations and the concentration of mercury Hgt and suspended matter TSM at $x = 0$ and d Hgt / dx and dTSM / dx at $x = O$ have therefore been included in the calibration. Values based upon the measurements are used as initial guesses. The initial guesses of the settling rates are found on the basis of sediment analyses. The Chlor Alkali plant started its production in 1950 and the settling rates found on the basis of the sediment profiles.

Figure 12 shows the validation results, where the measured and computed values of the mercury content of the tuna fish are compared. The agreement between model results and measured values is acceptable. As seen, the tuna fish will exceed a mercury concentration of 1 mg/kg already at a body weight of 350 g. A mercury concentration of more than 40 ppm (more than 40 times the maximum value recommended by WHO) has been found for tuna fish with a weight of several kg.

The model has been used to perform a series of scenarios considering a wide spectrum of mercury discharge reductions. Results of one of the simulations for the tuna fish is shown in Figure 13. The mercury concentration versus the weight is shown based on the assumption of a 90% reduction of the mercury discharge. Note that mercury concentration in tuna fish is reduced more than 10 times, which can be explained by the binding capacity of the sediment. The result shows that this reduction gives a satisfactory low mercury concentration in the tuna fish and these results should be used in environmental management, as a 90% reduction of the mercury concentration in waste water can be achieved by rather inexpensive solutions such as, for example, chemical precipitation.

Table 3 State variables of the Mex Bay Model

State variable	Unit	Comments
1. Salinity	%	Measured at all stations and in different depths
2. Hg - inorganic	µg/l	
3. Hg - prganic	µg/l	
4. Hg - total dissolved	µg/l	Hg - Total dissolved is the sum of Hg -inorganic and Hg- organic
5. Hg - particulate	µg/l	
6. Hg - total	µg/l	Hg-total is the sum of Hg - total dissolved and Hg-total particulate
7. Inorganic Hg - plankton	µg/kg	Wt. Weight Hg inorg. in plankton
8. Total Hg - plankton	µg/kg WW	
9. Inotganic Hg - Pelagic fish	µg/kg WW	5 different forms of Pelagic fish were examined
10. Total Hg - Pelagi fish	µg/kg WW	Measured in all five examined fish in muscle (flesh)
11. Inorganic Hg-benthic fish	µg/kg WW	Two species of benthic fish were examined
12. Total Hg-benthic fish	µg/kg WW	
13. Inorganic Hg-filter feed fish	µg/kg WW	Two species of filter feed fish were examined
14. Total Hg-filt.f. fish	µg/kg WW	Measured in muscle (flesh)
15. Inorganic Hg-Carnv. fish	µg/kg WW	Two species of carnivous fish were examined
16. Total Hg - Carnv. fish	µg/kg WW	Measuresd in muscle
17. Inorg. Hg - benthic invertebrates	µg/kg WW	Two species of benthic invertebrates were examined
18. Total Hg - Beth. invt.	µg/kg WW	
19. Suspended matter	µg/l	
20. Suspended matter C	µg C/l	Use of equation
21. Leachable Hg - sediment	µg/g DM	
22. Organic Hg - sed.	µg/g DM	
23. Total Hg - sed.	µg/g DM	
24-28 Hg f(weight(time)) in Pelagic -, Benthic-, Carnivous fish and Benthic Invertebrates		

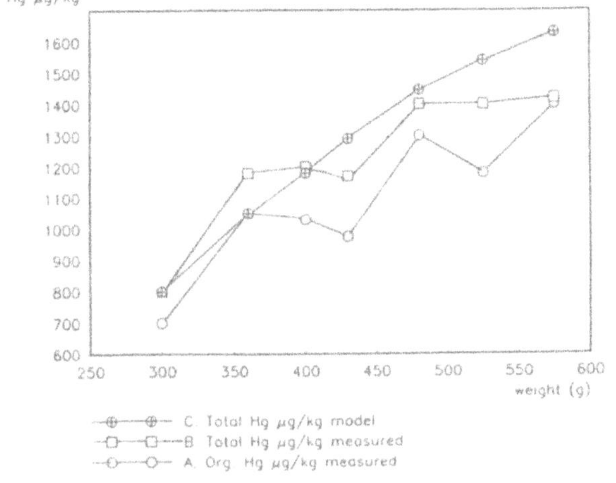

Figure 12.

Mercury concentration in tuna fish (µg/kg) as function of weight. A gives organic Hg in µg/kg (measured), B gives total Hg in µg/kg (measured) C shows the model results of total Hg in µg/kg.

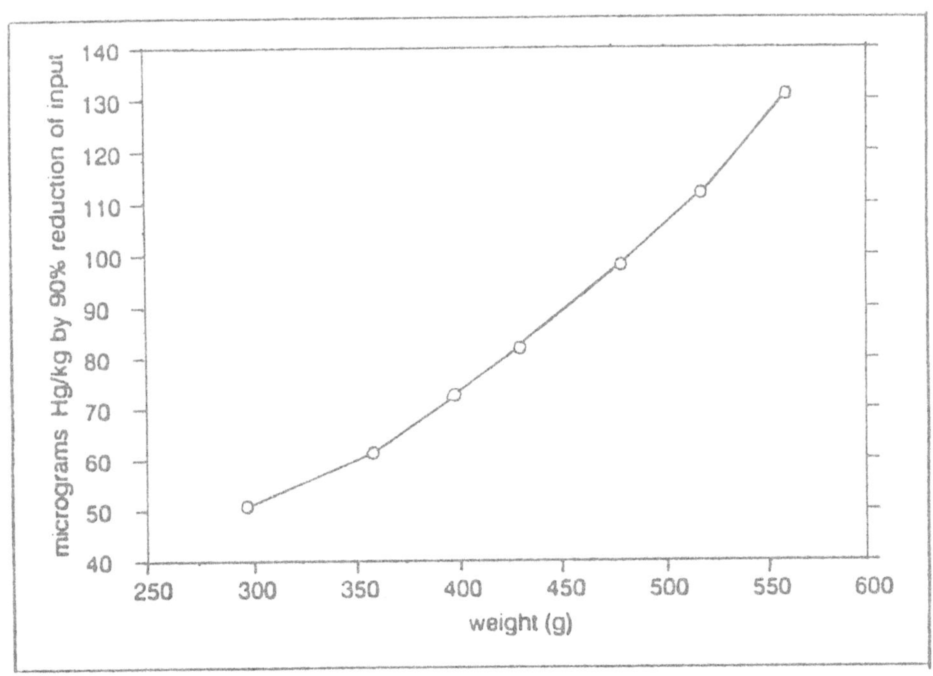

Figure 13. Mercury concentration in tuna (µg/kg) as function of weight by 90% reduction of mercury discharge.

The model applied in this case study is rather simple compared with the complex biological and hydrodynamic processes responsible for mercury concentrations of the fish species, which are the most central state variables. However, an acceptable agreement between measured values and model values has been found, although submodel 1 does not give an acceptable fit for the relationship between mercury concentration and the distance from the outlet, probably due to a too simple description of the hydrodynamics.

The model is an illustrative example of what can be achieved by using a simple model, resulting from considerations of where simplification can be made and what are the most essential processes and state variables. If the experience gained by development of this model is used to set up a procedure for development of a management model for the control of heavy metal pollution in aquatic ecosystems, the following procedure could be recommended:

1. Cover the hydrodynamics as simple as possible by considering a constant discharge. Space coordinates are used as independent variables.
2. Consider including the following physical-chemical processes: adsorption on suspended matter and sediment water exchange processes.
3. The concentration of heavy metals in the species with high level contamination can be determined by the use of concentration factors and bioaccumulation processes. If a description of the concentration as function of weight is required, the submodel 5 in this example should be applied.

6. SUMMARY AND CONCLUSIONS

Toxic substance models have been developed during the last 15 years. They may be divided into 6 classes with different characteristics, which are pertinent for their specific applications. Toxic substance models are in principle not different from other ecological/environmental models, but their specific use may determine different simplifications that often are significantly different from those made in eutrophication and BOD-models. This is mainly due to a more limited need for high accuracy in the predictions. It is of great importance in this context to state where in the system is the critical point for a given environmental issue.

Toxic substance models are increasingly used in environmental management mainly in three relations. First, models are used more and more often for a given environmental problem associated with toxic contamination of a specific ecosystem. The idea is to give recommendations, based upon model results, to which extent it is necessary to reduce the emission. This application is somewhat similar to the use of eutrophication and BOD-models. Second, models are increasingly used to assess the fate of chemicals. "Where and how much will the chemical do harm?" is a typical question which the model application attempts to answer. Last, toxic substance models are used to make risk assessments. The model attempts in this case to answer for instance: if we use this and that chemical in this and that context, what would be the risk that the chemical would contaminate human food, milk, vegetables, meat, fish etc.? As seen herein these are very urgent questions that toxic substance models are answering. Therefore, it is expected that ecotoxicological models will be used more frequently in the coming years in an environmental management context, particularly for aquatic ecosystems which have the longest tradition for modeling.

REFERENCES

Aoyama, I., Yos. Inoue and Yor. Inoue (1978) " Simulation analysis of the concentration process of trace heavy metals by aquatic organisms from the viewpoint of nutrition ecology," Water Research **12**:837-842

Bartell, S.M., Gardner, R.H. and O'Neill, R.V. (1984) " The fates of aromatics model," Ecol. Modeling, **22**:109-123

Breck, J.E., DeAngelis, D.L., Van Winkle, W. and Christensen, S.W. (1988) "Potential importance of spatial and temporal heterogeneity in pH, Al and Ca in allowing survival of a fish population: a model demonstration," Ecol. Modeling, **41**:1

Brown, D.S. and Flagg, E.W. (1981) "Empirical prediction of organic pollutant sorption in natural sediments," J. Environ. Qual., **10**: 382-386.

Edgington, D.H. and Callender, E. (1970) "Minor element geochemistry of Lake Michigan ferromanganese nodules," Earth Plant Sci. Lett. **8**:97-100.

Fagerstöm, T. and Aasell, B. (1973) "Methyl mercury accumulation in an aquatic food chain. A model and implications for research planning," Ambio **2**: 164-171

Gillett, J.W., *et al.* (1974) A conceptual model for the movement of pesticides through the environment. National Environmental Research Center, U.S. Environmental Protection Agency, Corvallis, OR Report EPA 660/3-74-024, pp. 79.

Halfon, E. (1986) "Modelling the fate of *Mirex* and *Lindane* in Lake Ontario, off the Niagara River mouth," Ecol. Modelling, **33**: 13

Jørgensen, S.E. (1984) "Parameter estimation in toxic substance models," Ecol. Modelling, **22**:1-13.

Jørgensen, S.E. and Gromiec, M. (1989) Mathematical Submodels of Water Quality Systems. Elsevier, Amsterdam.

Jørgensen, S.E. and Johnsen, I. (1989) Principles of Environmental Science and Technology, Studies in Environmental Science 33. Elsevier, Amsterdam

Jørgensen, S.E. (1990) Modelling in Ecotoxicology, Elseveier, Amsterdam, 340 pp.

Jørgensen, S.E. (1991) Modelling in Environmental Chemistry. Developments in Environmental Modelling, 17. Elsevier, Amsterdam, 505 pp.

Jørgensen, S.E., Nors-Nielsen, S. and Jørgensen, L.A. (1991) Handbook of Ecological Parameters and Ecotoxicology. Elsevier, Amsterdam, 1270 pp.

Jørgensen, S.E. (1994) Fundamentals of Ecological Modelling, 2nd. edition. Elsevier, Amsterdam, pp. 630.

Jørgensen, S.E. (1995) "The validity of Allometric Relationships in Toxic Substance Models," Ecol. Modelling, submitted

Lam, D.C.L. and Simons, T.J. (1976) "Computer model for toxicant spills in Lake Ontario." In Environmental Biogeochemistry vol. 2. Metals transfer and ecological mass balances. (J.O. Nriago, ed.) Ann Arbor Science Publ., Ann Arbor, MI, pp. 537-549

Larson, R.J. and Perry, R.L. (1981) "Use of the electrolytic respirometer to measure biodegradation in natural waters," Water Res., **15**: 697

Leung, D.K. (1978) Modelling the bioaccumulation of pesticides in fish. Center for the Ecological Modelling, Polytechnic Institute, Troy, NY Report 5.

Matthies, M., Brüggeman, R. and Münzer, B.(1992) Screening Assessment Model Systems SAMS: A Program of Simple Models for Exposure Assessment to Chemicals. Prepared for OECD Environment Directorate, Paris.

Miller, D.R. (1979) Models for total transport. Principles of Ecotoxicology Scope vol. 12. Ed. G.C. Butler, 1979. New York, NY, Wiley, pp. 71-90.

Morgan, M.G. (1984) "Uncertainty and quantitative assessment in risk management." In. J.V. Rodricks & R.G. Tardiff (eds.), Assessment and Management of Chemical Risks, Chapter 8, ACS Symposium Series 239. American Chemical Society, Washington, D.C

Nyholm, N., Nielsen, T.K. and Pedersen, K. (1984) " Modelling heavy metals transport in an arctic fjord system polluted from mine tailings." Ecol. Modelling **22**: 285-325

Orlob, G.T., Hrovrat, D. and Harrison, F. (1980) "Mathematical model for simulation of the fate of copper in a marine environment." American Chemical Society. Advances in Chemistry Series **189**: 195-212

Schwarzenbach, R.P. and Imboden, D.M. (1984) " Modelling concepts for hydrophobic pollutants in lakes," Ecol. Modelling, **22**:171-213

Seip, K.L. (1978) "Mathematical model for uptake of metals in benthic algae," Ecological Modelling **6**: 183-198.

Slovic, P., Fischoff, B. and Lichtenstein, S. (1982) "Facts and fares: Understanding perceived risk." In R.C. Schwing & W.A. Albers, Jr. (eds), Societal Risk Assessment: How Safe is Safe Enough? Plenum Press, New York.

Smith, J.H. *et.al.* (1977) Environmental Pathways of Selected Chemicals in Freshwater Systems. Part I, EPA 500/7-77-113.

Thomann, R.V., *et.al.* (1974) "A food chain model of cadmium in western Lake Erie," Water Research **8**: 841-851

Thomann, R.V. (1984) " Physico-chemical and ecological modelling the fate of toxic substances in natural water systems," Ecol. Modelling, **22**:145-186

CHAPTER 5

SEDIMENT TOXICITY AND EQUILIBRIUM PARTITIONING

DEVELOPMENT OF SEDIMENT QUALITY CRITERIA FOR TOXIC SUBSTANCES

Dominic M. DiToro[1,2]
Laura D. De Rosa[1]

1. INTRODUCTION

The toxicity of chemicals in sediments is strongly influenced by the extent to which the chemicals bind to the sediment. This modifies the chemical potential to which the organisms are subjected. As a consequence, different sediments will exhibit different degrees of toxicity for the same total quantity of chemical. These differences have been reconciled by relating organism response to the chemical concentration in the interstitial water of the sediments (Adams et al., 1985; Swartz, et al., 1985; Muir et al., 1985; Adams, 1987; Kemp and Swartz, 1988; Nebeker and Schuytema, 1988). The relevant sediment properties, therefore, are those which influence the distribution of chemical between the solid and aqueous phases.

This chapter presents a summary of the technical basis for establishing sediment quality criteria (SQC) for nonionic organic chemicals (DiToro et al., 1991) and metals by determining the distribution between the solid and aqueous phases. This methodology is termed equilibrium partitioning (EqP). A full development of EqP for nonionic organic chemicals is presented in the EPA technical basis document for deriving SQC (US EPA, 1993) and preparation of a report discussing the development of SQC for metals is underway for presentation to the EPA Science Advisory Board.

---

[1] HydroQual, Inc., Mahwah, New Jersey, USA

[2] Department of Env. Engineering and Sci., Manhattan College, Bronx, New York

NATO ASI Series, Partnership Sub-Series, 2. Environment – Vol. 3
Remediation and Management of Degraded River Basins
Edited by V. Novotny and L. Somlyódy

2. TOXICITY AND BIOAVAILABILITY OF NONIONIC ORGANIC CHEMICALS IN SEDIMENTS

Establishing SQC requires a determination of the extent of the bioavailability of sediment associated chemicals. It has frequently been observed that similar concentrations of a chemical, in units of mass of chemical per mass of sediment dry weight (e.g. micrograms chemical per gram sediment), can exhibit a range in toxicity in different sediments. Because the purpose of SQC is to establish chemical concentrations that apply to sediments of differing types, it is essential that the reasons for this varying bioavailability be understood and explicitly included in the criteria. Otherwise the criteria cannot be presumed to be applicable across sediments of differing properties.

The importance of this issue cannot be overemphasized. For example, if 1 $\mu g/g$ of kepone is the LC50 for an organism in one sediment and 35 $\mu g/g$ is the LC50 in another sediment, then unless the cause of this difference can be associated with some explicit sediment properties it is not possible to decide what would be the LC50 of a third sediment. The results of toxicity tests used to establish the toxicity of chemicals in sediments would not be generalizable to other sediments. Imagine the situation if the results of toxicity tests in water depended strongly on the particular water source - for example, Lake Superior versus well water. Until the source of the differences was understood, it would be fruitless to attempt to establish water quality criteria (WQC). It is for this reason that the understanding bioavailability is a principal focus in establishing sediment quality criteria.

The observations that provided the key insight to the problem of quantifying the bioavailability of chemicals in sediments were that the concentration-response curve for the biological effect of concern could be correlated not to the total sediment chemical concentration (micrograms chemical per gram sediment), but to the interstitial water (i.e., pore water) concentration (micrograms chemical per liter pore water). In addition, the effect concentrations found for the pore water were essentially equal to that found in water-only exposures. Organism mortality, growth rate, and bioaccumulation data are used to demonstrate this correlation, which is a critical part of the logic behind the EqP approach to developing SQC. For nonionic organic chemicals, it is shown that the concentration - response curves correlate equally well with the sediment chemical concentration on a sediment organic carbon basis (DiToro et al., 1991).

These observations can be rationalized by assuming that the pore water and sediment carbon are in equilibrium and that the concentrations are related by a partition coefficient, K_{oc}, as shown in Figure 1 (right). The name "Equilibrium Partitioning" (EqP) describes this assumption of partitioning equilibrium. The rationalization for the equality of water-only and sediment-exposure-effects concentrations on a pore water basis is that the sediment-pore water equilibrium system (Figure 1, right) provides the same exposure as a water-only exposure (Figure 1, left). The reason is that the chemical activity is the same in each system at equilibrium. These results do not imply that pore water or sediment organic carbon is the primary route of exposure because all exposure pathways are at equal chemical activity in an equilibrium experiment and the route of exposure cannot be determined. It should be pointed out that EqP assumptions are only approximately true, and, there fore, the predictions from the model have an inherent uncertainty. A discussion and quantification of uncertainty is found in the US EPA (1993) SQC derivation document.

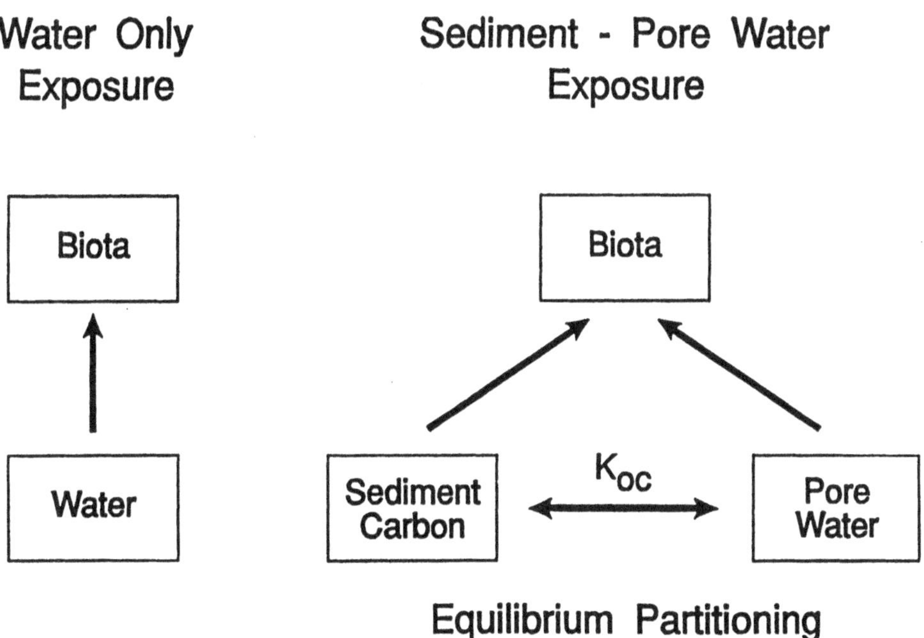

Figure 1. Diagram of the organism exposure routes for a water-only exposure (left) and a sediment exposure (right). *Equilibrium partitioning* refers to the assumption that an equilibrium exists between the chemical sorbed to the particulate sediment organic carbon and the pore water. The partition coefficient is K_{oc}.

2.1 Toxicity Experiments

A substantial amount of data has been assembled that addresses the relationship between toxicity and pore water concentration (DiToro et al., 1991; US EPA, 1993). One example of these data sets (Figure 2) are results from kepone experiments (Adams et al., 1985; Ziegenfus et al., 1986) where the biological response (mortality) is plotted versus the total sediment concentration in the top panel, versus the measured pore water concentration in the center panel and versus the sediment concentration, which is organic carbon normalized (microgram chemical per gram organic carbon) in the bottom panel. Table 1 summarizes the LC50 estimates and 95 percent confidence limits for these data on a total sediment, pore water and organic carbon basis, as well as the water-only values.

TABLE 1. KEPONE LC_{50} FOR SEDIMENT DRY WEIGHT AND SEDIMENT-ORGANIC CARBON NORMALIZATION AND FOR PORE-WATER AND WATER-ONLY EXPOSURES

foc (%)	Total Sediment (μg/g)	Pore Water (μg/L)	Organic Carbon (μg/g)	Water Only (μg/L)	Reference
0.90	09.0 (0.73-1.10)	29.9 (25.3-35.6)	1,000 (811-1,220)	26.4 (22.7-30.6)	Adams et al. 1985
1.50	6.9 (5.85-8.12)	31.3 (25.7-38.1)	460 (390-541)		
12.0	35.2 (30.6-40.5)	18.6 (15.7-21.9)	293 (255-337)		

The LC50 and the 95 percent confidence limits in parentheses are computed by the modified Spearman-Karber method (Hamilton et al., 1977).

The disparity among the sediments (Figure 2, top) are particularly dramatic. For the low organic carbon sediment (f_{oc} = 0.09 percent), the 50th percentile total kepone concentration for *Chironomus tentans* mortality (LC50) is < 1 μg/g. By contrast, the 1.5 percent organic carbon sediment LC50 is approximately 7.0 μg/g. The high organic carbon sediment (12 percent) exhibits still a higher LC50 value on a total sediment kepone concentration basis (35 μg/g). However, as shown in the center panel, essentially all the mortality data collapse into a single curve when the pore water concentrations are used as the correlating concentrations. On a pore water basis, the biological responses as measured by LC50 vary approximately less than a factor of two, whereas when they are evaluated on a total sediment kepone basis they exhibit an almost 40-fold range in kepone toxicity. The comparison between the pore water effects concentrations and the water-only results indicates that they are similar. The pore water LC50s are 19 to 30 μg/L, and the water-only exposure LC50 is 26 μg/L.

Acute Toxicity of Kepone

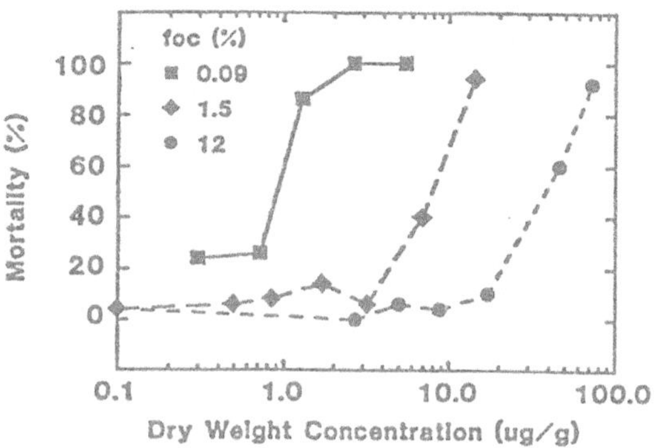

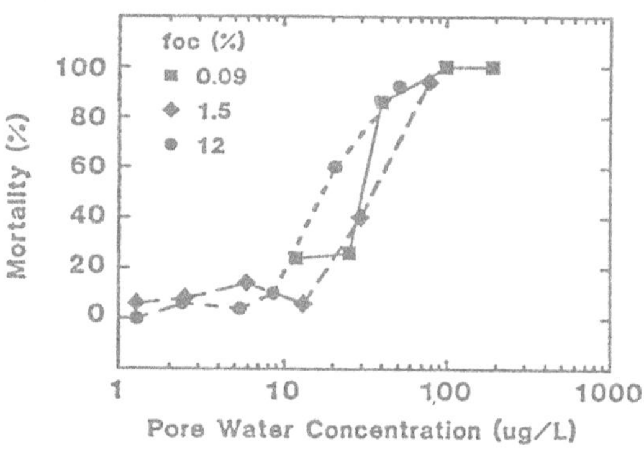

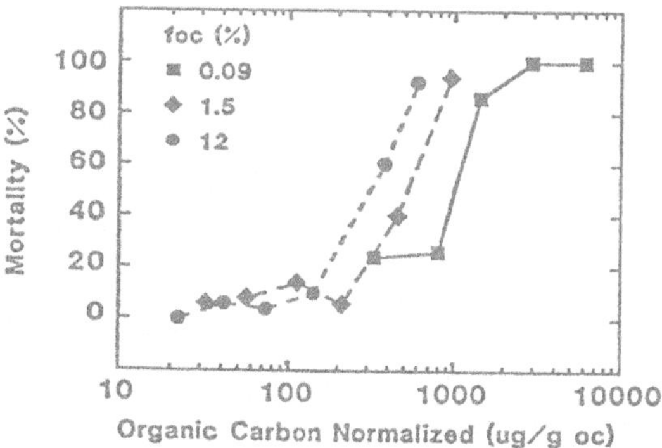

Figure 2. Comparison of percent mortality of *C. tentans* to Kepone concentration in bulk sediment (top), pore water (center) and bulk sediment using organic carbon normalization (bottom) for three sediments with varying organic carbon concentrations after Adams et al. (1985)

If a concentration-response curve correlates to pore water concentration, it should correlate equally well to organic carbon normalized total chemical concentration, independent of sediment properties. This is based on the partitioning formula $C_{s,oc} = K_{oc}\,C_d$ (Equation 16), discussed below which relates the free dissolved concentration to the organic carbon normalized particle concentration. This applies only to nonionic hydrophobic organic chemicals because the rationale is based on a partitioning theory for this class of chemicals. To demonstrate this relationship, the lower panel of Figure 2 presents the response versus the sediment concentration, which is organic carbon normalized. The variation of organic carbon normalized LC50s between sediments is less than a factor of two to three and is comparable to the variation in pore water LC50s (Table 1).

A more comprehensive comparison is presented in Figure 3, which also examines the use of the water-only LC50 to predict the pore water and sediment organic carbon LC50s. Figure 3 presents mortality data for various chemicals and sediments compared to pore water concentrations when normalized on a toxic unit basis. Three different sediments are tested for each chemical as indicated. Predicted pore water toxic units are the ratio of the measured pore water concentration to the LC50 obtained from water only toxicity tests. The EqP model predicts that the pore water LC50 will equal the water only LC50 which is obtained from a separate water only exposure toxicity test.

Define:

$$\textit{predicted pore water toxic unit} = \frac{(\textit{pore water concentration})}{(\textit{water only LC50})} \qquad (1)$$

Therefore, a toxic unit of one occurs when the pore water concentration equals the water-only LC50, at which point it would be predicted that 50 percent mortality would be observed. The correlation of observed mortality to predicted pore water toxic units in Figure 3 demonstrates (a) the efficacy of using pore water concentrations to remove sediment to sediment differences and (b) the applicability of the water-only effects concentration and, by implication, the validity of the EqP model. By contrast, it has been shown (DiToro et al.,1991; US EPA, 1993), that the mortality versus sediment chemical concentration on a dry weight basis varies dramatically from sediment to sediment.

The quality of the effects concentration on a pore water basis suggests that the route of exposure is via pore water. However, the quality of the effects concentration on a sediment organic carbon basis, which is demonstrated in Figure 2 and subsequently in Figure 5, suggests that the ingestion of sediment

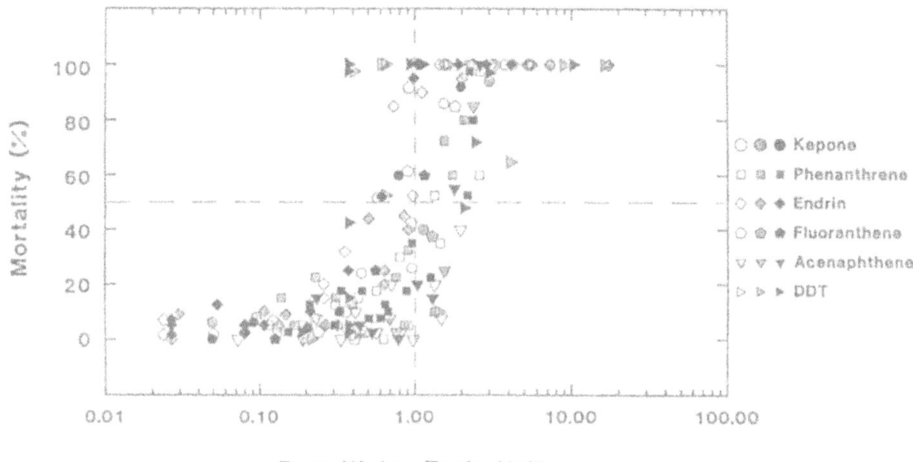

Figure 3. Mortality vs. predicted pore water toxic units for six chemicals up to three organic carbon content sediments per chemical. Sediment types are indicated by single hatching (lowest organic carbon content), cross hatching (intermediate organic carbon content) and filled symbols (highest organic carbon content). See US EPA (1993) for more detailed description of these data sets. Predicted pore water toxic units are the ratios of the pore water concentrations to the water only LC_{50} (Eq.1)

organic carbon is the primary route of exposure. It is important to realize that if the sediment and pore water are in equilibrium, then the effective exposure concentration is the same regardless of exposure route. Therefore, it is not possible to determine the primary route of exposure from equilibrated experiments.

Whatever the route of exposure, the correlation of toxicity to pore water suggests that if it were possible to either measure the pore water chemical concentration, or predict it from the total sediment concentration and the relevant sediment properties such as the sediment organic carbon concentration, then that concentration could be used to quantify the exposure concentration for an organism. Thus, the partitioning of chemicals between the solid and the liquid phase in a sediment becomes a necessary component for establishing SQC.

2.2 Sorption of Nonionic Organic Chemicals

For nonionic hydrophobic organic chemicals sorbing to natural soils and sediment particles, a number of empirical models have been suggested. See Karickhoff (1984) for an excellent review. The chemical property that indexes hydrophobicity is the octanol/water partition coefficient, K_{ow}. The important particle property is the weight fraction of organic carbon, f_{oc}. O'Connor and Connolly (1980) have shown that another important environmental variable appears to be the particle concentration itself.

In many experiments using particle suspensions, the partition coefficients have been observed to decrease as the particle concentration used in the experiment is increased (O'Connor and Connolly, 1980). Unfortunately very few experiments have been done on settled or undisturbed sediments. Therefore the correct interpretation of particle experiments is of critical importance. It is not uncommon for the partition coefficient to decrease by two to three orders of magnitude at high particle concentrations. If this partitioning behavior is characteristic of bedded sediments, then quite low partition coefficients would be appropriate. This would result in lower sediment chemical concentrations for SQC. However, if this phenomenon is an artifact or is due to a phenomenon that does not apply to bedded sediments, then a quite different partition coefficient would be used. The practical importance of this issue requires a detailed discussion of the particle concentration effect.

2.2.1 Particle concentration effect

For the reversible (or readily desorbable) component of sorption, a particle interaction model (PIM) has been proposed that accounts for the particle concentration effect and predicts the partition coefficient on nonionic hydrophobic chemicals over a range of nearly seven orders of magnitude with a $\log_{10}$ prediction standard error of 0.38 (DiToro, 1985). The reversible component partition coefficient K_p^*, is the ratio of reversibly bound chemical concentration, C_s (mg/kg dry weight), to the dissolved chemical concentration, C_d (mg/L):

$$C_s = K_p^* C_d \tag{2}$$

The PIM model for K_p^* is:

$$K_p^* = \frac{f_{oc} K_{oc}}{1 + m f_{oc} K_{oc} / \nu_x} \tag{3}$$

where:

K_p^*	=	reversible component partition coefficient (L/kg dry weight)
K_{oc}	=	particle organic carbon partition coefficient (L/kg organic carbon)
f_{oc}	=	particle organic carbon weight fraction (kg organic carbon/kg dry weight)
m	=	particle concentration in the suspension (kg dry weight/L)
v_x	=	1.4, an empirical constant (unitless).

The regression of K_{oc} to the octanol/water coefficient, K_{ow}, yields

$$\log_{10}K_{oc} = 0.00028 + 0.983\log_{10}K_{ow} \qquad (4)$$

so that K_{oc} approximately equals K_{ow}.

A number of explanations have been offered for the particle concentration effect. The most popular is to posit the existence of an additional third sorbing phase or complexing component that is associated with the particles but is inadvertently measured as part of the dissolved chemical concentration due to experimental limitations. Colloidal particles that remain in solution after particle separation (Benes and Majer, 1980; Gschwend and Wu, 1985) and dissolved ligands or macromolecules that desorb from the particles and remain in solution (Carter and Suffet, 1983; Voice et al., 1983; Curl and Keolelan, 1984; Nelson et al., 1985) have been suggested. It has also been suggested that increasing particle concentration increases the degree of particle aggregation, decreasing the surface area and hence the partition coefficient (Karickhoff and Morris, 1985). The effect has also been attributed to kinetic effects (Karickhoff, 1984).

Sorption by nonseparated particles or complexing by dissolved organic carbon can produce an apparent decrease in partition coefficient with increasing particle concentration if the operational method of measuring dissolved chemical concentration does not properly discriminate the truly dissolved or free chemical concentration from the complexed or colloidally sorbed portion. However, the question is not whether improperly measured dissolved concentrations can lead to an apparent decrease in partition coefficient with increasing particle concentrations. The questions is whether these third-phase models explain all (or most) of the observed partition coefficient - particle concentration relationships.

An alternate possibility is that the particle concentration effect is a distinct phenomena that is a ubiquitous feature of aqueous-phase particle sorption. A number of experiments have been designed to explicitly exclude possible third-phase interferences. Both the resuspension experiment for polychlorinated biphenyls (PCBs) (DiToro and Horzempa, 1983) and metals (DiToro et al., 1986; McIlroy et al., 1986) in which particles are resuspended into a reduced volume of supernatant and the dilution experiment (DiToro and Horzempa, 1083) in which the particle suspension is diluted with supernatant from a parallel vessel display particle concentration effects. It is difficult to see how third-phase models can account for these results because the concentration of the colloidal particles is constant while the concentration of the sediment particles varies substantially.

The model (Equation 3) is based on the hypothesis that particle concentration effects are due to an additional desorption reaction induced by particle-particle interactions (DiToro, 1985). Mackay and Powers (1987) suggested that actual particle collisions were responsible. This interpretation relates v_x to the collision efficiency for desorption and demonstrates that it is independent of the chemical and particle properties, a fact that has been experimentally observed (DiToro, 1985; DiToro et al., 1986).

It is not necessary to decide which of these mechanisms is responsible for the effect if all the possible interpretations yield the same result for sediment/pore water partitioning. Particle interaction models would predict that $K_{oc} \approx K_{ow}$ because the particles are stationary in sediments. Third-phase models would also relate the free (i.e., uncomplexed) dissolved chemical concentration to particulate concentration via the same equation. As for kinetic effects, the equilibrium concentration is again given by the relationship $K_{oc} \approx K_{ow}$. Thus there is unanimity on the proper partition coefficient to be used in order to relate the free dissolved chemical concentration to the sediment concentration, that is, $K_{oc} \approx K_{ow}$.

2.2.2 Organic carbon fraction

The unifying parameter that permits the development of SQC for nonionic hydrophobic organic chemicals that are applicable to a broad range of sediment types is the organic carbon content of the sediments. This can be shown as follows. The sediment/pore water partition coefficient, K_p, is given by

$$K_p = f_{oc} K_{oc} \tag{5}$$

and the solid phase concentration is given by

$$C_s = f_{oc} K_{oc} C_d \tag{6}$$

where C_s is the concentration on sediment particles. An important observation can be made that leads to the idea of organic carbon normalization. Equation 5 indicates that the partition coefficient for any nonionic organic chemical is linear in the organic carbon fraction, f_{oc}. DiToro et al. (1991), US EPA (1993) and DiToro (1985) presented data to support the linearity of partitioning to a value of f_{oc} = 0.2 percent. This result and the toxicity experiments suggest that for f_{oc} > 0.2 percent, organic carbon normalization is valid.

As a consequence of the linear relationship of C_s and f_{oc}, the relationship between sediment concentration C_s, and free dissolved concentration, C_d, can be expressed as

$$\frac{C_s}{f_{oc}} = K_{oc} C_d \qquad (7)$$

If we define

$$C_{s,oc} = \frac{C_s}{f_{oc}} \qquad (8)$$

as the organic carbon normalized sediment concentration (mg chemical/kg organic carbon), then from Equation (7):

$$C_{s,oc} = K_{oc} C_d \qquad (9)$$

Therefore, for a specific chemical with a specific K_{oc}, the organic carbon normalized total sediment concentration, $C_{s,oc}$, is proportional to the dissolved free concentration, C_d, for any sediment with f_{oc} > 0.2 percent. Karickhoff (1984) has shown that this latter qualification is judged to be necessary because at f_{oc} < 0.2 percent the other factors that influence partitioning (e.g., particle size and sorption to nonorganic mineral fractions) become relatively more important. Using the proportional relationship given by Equation 9, the concentration of free dissolved chemical can be predicted from the normalized sediment concentration and K_{oc}. The free concentration is of concern as it is the form that is bioavailable.

A verification of Equation 9 is from data collected during sediment toxicity tests in the laboratory. These tests yield sediment ($C_{s,oc}$) and pore water (C_d) chemical concentrations at several dosages bounding an experimentally estimated toxic concentration for the test organism. The organic content of the sediments is measured also. Sediment toxicity tests are done under quiescent conditions and sediment and pore water are in equilibrium. The results of these tests can be used to compute the organic carbon partition coefficient K_{oc}. Estimates of K_{oc} from laboratory measurements of K_{ow} are then compared to partitioning in the sediment toxicity tests. Sediment toxicity tests and K_{ow} measurements are available for four chemicals: endrin (Kemp and Schwartz, 1986; Nebeker et al., 1989; Stehly, 1991), acenaphthene (Swartz, 1991), phenanthrene (Swartz, 1991), and fluoranthene (Swartz et al, 1990; DeWitt et al. 1992). Sediment toxicity tests were performed as part of the development of SQC. Mortality results for these tests are presented in Figure 3.

Figure 4 shows organic carbon normalized sorption isotherms for acenaphthene, endrin, phenanthrene and fluoranthene, where the sediment concentration (ug/g OC) is plotted versus pore water concentration (ug/L). These tests represent freshwater and marine sediments having a range of organic carbon of 0.07 to 11.0 percent. In each panel, the line corresponds to Equation 9 where K_{oc} is derived from K_{ow} measurements in the laboratory. A full discussion of laboratory K_{ow} measurements is presented in the EPA (1993) SQC development document. In each of the panels the toxicity test data are in agreement with the line computed from experimentally determined K_{ow}. Hence we arrive at the following important conclusion: For a specific chemical having a specific K_{oc}, the organic carbon normalized sediment concentration, SQC_{oc}, is independent of sediment properties.

As discussed above, hydrophobic chemicals also tend to partition to colloidal-sized organic carbon particles that are commonly referred to as dissolved organic carbon, or DOC. DiToro et al. (1991) have documented that although DOC affects the apparent pore water concentrations of highly hydrophobic chemicals, the DOC-bound fraction of the chemical appears not to be bioavailable.

Therefore, we expect that toxicity in sediment can be predicted from the water-only effects concentration and the K_{oc} of the chemical. The utility of these ideas can be tested with the same mortality data as these in Figure 3 but restricted to nonionic organic chemicals for which organic carbon normalization applies. The concept of sediment toxic units is useful in this regard. These are computed as the ratio of the organic carbon-normalized sediment concentrations, C_s/f_{oc}, and the predicted sediment LC50 using K_{oc} and the water-only LC50. That is:

$$\textit{predicted sediment toxic unit} = \frac{C_s/f_{oc}}{Koc\ (\textit{water only LC50})} \tag{10}$$

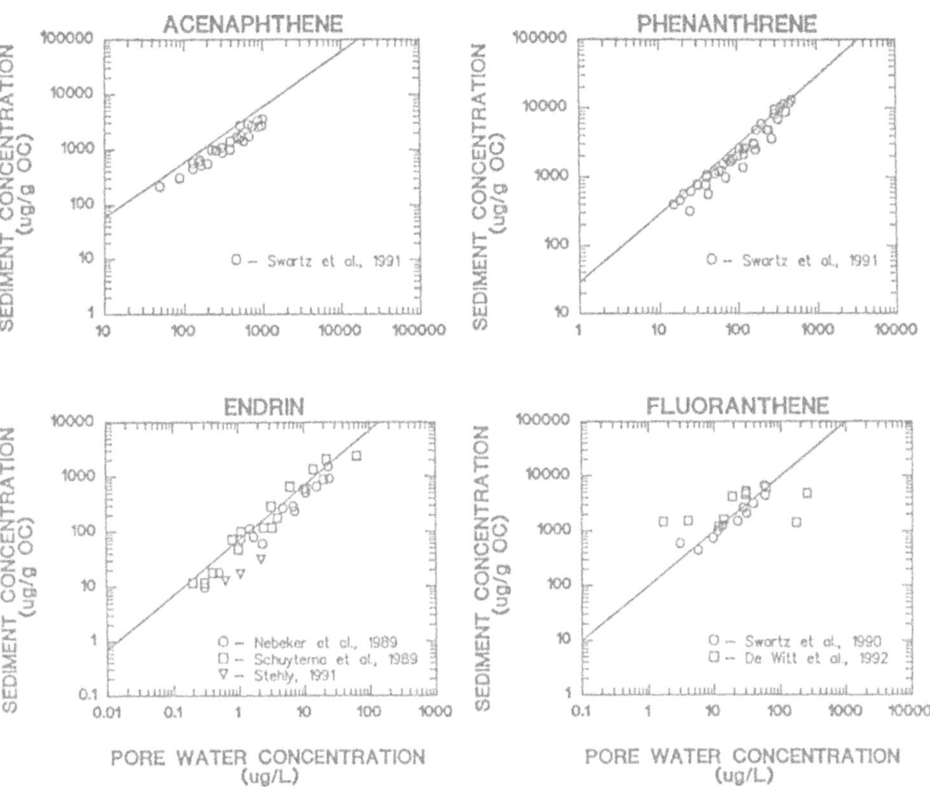

Figure 4. Comparison of organic carbon partition coefficient (K_{oc}) observed in toxicity tests (symbols) to K_{oc} derived from laboratory K_{ow} and Equation 4 (solid line). Symbols are sediment concentration, ordinate, vs. pore water concentration, abscissa. Solid line is $C_{S,OC}=K_{OC}*C_d$, where $\log_{10}K_{OC}$ is 3.76 for acenaphthene, 4.84 for endrin, 4.46 for phenanthrene, and 5.00 for fluoranthene. These $\log_{10}K_{OC}$ values are estimated from $\log_{10}K_{OC}$ values measured at the U.S. EPA Environmental research Laboratory at Athens, Georgia (US EPA, 1993). Data sources as indicated.

Figure 5 presents the percent mortality versus predicted sediment toxic units. The correlation is similar to that obtained using the pore water concentrations in Figure 3. The cadmium data are not included because partitioning is not determined solely by sediment organic carbon. The predicted sediment toxic units for each chemical follow a similar concentration-response curve independent of sediment type. The data demonstrate that 50 percent mortality occurs at about one sediment toxic unit, independent of chemical, species or organism or sediment type, as expected if the EqP assumptions are correct.

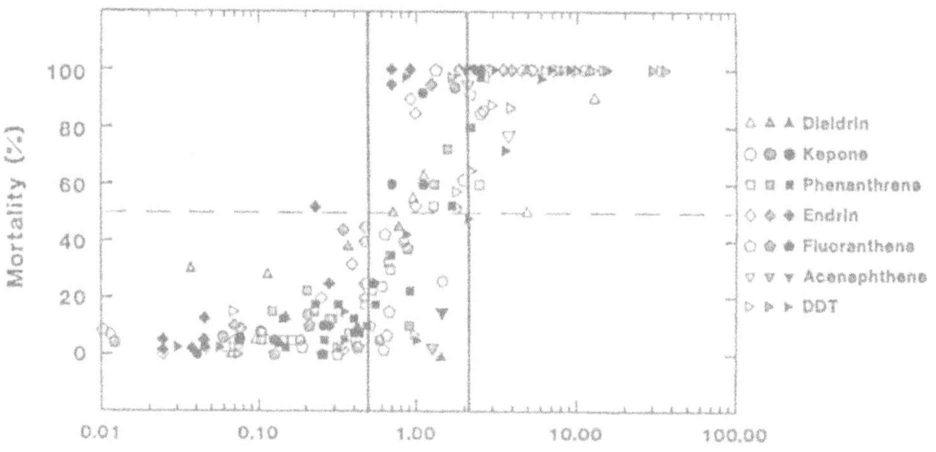

Figure 5. Mortality vs. predicted sediment toxicity units for seven chemicals and up to three percent organic carbon content sediments per chemical. Sediment types are indicated by single hatching (lowest organic carbon content), cross hatching (intermediate organic carbon content) and filled symbols (highest organic carbon content). See US EPA (1993) for more detailed description of these data sets. Predicted sediment toxic units are the ratio of the organic carbon-normalized sediment chemical concentration to the predicted LC_{50} (Eq. 10). K_{OC} values are computed from K_{OW} and Equation 4. K_{OW} for endrin (4.85), fluoranthene (5.00), dieldrin (5.25), phenanthene (4.46), and acenaphthene (3.76) were measured by the U.S. EPA Research Laboratory, Athens, Georgia. Methods are presented in US EPA (1993). K_{OW} for DDT (5.84) is the geometric mean of the reported values in the Log P database (Leo and Hansch, 1986). The kepone K_{OC} is the log mean of the ration of organic carbon-normalized kepone concentrations to pore water - kepone concentration from the toxicity data set published by Adams et al. (1985).

If the assumptions of EqP were exactly true and there were no experimental variability or measurement error, then the data in Figures 3 and 5 should all predict 50 percent mortality at one toxic unit. There is an uncertainty of approximately a factor of two in the results as indicated by the error bars. This uncertainty associated with sediment quality criteria was obtained from a quantitative estimate of the degree to which the data in Figure 5 support the assumptions of EqP (DiToro et al., 1991). This variation reflects inherent variability in these experiments as well as phenomena that have not been accounted for in the EqP model. This appears to be the limit of the accuracy and precision to be expected.

2.3 Effects Concentration

The development of SQC requires an effects concentration for benthic organisms. Because many of the organisms used to establish the water quality criteria (WQC) are benthic, perhaps the WQC are adequate estimates of the effects concentrations for benthic organisms. To examine this possibility, the acute toxicity data base, which is used to establish the WQC is segregated into benthic and water column species, and the relative sensitivities of each group are compared. The data are from the 40 freshwater and 30 saltwater U.S. Environmental Protection Agency (EPA) criteria documents. If it were true that benthic organisms are as sensitive as water column organisms - and the evidence to be presented appears to support this supposition - then SQC could be established using the final chronic value (FCV) from these WQC documents as the effects concentration for benthic organisms. The apparent equality between the effects concentration as measured in pore water and in water-only exposures (Figure 3) or as predicted from organic carbon normalization (Figure 5) supports using an effects concentration derived from water only exposures.

This use of WQC assumes that (a) the sensitivities of benthic species and species tested to derive WQC predominantly water column species, are similar and (b) the levels of protection afforded by WQC are appropriate for benthic organisms. This section examines the assumption of similarity of sensitivity using a comparative toxicological examination of the acute sensitivities of benthic and water column species. A comparison of the FCVs and the chronic sensitivities of benthic saltwater species in a series of sediment colonization experiments was done by DiToro et al. (1991) and US EPA (1993). Although there is considerable scatter, these results, a more detailed analysis of all the acute toxicity data, and the results of benthic colonization experiments support the contention of equal sensitivity (US EPA, 1993).

The relative acute sensitivities of benthic and water column species are examined by using LC50s for freshwater and saltwater species from draft or published WQC documents that contain minimum data base requirements for calculation of final acute values. These data sets are selected because exposures were via water, durations were similar, and data and test conditions have been scrutinized by reviewing the original references. Each life state of the tested species was classified according to habitat. Habitats were based on degree of association with sediment.

The relative acute sensitivities of the most sensitive benthic and water column species were examined by comparing the lowest acute LC50 concentration for the benthic and water column organisms, using acute values from the 40 freshwater and the 30 saltwater WQC documents. When benthic species were defined as only infaunal organisms and water column species were defined as all others, the water column species were typically the most sensitive. The results are cross-plotted on Figure 6 (left). The line represents perfect agreement. In most instances where acute values for saltwater benthic and water column species are identical, it is because penaeid shrimp are most sensitive to insecticides and are classified as both infaunal (benthic) and epibenthic (water column). Unfortunately data on the sensitivities of benthic infaunal species are limited. Therefore, it is probably premature to conclude from the existing data that infaunal species are more tolerant than water column species.

A similar examination of the most sensitive benthic and water column species, where the definition of benthic includes both infaunal and epibenthic species is based on more data and suggests a similarity in sensitivity (Figure 6, right). The variability of these data is high, suggesting that for some chemicals, benthic and water column species may differ in sensitivity, that additional testing would be desirable, or that this approach to examining species sensitivity is not sufficiently rigorous.

Frequency distributions of the sensitivities of all species to all chemicals indicate that infaunal species may be relatively insensitive but that infaunal and epibenthic species appear almost evenly distributed among both sensitive and insensitive species overall (US EPA, 1993).

2.4 Example Calculations

The calculation procedure for establishing SQC is as follows. If FCV (mg/L) is the final chronic WQC for the chemical of interest, then the SQC (mg/kg sediment) are computed using the partition coefficient K_p (L/kg sediment) between sediment and pore water:

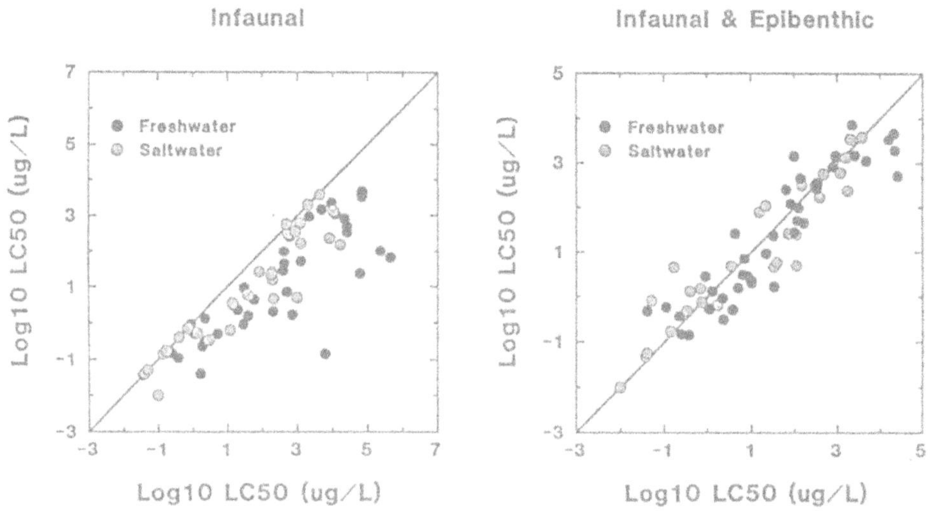

Figure 6. Comparison of LC_{50} acute values from EPA WQC or draft criteria documents for the most sensitive benthic (absisa) and water column (ordinate) species. Benthic species are defined as infaunal species (left panel) or infaunal and epibenthic species (right panel).

$$SQC = K_p \, FCV \tag{11}$$

This is the fundamental equation from which SQC are generated. The partition coefficient, K_p, the ratio of sediment concentration, C_s, to pore water concentration, C_d is given by

$$K_p = \frac{C_s}{C_d} = f_{oc} \, K_{oc} \tag{12}$$

where K_{oc} is the partition coefficient for sediment organic carbon.

Using Equations 11 and 12, an SQC is calculated from

$$SQC = f_{oc} \, K_{oc} \, FCV \tag{13}$$

This equation is linear in the organic carbon fraction, f_{oc}. As a consequence, the relationship can be expressed as

$$\frac{SQC}{f_{oc}} = K_{oc}\, FCV \tag{14}$$

If we define

$$SQC_{oc} = \frac{SQC}{f_{oc}} \tag{15}$$

as the organic carbon normalized SQC concentration (microgram chemical per kilogram or organic carbon), then

$$SQC_{oc} = K_{oc}\, FCV \tag{16}$$

Equation 16 can be used to compute SQC_{oc} for a range of K_{ow}s and FCVs. The results for several chemicals are shown in Figure 7 in the form of a nomograph. The diagonal lines are for constant FCVs as indicated. The abscissa is $\log_{10} K_{ow}$. For example, if a chemical has an FCV of 1.0 mg/L and a $\log_{10} K_{ow}$ of 4, so that $K_{ow} = 10^4$, the $\log_{10} SQC_{oc}$ is approximately 1 and the SQC $= 10^1 = 10.0$ μg chemical/g organic carbon.

As can be seen, the relationships between SQC_{oc} and the parameters that determine its magnitude, K_{ow} and FCV, are essentially linear on a log-log basis. For a constant FCV, a 10-fold increase in K_{ow} (one log unit) increases the SQC_{oc} by approximately 10-fold (one log unit) because K_{oc} also increases approximately 10-fold. Thus, chemicals with similar FCVs will have larger SQC_{oc}s if their K_{ow}s are larger.

The FCVs that are available for nonionic organic insecticides range from approximately 0.01 μg/L to 0.3 μg/L, a factor of 30. The SQC_{oc}s range from approximately 0.01 μg/g organic carbon to in excess of 10 μg/g organic carbon, a factor of over 1,000. This increased range in values occurs because the K_{ow}s of these chemicals span over two orders of magnitude. Hence the most stringent SQC_{oc} in this example is for chlordane, a chemical with the lowest K_{ow} among the chemicals with an FCV of approximately 0.01 μg/L.

By contrast, the PAHs included in this example have a range of FCVs and K_{ow}s of approximately one-half order of magnitude. But these values vary inversely: The chemical with the larger FCV has a smaller K_{ow}. The result is that the SQC_{oc}s are approximately the same, 200 μg/g organic carbon. Classes of chemicals for which the effects concentrations decrease logarithmically with increasing K_{ow}s, for example, chemicals that are narcotics (Abernethy and Mackay, 1987), will have SQC that are more nearly constant.

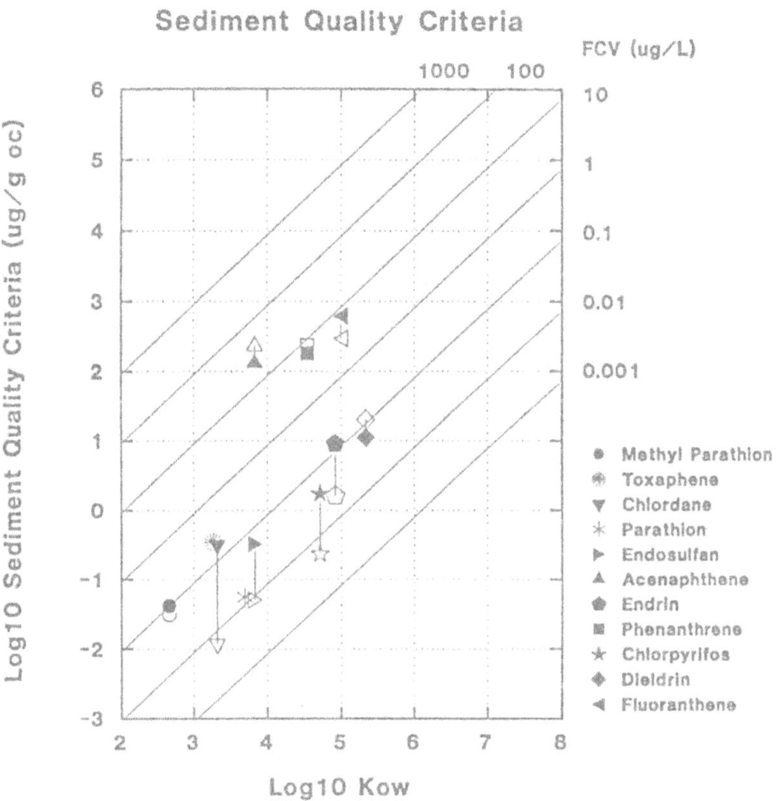

Figure 7. Log_{10} SQC vs. $log_{10}K_{OW}$. The diagonal lines indicate the FCV values. The criteria are computed from Equation 16. K_{OC} is obtained from K_{OW} and Equation 4. the symbols indicate SQC_{OC} for the freshwater (filled) and saltwater (hatched) criteria for the listed chemicals. The vertical line connects symbols for the same chemical. The FCV for methyl parathion, toxaphene, chlordane, parathion, endosulfan and chlorpyrifos are from EPA Water Quality Criteria or from draft criteria documents. The FCVs for acenaphthene, endrin, phenanthene, dieldrin and fluoranthene were computed as part of the development of SQC as were the K_{OW}s (US EPA, 1993). The K_{OW}s for methyl parathion, toxaphene, chlordane, parathion, endosulfan and chlorpyrifos are the log mean of the values reported in the Log P database (Leo and Hansch, 1988).

The final validation of SQC will come from field studies that are designed to evaluate the extent to which biological effects can be predicted from SQC. Sediment quality criteria can possibly be validated more easily than WQC because determining the organism exposure is more straightforward. The benthic population exposure is quantified by the organic carbon normalized sediment concentration.

Landrum (1989) has suggested that the kinetics of PAH desorption from sediments control the chemical body burden of a benthic amphipod. The extent to which kinetics can be important in field situations is unknown at present, and field studies would be an important component in examining this question. In addition, more laboratory sediment toxicity tests, particularly chronic tests involving multiple sediments, would also be helpful. In a typical practical application of SQC mixtures of chemicals are involved. The extension of EqP methodology to mixtures would be of great practical value. Initial experiments by Swartz et al. (1988) indicate that it should be possible.

The EqP method is presently restricted to computing effects-based criteria for the protection of benthic organisms. The direct extension of this methodology for computing sediment criteria that are protective of human health, wildlife, and marketability of fish and shellfish requires that the equilibrium assumption be extended to the water column and to water column organisms. This is, in general, an untenable assumption. Water column concentrations can be much lower than pore water concentrations if sufficient dilution flow is present. Conversely, upper-trophic-level organisms are at concentrations well above equilibrium values (Connolly and Pedersen, 1988). Hence, the application of the final residue values from the WQC for the computation of SQC, as was done for certain interim criteria (Cowan and DiToro, 1988), is not technically justifiable. At present, organism lipid-to-sediment organic carbon ratios might be useful in estimating the concentration of contaminants in benthic species, for which the assumption of equilibrium is reasonable. However, a site- specific investigation, for example, by Connolly (1991), appears to be the only available method for performing an evaluation of the effect of contaminated sediments on the body burdens of upper-trophic-level organisms.

3. TOXICITY AND BIOAVAILABILITY OF METALS IN SEDIMENTS

The varying toxicity of non-ionic organic chemicals in different sediments has been found to be related to the organic carbon content of the sediments (Adams et al., 1985; Nebeker and Schuytema, 1988; DiToro et al., 1991; and US EPA, 1993). This is due to the importance of sediment organic carbon in determining the extent of sorption of non-ionic organic chemicals to sediments. The analogous sediment properties for metals would be the phases that influence their partitioning behavior. It has been speculated that the oxides of iron and manganese as well as organic carbon would be relevant in determining the toxicity of metals in sediments (US EPA, 1989).

The primacy of the acid volatile sulfide (AVS) phase - the solid phase sediment sulfides that are soluble in cold acid - in determining the toxicity of cadmium in sediments has recently been established by DiToro et al. (1990, 1992). The results of toxicity tests using three sediments with differing AVS concentrations indicate that the LC50s are quite different on a dry weight basis. However the interstitial water and water only LC50s are similar. The problem is to determine what sediment parameter is controlling the cadmium activity.

Experimental cadmium titration of iron sulfide and natural sediments indicate that cadmium can react with the solid phase AVS to form cadmium sulfide precipitate. If the quantity of AVS in a sediment exceeds the quantity of added cadmium, the concentration of cadmium in the interstitial water is non-detectable and no mortality is observed. The reason is that the AVS is sufficiently reactive so that the added cadmium precipitates as cadmium sulfide which is quite insoluble so that the interstitial water cadmium concentration is low. In addition CdS itself apparently is not bioavailable. As long as excess AVS is present no free cadmium exists. However, if the added cadmium exceeds the AVS, free cadmium is measured in the interstitial water and amphipod mortality occurs. The presentation that follows gives the experimental evidence that leads to these conclusions.

3.1 Cadmium Toxicity

The toxicity of cadmium to *Ampelisca* in Long Island Sound sediment, *Rhepoxynius hudsoni* in Ninigret Pond sediment and *Ampelisca* in an equal parts mixture of the two sediments, is shown in Figure 8. Mean control mortalities were 5.0, 1.7, and 16.7 percent, respectively. The curves are log-logistic concentration response functions fit to the data simultaneously using the same slope parameter. They are included as an aid in visualizing the data. The LC50 range from 290 to 2,850 μg/g on a sediment dry weight basis. These two organisms have nearly the same 96 hour cadmium activity LC50s in water-only exposures: 17.0 μg Cd^{2+}/L, for *Rhepoxynius* and 32.0 μg Cd^{2+} /L for *Ampelisca*. Hence the differences in the cadmium toxicity are likely to be attributable to sediment properties affecting bioavailability.

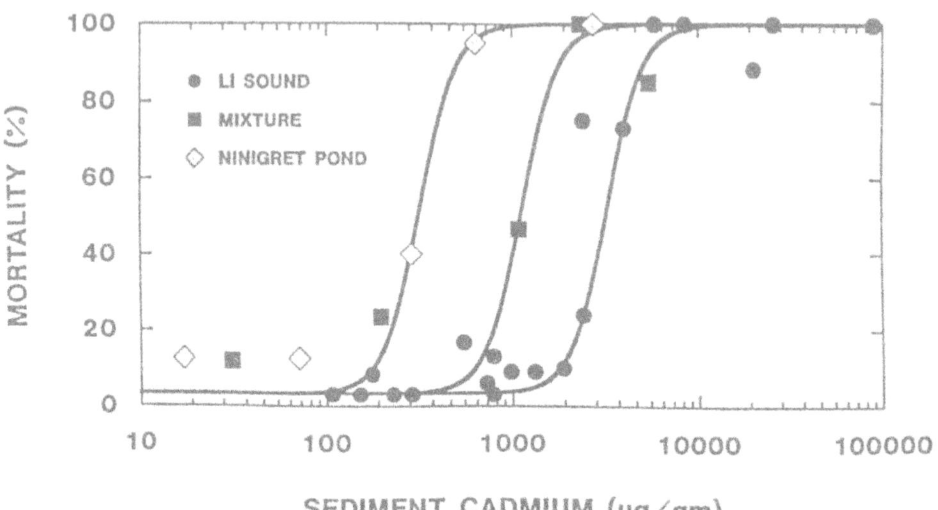

Figure 8. Toxicity test results for Long Island Sound sediments (*Ampelisca*), Ninigret Pond (*Rhepoxynius hudsoni*), and the 50/50 percent mixture of the two sediments (*Ampelisca*) after DiToro et al. (1990). Cadmium concentrations on a dry sediment weight basis.

3.1.1 Metal sulfides and cadmium titration

The importance of sulfide in the control of metal concentrations in the interstitial water of marine sediments is well documented (Morse et al., 1987; Boulegue et al., 1982; Emerson et al., 1983; and Davies-Colley, 1985). Metal sulfides are very insoluble and the equilibrium interstitial water metal concentrations in their presence are small. It is possible that the interstitial water sulfide concentration in the sediments used for these toxicity tests was initially high enough so that as cadmium was added to the sediment, cadmium sulfide was precipitating following the reaction:

$$Cd^{2+} + S^{2-} \rightarrow CdS(s) \tag{17}$$

However direct measurements of the interstitial water sulfide activity, $\{S^{2-}\}$, with a sulfide electrode failed to detect any free sulfide in the unspiked sediments. This was a most puzzling result since a bright yellow cadmium sulfide precipitate was clearly forming as cadmium was added to the sediment.

The lack of significant quantity of dissolved sulfide in the interstitial water and the evident formation of solid phase cadmium sulphide suggested the following possibility. Morse et al. (1987) documented that the majority of the sulfide in sediments is in the form of solid phase iron sulfides. Perhaps the source of the sulphide is this solid phase sulfide initially present. As cadmium is added to the sediment it causes the solid phase iron sulfide to dissolve releasing sulfide which is available for the formation of cadmium sulfide. The plausibility of this mechanism is examined below.

3.2 Solubility Relationships and Displacement Reactions

The majority of sulfide in sediments is in the form of iron monosulfides (mackinawite and greigite) and iron bisulfide (pyrite) of which the former are the most reactive. These sulfides can be partitioned into three broad classes which reflect the techniques used for quantification (Morse et al., 1987; Berner, 1967; and Goldhaber and Kaplan, 1974). The most labile fraction, acid volatile sulfide (AVS), is associated with the more soluble iron and manganese monosulfides. The more resistant sulfide mineral phase, iron pyrite, is not soluble in the cold acid extraction used to measure AVS. Neither is the third compartment, organic sulfide associated with the organic matter in sediments (Landers et al., 1983).

Iron monosulfide, FeS(s), is in equilibrium with aqueous phase sulfide via the reaction:

$$FeS(s) \leftrightarrow Fe^{2+} + S^{2-} \tag{18}$$

If cadmium is added to the aqueous phase, the result is:

$$Cd^{2+} + FeS(s) \leftrightarrow Cd^{2+} + Fe^{2} + S^{2-} \tag{19}$$

As the cadmium concentration increases, $[Cd^{2+}][S^{2-}]$ will exceed the solubility product of cadmium sulfide and CdS(s) will start to form. Since cadmium sulfide is more insoluble than iron monosulfide, FeS(s) should start to dissolve in response to the lowered sulfide concentration in the interstitial water. The overall reaction is:

$$Cd^{2+} + FeS(s) \leftrightarrow CdS(s) + Fe^{2+} \tag{20}$$

The iron in FeS(s) is displaced by cadmium to form soluble iron and solid cadmium sulfide, CdS(s). The consequence of this replacement reaction can be seen using an analysis of the M(II)-Fe(II)-S(II) system with both MS(s) and FeS(s) present as shown by DiToro et al (1992). M(II) represents any metal that forms a sulfide that is more insoluble than FeS. If the added metal, $[M]_A$, is less than the AVS present in the sediment then the ratio of metal activity to total metal in the sediment-interstitial water system is less than the ratio of the MS to FeS solubility products:

$$\{M^{2+}\} / [M]_A \ < \ K_{MS} / K_{FeS} \tag{21}$$

This is a general result that is independent of the details of the interstitial water chemistry. In particular it is independent of the Fe^{2+} activity. Of course the actual value of the ratio $\{M^{+2}\}/[M]_A$ depends on aqueous speciation, as indicated by Equation 20. However, the ratio is still less than the ratio of the sulfide solubility products.

The sulfide solubility products and the ratios are listed in Table 2. The ratio of cadmium activity to total cadmium is less than $10^{-10.5}$. For nickel the ratio is less than $10^{-5.6}$. By inference this reduction in metal activity will occur for any other metal that forms a sulfide that is significantly more insoluble than iron monosulfide. The ratios for the other metals in Table 2, Zn, Pb, Cu, and Hg, indicate that metal activity for these metals will be very small in the presence of excess AVS.

3.3 Sediment Titration Results

A titration procedure was used to evaluate the behavior of sediments taken from four quite different marine environments: the Long Island sound and Ninigret Pond sediments used in the toxicity tests; and sediments from Black Rock Harbor and the Hudson River (Figure 9). The binding capacity for cadmium is estimated by extrapolating a straight line to fit to the dissolved cadmium data. The equation is:

$$[\sum Cd(aq)] \ = \ \max\{O, m([Cd]_A \ - \ [CD]_B)\} \tag{22}$$

where $[\sum Cd(aq)]$ is the total dissolved cadmium, $[Cd]_A$ is the cadmium added, $[Cd]_B$ is the bound cadmium, and m is the slope of the straight line. The sediments exhibit quite different binding capacities for cadmium, listed in Table 3, ranging from approximately 1 μmol/g to more than 100 μmol/g.

TABLE 2 METAL SULFIDE SOLUBILITY PRODUCTS*			
Metal Sulfide	log $K_{sp,2}$	log K_{sp}	Log (K_{MS}/K_{FeS})
FeS	-3.64	-22.39	
NiS	-9.23	-27.98	-5.59
ZnS	-9.64	-28.39	-6.00
CdS	-14.10	-32.85	-10.46
PbS	-14.67	-33.42	-18.55
CuS	-22.19	-40.94	-34.86
HgS	-38.50	-57.25	-34.86

* Solubility products, $K_{sp,2}$ for the reaction $M^{2+} + HS^- \leftrightarrow MS(s) + H^+$ for CdS (greenockite), FeS (mackinawite), and NiS (millerite) from Emerson et al. (1983). Solubility products for CuS (covellite), HgS (metacinnabar), PbS (galena)., and ZnS (wurzite), and $pK_2 = 18.57$ for the reaction $HS^- \leftrightarrow H^+ + S^{2-}$ from Schoonen and Barnes (1988). K_{sp} for the reaction $M^{2+} + S^{2-} \leftrightarrow MS(s)$ is computed from log $K_{sp,2}$ and pK_2

TABLE 3. CADMIUM BINDING CAPACITY AND AVS OF SEDIMENTS			
Sediment	Initial AVS (μmol/g)[a]	Final AVS (μmol/g)[b]	Cd Binding Capacity (μmol/g)
Black Rock Harbor	175 (41)	-	114 (12.1)
Hudson River	12.6 (2.8)	-	8.58 (2.95)
LI Sound(c)	15.9 (3.3)	13.9 (6.43)	4.57 (2.52)
Mixture[c]	5.45 (-)	3.23 (1.18)	-
Ninigret Pond[c]	2.34 (0.73)	0.28 (0.12)	1.12 (0.42)

[a] Average (Standard Deviation) AVS of separated measurements of the stocl
[b] Average (Standard Deviation) AVS after the sediment toxicity experiment
[c] From the three sediment experiments

The possibility that acid volatile sulfide is a direct measure of the solid phase sulfide that reacts with cadmium can be examined in Table 3 which, along with the sediment binding capacity for cadmium, lists the measured AVS for each sediment. The sediment cadmium binding capacity appears to be somewhat less than the initial AVS for the sediments tested. However, a comparison between the initial AVS of the sediments and that remaining after the cadmium titration is completed, Table 3, suggests that some AVS is lost during titration. In any case the covariation of sediment binding capacity and AVS is clear in the data in Table 3. This suggests that AVS is the proper quantification of the solid phase sulfides that can be dissolved by cadmium.

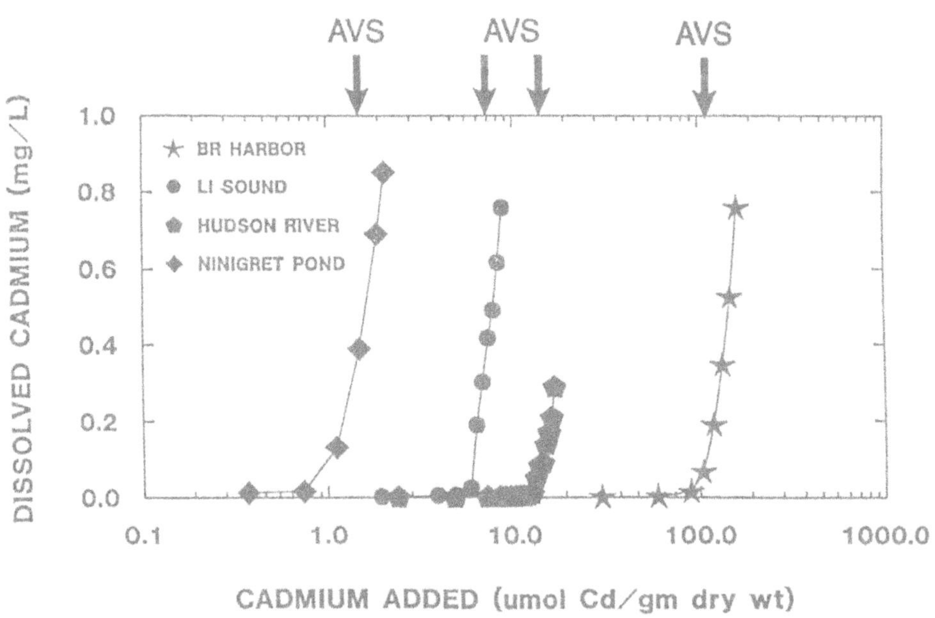

Figure 9. Cadmium titration of sediments: Black Rock Harbor, Long Island Sound, Hudson River, Ninigret Pond. Cadmium added per unit dry weight of sediment vs. total dissolved cadmium. From DiToro et al. (1990)

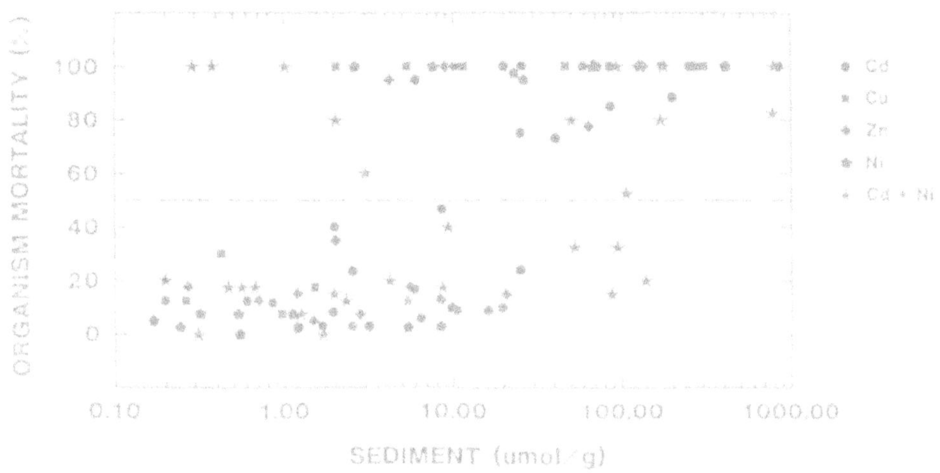

Figure 10. Organism mortality vs. sediment metal concentration for various sediment toxicity studies. Metals represented are cadmium, copper, zinc and nickel.

3.4 Sediment Toxicity and AVS Normalization

The three sediment toxicity experiment illustrated in Figure 8 was designed to test the utility of AVS as a predictor of the cadmium binding capacity of sediments and therefore a predictor of the concentration of cadmium that would cause sediment toxicity. These data along with other data from sediment metals toxicity tests are plotted on a dry weight basis in Figure 10. The metals represented here are cadmium, copper zinc and nickel. These divalent metals were chosen because they have lower sulfide solubility parameters (Table 3) than FeS and will form metal sulfides in the presence of AVS. No correlation of sediment concentration to organism mortality is evidenced.

AVS is extracted from the sediments using cold hydrochloric acid. The metals concentration that is simultaneously extracted is termed the simultaneously extracted metal or SEM. Figure 11 presents the sediment SEM/AVS ratio versus mortality. As the ratio approaches 1 mortality increases. These results suggest that if the ratio of SEM/AVS = 1 is used to discriminate toxic from nontoxic sediments (greater or less than 50 percent mortality, respectively), then in general the data are correctly classified as nontoxic (bottom left quadrant) and toxic (top right quadrant). The data that are incorrectly classified as toxic (bottom right quadrant) follows from the assumption that metal activity will invariably be high enough to cause toxicity if SEM/AVS > 1. It is possible that other ligands, associated with sediment sorption, for example, are reducing the metal activity below that which is lethal to the text organisms. Also, less sensitive organisms can tolerate the increased metal activity even if SEM/AVS > 1. For organisms that are present when SEM/AVS > 1, preliminary data by Ankley et al. (1991) and Striplin (1990) suggests that the extent to which metals bioaccumulate is strongly influenced by the AVS concentration .

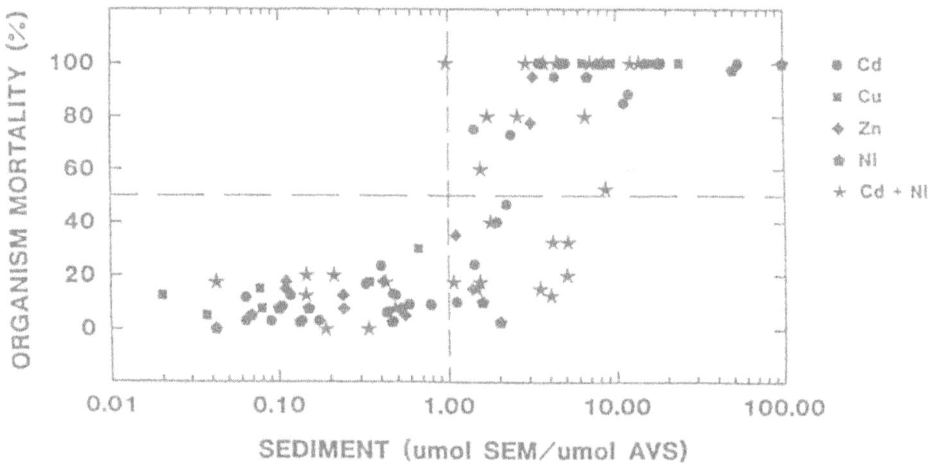

Figure 11. Organisms mortality vs. the molar ratio of SEM to AVS of the sediment for various sediment toxicity studies. Metals represented are cadmium, copper, zinc, and nickel.

If a more restrictive interpretation is adopted and the criterion SEM/AVS < 1 is used only to predict when a sediment is *not* acutely toxic, then all experiments are correctly classified. This is directly attributable to the excess AVS in the sediment, which assures that the metal activity in the sediment-interstitial water system is below the lethal metal activity for the organisms tested.

It is possible that these results are due to a covariation of AVS with other sediment properties, for example, organic carbon or iron content, which are actually controlling the metal bioavailability. However, the fact that the boundary occurs at SEM/AVS = 1, which is based on the supposition that AVS is the controlling sediment property, strongly argues that the relationship is casual, rather than correlative.

It should be noted that if the AVS concentration is effectively zero, as it would be in fully aerobic sediments, then other sediment properties would control the metal activity. This does not contradict the assertion that the SEM/AVS molar ratio of less than 1 predicts an absence to toxicity since in this case the molar ratio would be very large. The prediction is not much help but it is still correct. However, even a small AVS concentration, [AVS] $\sim$ 0.1 μmol/g, can sequester a significant quantity of metal and should be taken into account in determining the potential for metal toxicity for these sediments.

Additionally, it is important to realize that "anaerobic" and "aerobic" sediments are not precise classifications. Sediments are characterized by an aerobic layer underlaid by an anaerobic layer. Most benthic aerobic organisms survive in sediments that are underlaid with completely anaerobic sediments, which are characterized by significant AVS concentrations. Experiments reported by DiToro et al. (1992) suggest that the presence of AVS in the anaerobic layer is sufficient to reduce the metal activity to which the animals are exposed.

However, we have not examined the extent to which an aerobic layer depth is sufficient to mitigate the influence of the lower layer AVS. Thus it is possible to imagine a situation where AVS at depth (> 10 cm) might not control metal activity in a completely aerobic overlying layer--the top 10 cm for example--where the animals are exposed. It seems likely that even in this situation the presence of a sink of metals at depth would reduce the activity in the entire sediment to below toxic levels. The reasoning is that the diffusional transport of metal in the interstitial water would bring metals from a presumably higher concentration in the aerobic layer interstitial water to the lower concentration in the anaerobic layer. This would eventually deplete the aerobic layer of metals and establish a uniformly low metal activity in the interstitial water--sediment system. Thus, even in this case, we would expect that excess AVS would predict the absence of acute toxicity.

We believe that the data demonstrate that, it is possible to predict when a sediment will *not* be acutely toxic due to toxic metals that form sulfides that are significantly less soluble than FeS. The criterion is that the molar sum of simultaneously extracted Cd, Cu, Ni, Pb, and Zn is less than the molar acid volatile sulfide concentration:

$$\frac{[SEM]_{Cd} + [SEM]_{Cu} + [SEM]_{Ni} + [SEM]_{Pb} + [SEM]_{Zn}}{[AVS]} < 1 \qquad (23)$$

It should be noted that in order to apply this relationship it is necessary to measure all the toxic simultaneously extracted metals that are present in amounts that contribute significantly to the molar SEM sum, typically Cd, Cu, Ni, Pb, and Zn. Failing to do this could lead to an incorrect prediction of the lack of acute toxicity, i.e., the sum of the measured SEMs to AVS ratio is less than 1, when in fact the unmeasured metal could increase the SEM concentration so that the ratio exceeds 1 and toxicity would be possible. Thus, the acute toxicities of these metals are interrelated, with each contributing to the AVS that is bound by toxic metal. If excess AVS remains, no acute toxicity is expected. If excess metal remains, then the most soluble, Ni and Zn, will appear as free metal and toxicity is possible.

REFERENCES

Abernethy, S. and Mackay, D. (1987) " A discussion of correlations for narcosis in aquatic species." In K.L. Kaiser, ed. QSAR in Environmental Toxicology II. D. Reidel Publishing Company, Dordrecht, The Netherlands. 1-16.

Adams, W.J. Kimerle, R.A. and Mosher, R.G. (1985) "Aquatic safety assessment of chemicals sorbed to sediment." In R.D. Cardwell, R. Purdy and R.C. Bahner, eds., Aquatic Toxicology and Hazard Assessment: Seventh Symposium. American Society for Testing and Materials, Philadelphia, Pennsylvania. pp. 429-453.

Adams, W.J. (1987) "Bioavailability of neutral lipophilic organic chemicals contained in sediments: A review." In K.L. Dickson, A.W. Maki and W.A. Brungs, eds., Fate and Effects of Sediment-Bound Chemicals in Aquatic Systems. Pergamon Press, New York, New York, pp. 219-244.

Ankley, G.T., Phipps, G.L., Leonard, E.N., Benoit, D.A., Mattson, V.R., Kosian, P.A., Cotter, A.M., Dierkes, J.R., Hansen, D.J. and Mahony, J.D. (1991) "Acid volatile sulfide as a factor mediating cadmium and nickel bioavailability in contaminated sediments," Environ. Toxicol. Chem. **10**:1299-1307.

Benes, P. and Majer, V. (1980) Trace Chemistry of Aqueous Solutions. Elsevier, New York, New York.

Berner, R.A. (1967) "Termodynamic stability of sedimentary iron sulfides," Am. J. Sci. **265**:773-785.

Boulegue, J., Lord, III, C.J. and Church, T.M. (1982) "Sulfur speciation and associated trace metals (Fe, Cu) in the pore waters of Great Marsh, Delaware," Geochim. Cosmochim. Acta. 46:453-464.

Carter, C.W. and Suffett, I.H. (1983) "Interactions between dissolved humic and fulvic acids and pollutants in aquatic environments." In R.L. Swann and A Eschenroeder, eds., Fate of Chemicals in the Environment. ACS Symposium Series 225, American Chemical Society, Washington, D.C. pp. 215-230.

Connolly, J.P. and Pederson, C.J. (1988) "A thermodynamic-based evaluation of organic chemical accumulation in aquatic organisms," Environ. Sci. Technol. **22**:99-103.

Connolly, J.P. (1991) "Application of a food chain model to polychlorinated biphenyl contamination of the Lobster and Winter Flounder food chains in New Bedford Harbor," Environ. Sci. Technol. **25**(4):760-770.

Cowan, C.E. and Di Toro, D.M. (1988) Interim sediment criteria values for nonpolar hydrophobic compounds. U.S. Environmental Protection Agency, Office of Water Regulations and Standards Division, Washington, D.C.

Curl, R.L. and Keolelan, G.A. (1984) "Implicit-adsorbate model for apparent anomalies with organic adsorption on natural adsorbents," Environ. Sci. Technol. 18:916-922

Davies-Colley, R.J., Nelson, P.O. and Williamson, K.J. (1985) " Sulfide control of cadmium and copper concentrations in anaerobic estuarine sediments," Mar. Chem. 16:173-186

DeWitt, T.H., Ozretich, R.J., Swartz, R.C., Lamberson, J.O., Schults, D.W., Ditsworth, G.R., Jones, J.K.P., Hoselton, L. and Smith, L.M. (1992) " The influence of organic matter quality on the toxicity and partitioning of sediment associated fluoranthene," Environ. Tox. Chem. **11**:197-208.

Di Toro, D.M. (1985) "A particle interaction model of reversible organic chemical sorption," Chemosphere **14**:1503-1538.

Di Toro, D.M. and Horzempa, L. (1983) "Reversible and resistant component model of hexachlorobiphenyl adsorption-desorption: Resuspension and dilution." In d. Mackay, S. Paterson, S.J. Eisenreich and M.S. Simmons, eds., Physical Behavior of PCB's in the Great Lakes. Ann Arbor Science, Ann Arbor, Michigan, pp. 89-114.

Di Toro, D.M., Mahony, J.D., Kirchgraber, P.R., O'Byrne, A.L., Paquale, L.R. and Piccirilli, D.C. (1986) "Effects of nonreversibility, particle concentration, and ionic strength on heavy metal sorption," Environ. Sci. Technol. **20**:55.

Di Toro, D.M., Mahony, J.D., Hansen, D.J., Scott K.J., Hicks, M.B., Mayr, S.M. and Redmond, M.S. (1990) "Toxicity of cadmium in sediments: The role of acid volatile sulfide," Environ. Tox. Chem. **9**:1487-1502.

Di Toro, D.M. Zarba, C., Hansen, D.J., Swartz, R.C., Cowan, C.E., Allen, H.E., Thomas, N.A., Paquin, P.R., and Berry, W.J. (1991) "Technical basis for establishing sediment quality criteria for non-ionic organic chemicals using equilibrium partitioning," Environ. Toxicol. Chem. **10**:1541-1583.

Di Toro, D.M., Mahony, J.D., Hansen, D.J., Scott, K.J., Carlson, A.R., and Ankley, G.T. (1992) "Acid volatile sulfide predicts the acute toxicity of cadmium and nickel in sediments," Environ. Sci. Technol., **26**(1): 96:101.

Emerson, S., Jacobs, L. and Tebo, B. (1983) "The behavior of trace metals in marine anoxic waters: Solubilities at the oxygen-hydrogen sulfide interface." In C.S. Wong, E. Boule, K.W. Bruland and J.D. Burton, eds., Trace Metals in Sea Water. Plenum Press, New York, New York, pp. 579-608.

Goldhaber, M.B. and Kaplan, I.R. (1974) "The Sulfur Cycle." In E.D. Goldberg, ed., The Sea, Volume 5. Marine Chemistry. J. Wiley & Sons, New York, New York, pp. 569-655.

Gschwend, P.M. and Wu, S. (1985) "On the constancy of sediment-water partition coefficients of hydrophobic organic pollutants," Environ. Sci. Technol. **19**:90-96.

Hamilton, M.A., Russo, R.C. and Thurston, R.V. (1977) "Trimmed spearman-Karber method of estimating median lethal concentrations in toxicity bioassays," Environ. Sci. Technol. **11**:714-719.

Karickhoff, S.W. (1984) "Organic pollutant sorption in aquatic systems," J. Hydraul. Div., ASCE **110**:707-735.

Karickhoff, S.W. and Morris, K.R. (1985) "Sorption dynamics of hydrophobic pollutants in sediment suspensions," Environ. Toxicol. Chem. 4:469-479.

Kemp, P.F. and Swartz, R.C. (1986) "Acute toxicity of interstitial and particle-bound cadmium to a marine infaunal amphipod," Mar. Envir. Res. **26**:135-153.

Kemp, P.F., and Swartz, R.C. (1988) "Acute toxicity of interstitial and particle-bound cadmium to a marine infaunal amphipod," Mar. Environ. Res. **26**:135-153.

Landers, D.H., David, M.B. and Mitchell, M.J. (1983) "Analysis of organic and inorganic sulfur constituents in sediments, soils and water," Intern. J. Environ. Anal. Chem. **14**:245:256.

Landrum, P.F. (1989) "Bioavailability and toxicokinetics of polycyclic aromatic hydrocarbons sorbed to sediments for the amphipod Pontoporeia hoyi," Environ. Sci. Technol. **23**:588-595.

Leo, A. and Hansch, C., eds. (1986) Log(P) Database and Related Parameters. Pomona College, Claremont, California.

Mackay, D. and Powers, B. (1987) "Sorption of hydrophobic chemicals from water: A hypothesis for the mechanism of the particle concentration effect," Chemosphere **16**:745-757.

McIlroy, L.M., DePinto, J.V., Young, T.C. and Martin, S.C. (1986) "Partitioning of heavy metals to suspend solids in the Flint River, Michigan," Environ. Toxicol. Chem. **5**:609-623.

Morse, J.W., Millero, F.J. Cornwell, J.C. and Rickard, D. (1987) "The chemistry of the hydrogen sulfide and iron sulfide systems in natural waters," Earth Sci. Rev. **24**:1-42.

Muir, D.C.G., Rawn, G.P., Townsend, B.E., Lockhart, W.L. and Greenhalgh, R. (1985) "Bioconcentration of cypermethrin, deltamethrin, fenvalerate, and permethrin by *Chironomus tentans* larvae in sediment and water," Environ. Tox. Chem. **4**:51.

Nebeker, A. and Schuytema, G. (1988) DDT/Endrin Results. Evaluation of Carbon Normalization Theory. U.S. Environmental Protection Agency Environmental Research Laboratory, Corvallis, Oregon.

Nebeker, A.V., Schuytema, G.S, Griffis, W.L., Barbitta, J.A. and Carey, L.A. (1989) "Effect of sediment organic carbon on survival of Hyalella azteca exposed to DDT and endrin," Environ. Toxicol. Chem. **8**:705-718.

Nelson, D.M., Penrose, D.M., Karttunen, J.O. and Mehlhaff (1985) "Effects of dissolved organic carbon on the adsorption properties of plutonium in natural waters," Environ. Sci. Technol. **19**:127-131.

O'Connor, D.J. and Connolly, J. (1980) "The effect of concentration of adsorbing solids on the partition coefficient," Water Resour. **14**:1517-1523.

Schoonen, M.A.A. and Barnes, H.L. (1988) "An approximation of the second dissociation constant for H_2S," Geochim. Cosmochim. Acta **52**:649-654

Stehly, G.R. (1991) memorandum to W.J. Berry, April 25, 1993.

Striplin, B.D. (1990) Skagway Harbor Field Investigation. Tetra Tech, Inc. Bellevue Washington

Swartz, R.C. (1991) Acenaphthene and phenanthrene files. memorandum to David Hansen, June 26, 1991.

Swartz, R.C., Ditsworth, G.R., Shults, D.W. and Lamberson, J.O. (1985) "Sediment toxicity to a marine infaunal amphipod: Cadmium and its interaction with sewage sludge," Mar. Environ. Res. **18**:133-153.

Swartz, R.C., Kemp, P.F., Schults, D.W. and lamberson, J.O. (1988) "Effects of mixtures of sediment contaminants on the marine infaunal amphipod, Rhepoxynius abronius," Environ. Toxicol. Chem. 7:1013-1020.

Swartz, R.C., Schults, D.W., DeWitt, T.H., Ditsworth, G.R. and Lamberson, J.O. (1990) "Toxicity of fluoranthene in sediment to marine amphipods: A test of the equilibrium partitioning approach to sediment quality criteria," Environ. Toxicol. Chem. **9**:1071-1080.

U.S. Environmental Protection Agency (1989) Briefing report to the EPA science advisory board on the equilibrium partioning approach to generating sediment quality criteria. EPA 440/5-89-002. U.S. Environmental Protection Agency, Office of Water Regulations and Standards, Criteria and Standards Division, Washington, D.C.

U.S. Environmental Protection Agency (1993) Proposed technical basis for deriving sediment quality criteria for nonionic organic contaminants for the protection of benthic organisms by using equilibrium partitioning. EPA 822-R-93-011. U.S. Environmental Protection Agency, Office of Water. Washington, D.C.

Voice, T.C., Rice, C.P., and Weber, Jr., W.J.(1983) "Effect of solids concentration on the sorptive partitioning of hydrophobic pollutants in aquatic systems," Environ. Sci. Technol. **17**:513-518.

Ziegenfuss, P.S., Renaudette, P.S. and Adams, W.J. (1986) "Methodology for assessing the acute toxicity of chemicals sorbed to sediments: Testing the equilibrium partitioning theory." In T.J. Poston and R. Prudy eds., Aquatic Toxicology and Environmental Fate Ninth Volume. STP 921. American Society for Testing and Materials, Philadelphia, Pennsylvania, pp. 479-493.

CHAPTER 6

CONTAMINATED SEDIMENTS AND REMEDIATION - GEOCHEMICAL PERSPECTIVE

Ulrich Förstner[1]

1. INTRODUCTION

Sediments are both carriers and potential sources of contaminants in aquatic systems and these materials may also affect groundwater quality and agricultural products when disposed on land. Contaminants are not necessarily fixed permanently by sediment, but may be recycled via biological and chemical agents both within the sediment compartments (pore water) and water column. Bioaccumulation and food chain transfers may be strongly affected by sediment associated pollutants. Benthic organisms in particular, have direct contact with sediment and the contaminant level in the sediment may have greater impact on their survival than that in aqueous solution.

Similarly to other waste materials, management of contaminated sediments requires a holistic approach. In addition to common predictive techniques for estimating losses, the interactive nature of various parameters has to be recognized. This means that particular emphasis should be given to the evaluation of the driving forces of geochemical processes and to the capacity controlling properties in contaminated sediments. Assessment of these parameters and their long-term borderline conditions should be an integrated part of the wider management scheme, i.e., the analytical and experimental parameters should always be related to potential remediation options for a specific sediment problem.

In modern sediment research on contaminants, five aspects are discussed, which in an overlapping succession also reflects development of knowledge on particle-associated pollutants during the past twenty-five years: (i) identification of sources and their distribution; (ii) evaluation of solid/solution relations; (iii) study of transfer mechanisms to biological systems; (iv) assessment of environmental

[1] Technical University of Hamburg, Germany

NATO ASI Series, Partnership Sub-Series, 2. Environment – Vol. 3
Remediation and Management of Degraded River Basins
Edited by V. Novotny and L. Somlyódy

impact; and (v) selection and further development of remedial measures, in particular, of dredged materials. In practical aspects (i), (iv) and (v) are of particular relevance and recent developments will be treated in this review.

2. IDENTIFICATION OF SOURCES AND TEMPORAL DEVELOPMENTS

A program of sediment studies will typically consist of a series of objectives of increasing complexity, each drawing part of its information from the preceding data base (Golterman et al., 1983):

- *Preliminary site characterization* - Low density sampling with limited analytical requirements to provide a general characterization of an area for which little or no previous information exists.

- *Identification of anomalies* - More detailed sampling and analysis designed to establish the presence and extent of anomalies.

- *Establishment of reference* - To create reference points in the form of some measured parameters for future comparison.

- *Identification of time change* - To show trends and variations of sediment data over time, using sediment cores or other repetitive sediment sampling.

- *Calculation of mass balances* - To account for the addition and subtraction of sediment-related components within an aquatic environment (a complex study) by means of accurate and representative sampling and analysis.

- *Process studies* - Specialized sampling to improve state of knowledge about aquatic systems, for example, by supplementary laboratory experiments.

Program objectives largely control the type, density, and frequency of sediment sampling and associated analyses, whereas the type of environment (rivers, lakes, estuaries, etc.) largely controls the locations and logistics of sampling. Logistic factors include (Golterman et al., 1983):

- local availability of sampling platform or vessel
- available time
- suitability of survey system to locate sample position
- availability of trained personnel and support staff
- availability of equipment
- storage and security
- transport systems
- capability of follow-up investigations and monitoring

For complex surveys there are numerous types of sampling patterns to choose, e.g., spot samples, square grids (including nested and rotated grids), parallel line and traverse line grids (with equal or non-equal sampling), ray grids or concentrated arc sampling, each of which offers some particular advantage (Golterman et al., 1983).

The suitability of corrers and bottom samplers has been tested during equipment trials by Sly (1969.) For source reconnaissance analysis, fine - to medium - grained bottom deposits from a depth of 15 to 20 cm can be collected, for example, with an Ekman grab sample. In environments with a relatively uniform sedimentation, for example, in lakes and marine coastal basins where deposits are fine-grained and deposit at a rate of 1 to 5 mm/year, a better sampling procedure involves taking vertical profiles with a gravity or valve corer.

The study of dated sediments cores has proven particularly useful as it provides an historical record of various influences on aquatic systems by indicating both natural background levels and man-induced accumulation of elements over an extended period of time. Various approaches to dating of sediment profiles have been used but the isotopic techniques, using ^{210}Pb, ^{137}Cs, and $^{238-240}Pu$, have produced more unambiguous results and, therefore, have been most successful. "Historical monitoring" by sediment studies has been summarized by Alderton (1985).

3. SEDIMENT QUALITY CRITERIA

Reasons for development of sediment quality criteria have been formulated in the mid 1980s as follows (Förstner et al., 1990): (1) sediment integrate contaminant concentrations over time; (2) long-term perspectives of water management need "*integrated strategies*" in which sediment-associated pollutants have to be considered as well (generally, the contaminant levels in the sediment have greater impact on survival of benthic organisms than do aqueous concentrations); (3) management plans have increasingly been based on the *assimilative capacity* of certain receiving systems, which requires knowledge of the properties of sediment components as the major sink; (4) permission for *dredging activities* and *deposition of dredged materials* have to be based on standardized sampling protocols and test procedures. In this context, the conventions for the marine environment held in Oslo and London, should be mentioned.

The United States Environmental Protection Agency was first to develop standard procedures for the assessment of environmental impact of sediment-bound pollutants (see also the preceding chapter by DiToro and Da Rosa). Further discussions led to the differentiation of biological and chemical-numerical approaches. Biological criteria integrate sediment characteristics and pollutant loads, but they do not generally indicate the cause of effect. With respect to chemical criteria there is no immediate indication on biological effect; their major advantage lie in their easy application and applicability to modeling approaches.

3.1 Biological Criteria

Biological criteria have been developed and are already applied in various areas (Anderson et al., 1987; Chapman et al., 1987):

- *Field biological surveys* - conduct on-site studies of biota to evaluate possible impact at the site (biological response pass/fail).

- *Bioassay of spiked sediment* - estimate effect/no effect of the sediment concentration for a specific chemical (numerical) criterion. They may be required for clearance of a new chemical.

- *Tissue action level* - link sediment concentration to safe tissue concentration (for example, FDA action level or body burden-response data) through application of equilibrium or kinetic models (numerical criterion).

- *Aqueous toxicity data* - apply toxicity data from typical water-column bioassays to sediments through direct measurements of pore water concentrations or estimation of pore water from sediment concentrations through application of equilibrium models (see preceding chapter by DiToro and Da Rosa). This may be required in evaluation of new chemicals.

Biological approaches to development and application of sediment quality criteria exhibit a common basis in the study of damaging impacts from contaminated sediments on organisms. The biological parameters "bioaccumulation," "toxicity," and "mutagenicity" have to be considered separately in any case. Bioassays like field surveys are empirical considerations which cannot provide numerical criteria to be transferred to different situations.

Generally, it is difficult to establish a clear cause-and-effect relationship between acute and chronic toxic effects on biota and the occurrence of specific pollutants in sediments. One major limitation is that not all sediment-associated chemicals can presently be identified, thus, unidentified compounds cannot be ruled out as principal ecological factors.

Relatively simple and implementable liquid, suspended particulate and solid-phase bioassays have been carried out for assessing the short-term impact of dredging and disposal operations on aquatic organisms (Ahlf and Munawar, 1988, Ahlf et al. 1992). Standardized tests are characterized by their lack of variability, but essential information (e.g., lethality, alterations of growth rate) can only be obtained with single species tests. The influence of the main environmental variables on the interactions of suspended particulates, or in-situ sediment contaminants and organisms, should also be determined under simulated field conditions. In particular due to recent developments, benthic bioassay procedures are important in evaluating the relationship between laboratory and field impacts (Reynoldson, 1987).

With restriction to the effects on benthic communities, the sediment quality "triad" by Chapman (1986) combines chemistry and sediment bioassays data with *in-situ* studies. Chemistry and bioassay estimates are based on laboratory measurements of resident organism histopathology, benthic community structure and bioaccumulation/metabolism. Areas where the three facets of the triad show the greatest overlap (in terms of positive or negative results) provide the strongest data for determining numerical sediment criteria.

3.2 Chemical-Numerical Quality Approaches

Numerical approaches for the assessment of environmental impact of sediment-associated metals are based on (1) accumulation, (2) pore water concentrations, (3) solid/liquid equilibrium partition (both sediment/water and organism/water) and (4) elutriation properties of contaminants:

Background (Ecoregional) Approach: Compares the actual data with reference sites comprising natural or insignificant pollutant concentrations. Particularly useful are samples from deeper layers of the sediment sequence at a given site, for example, from drill holes, since the material is derived from the same catchment and usually is similar in its substrate composition. Nevertheless, standardization with respect to grain size distribution is indispensable.

Pore water approach: Based on the experience that the composition of interstitial waters is the most sensitive indicator of reactions that take place between the pollutants on particles and the aqueous phase which interact. There is the advantage of direct recovery and analysis of water-borne constituents. But there are several disadvantages, mainly arising from sampling and sample preparation, which need considerable precaution such as exclusion of oxygen.

Equilibrium Approaches: These approaches are related to the broad toxicological basis of food and water quality data which is a very important advantage. On the other hand there are the effects of sample preparation, e.g., drying procedures; the separation procedures, for example, filtration or centrifugation. There are strong effects of grain size composition and the influence of suspended matter concentration in an aquatic system, which is even more important if the kinetics of sorption and desorption are too slow for equilibrium to be achieved in a given time of interaction. Unlike non-ionic organic chemicals, K_D -values of metals are not only correlated to organic substances but also to other sorption-active surfaces, therefore, the equilibrium partition approach exhibits strong limitations for metallic elements.

Remobilization: Short-term effects may be studied from water/sediment-suspensions, medium-term effects from experiments using tanks, and long-term effects by applying chemical extractions, either single or in sequence. In connection with problems arising from disposal of solid wastes, particularly of dredged materials, the chemical extraction sequence has been applied which is designed to differentiate between the exchangeable, carbonate, reducible (hydrous Fe/Mn oxides), oxidizable (sulfides and organic phases) and residual fractions. The undisputed advantage of this approach with respect to the estimation of long-term effects on metal mobilization lies in the fact that rearrangements of specific solid "phases" can be evaluated prior to the actual remobilization of certain proportions of an element into the dissolved phase.

The approaches to sediment quality assessment described herein so far can be separated into two groups, i.e., either as a "quality standard" or with regard to an "indicator" function. The former group would comprise, for example,"pore water," "sediment/water equilibrium," and "sediment/organism equilibrium" approaches which are based on toxicological relevant standards. The approaches of the latter group consisting of the "elution" and "background (ecoregional)" approaches will, disregarding their numerical character, primarily only provide qualitative indicators on a certain status of the extent of sediment pollution. However, it seems to be advantageous to have such a relatively simple assessment, which may then be extended with more complicated procedures, including biological criteria, into an integrated ecological evaluation.

3.3 Geochemical Characterization of the Potential Mobility of the Pollutant

In management of dredged materials, two different target areas for combined matrix/pollutant criteria can be distinguished: (1) sediment resuspension, and (2) dredged material disposal. With respect to "resuspension of aquatic sediments", which involves more short-term effects than disposal of dredged materials, special emphasis should be placed on the factor of "available species of a certain pollutant." The environmental impact of sediment deposits is influenced by internal chemical conditions rather than by the concentration and extractability of a pollutant. Priority, therefore, should be given to the optimization of long-term chemical stability (see Section 4.2 of this chapter on "geochemical engineering").

Regarding the potential release of contaminants, particularly metals, from sediments, changing pH and redox conditions are very important. Table 1 shows data of pore water analyses from dredged material disposal sites in Hamburg Harbor. Recent deposits are characterized by low concentrations of nitrate, cadmium and zinc; when these low-buffered sediments are oxidized during a time period of a few months to years, the concentrations of ammonia and iron in pore water typically decrease, whereas those of cadmium and zinc increase (consequently, these metals are easily taken up by agricultural crops).

Direct assessment of pH-changes resulting from the oxidation of anoxic sediment constituents can be performed by aeration of sediment suspensions with air or oxygen and subsequent determination of the pH difference between the original sample and oxidized material. The greater the difference the higher is the short-term mobilization potential of metals, e.g., during dredging, resuspension and other processes by which anoxic sediments get into contact with oxygenated water or - following land deposition of dredged materials - with atmospheric oxygen.

Table 1 Mobilization of metals and nitrogen compounds from dredged materials after land depositions (Maaß et al., 1985 and other authors).

Element or Compound	Reduced Water	Oxidized Water
Ammonia	125 mg/l	< 3 mg/l
Iron	80 mg/l	< 3 mg/l
Nitrate	< 3 mg/l	120 mg/l
Zinc	< 10 μg/l	5000 μg/l
Cadmium	<0.5 μg/l	80 μg/l

Periodical redox processes can spur an increase or decrease in "acid producing potential" (APP) or pH in a sediment system. In a *closed* system, periodical redox processes can lead to the change or transfer between APP(s) and APP(aq) but the total APP of the system does not change. The processes are reversible. The hydrogen ions produced in oxidation will be consumed by subsequent reduction. In contrast, in an *open* system, the total APP of the system will change depending on the properties of the system and reaction processes. Under certain conditions total APP of the system increases, while under other conditions total APP in the system decreases. Some processes are irreversible. The components producing or consuming H^+ leave the system and cause the change in APP(s), APP(aq) and permanent ANC (acid neutralization capacity) (Van Breemen, 1987).

Within certain categories of acid producing and consuming capacities, guideline values for individual metals should be based on elutriate data, preferentially at pH 4, for better comparison with other solid matrices, for example, with Swiss Ordinances for Waste Materials contained in Anonymous (1990). These guidelines include two categories of limit values: "Inert/Construction Materials" and "Residual Material Deposits", the latter mainly comprising of pretreated products of municipal solid waste incineration.

Regarding prediction of long-term effects of sediment bound metals, chemical extraction procedures are of limited value because they usually involve neither reaction-mechanistic or kinetic considerations. This shortcoming can be avoided, for example, by an experimental approach- procedure which was originally used by Patrick et al. (1973) and Herms and Brümmer (1978). In these procedures sediments can be treated in a circulation system under controlled intensification of significant release controlling parameters such as pH, redox-potential, and temperature. Our experimental design at the University of Hamburg (Schoer and Förstner, 1987) includes an ion-exchange system for extracting and analyzing the released metals at an adequate frequency, and compares sequential extraction results before and after treatment of the sample in the circulation apparatus. Individual metal species are released at different time intervals. Taking into account both element contents released during 10 weeks experiments (equivalent to several thousand years of solid/water interaction) and those extrapolated from extraction "pools", concentrations can be calculated for different scenarios.

In the approach by Obermann and Cremer (1992) outlined on Figure 1, pH is automatically adjusted to 4 during the time period of 24 hours. In addition to the release rates of metals which can be determined from samples taken at different time intervals, the sum curve of acid consumption provides information on the potential changes of the matrix composition during acidification and availability of the buffering capacity at different time scales. To complement the relationship a diagram of discrete acid additions should also be recorded and plotted.

The above discussion has shown that potential mobility, bioavailability and toxicity of metals in sediments strongly depend on redox conditions (see also Chapter by Salomons). From earlier chemical extraction procedures, as proposed by Tessier et al. (1979) to more recent SEM/AVS[2] ratio determinations (DiToro et al., 1990, 1992, also Chapter by DiToro and Da Rosa), metal speciation is a major characteristic describing its potential mobility in sedimentary environments. The inherent link between the parameters to describe "metal speciation" is the characterization of chemical stability and bioavailability of metal compounds in sediments under certain redox conditions. In a sulfidic anoxic environment, even if the sediment is heavily polluted by metals, organisms are considered to be still safe due to strong fixation of metal ions by $S^{=}$ or HS^{-} which are a source of Acid Volatile Sulfide (AVS). Such heavily polluted sediments, however, can behave as a time bomb which is triggered by only one

[2] SEM = Simultaneously Extractable Metals, and AVS = Acid Volatile Sulfide refer to an extraction procedure for metals and sulfides. See Chapter by DiToro and Da Rosa and pertinent references in the chapter.

factor - a redox increase to a critical point, for example, by exposure to oxygen-rich overlying water or directly to the air. Once this situation occurs (a possible pathway is oxygen transfer via plant roots), toxic metals in sediments will be released to water phase or transformed into more bioavailable species (see Calmano et al., 1993).

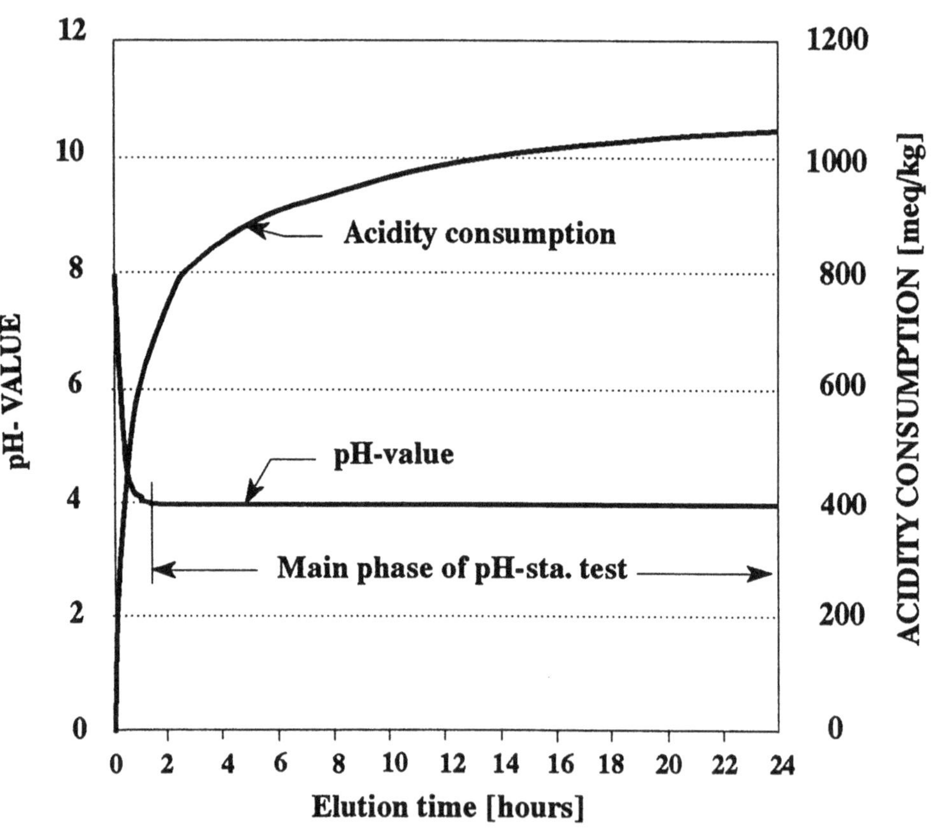

Figure 1. Patterns of a suspension titration (after Obermann and Cremer, 1992). The initial phase during which the target pH-value is not attained, is followed by a 50-fold main phase with highly stable pH conditions.

At the moment, research on long-term effects of redox variations on metal behavior in sediments is mostly based on thermodynamic considerations. Future research should emphasize studies on the kinetics of transformation of metal species, hydrogen ion production and metals release as affected by changing redox conditions. Additional important aspects involve the bridging of gaps between numerical criteria approaches, as reflected by matrix composition and metal mobility, and biological approaches. Assessment for bioavailability of metal species is still based on results of water phase elutriate, mostly using acute toxicity data (Ankley et al., 1991). Initial data on solid phase bioassays suggest good perspectives for future applications in sediment quality assessment.

4. REMEDIATION PROCEDURES FOR DREDGED SEDIMENTS

Dredging is a human activity which is comparable in magnitude with highly dynamic natural geomorphologic processes such as landslide. In the harbors around the North Sea, approximately 100 million m^3 of sediment has to be dredged annually - about 10 times the average annual discharge of the Rhine River. Typical problems associated with these activities and sediment are;

- increasing volummes of dredged materials,
- high concentrations of toxic substances, and
- the fact that these problems have been concentrated at the mouth of large rivers and in coastal areas.

In the Rotterdam Harbor located at the mouth of the Rhine River, the volume of sediment which has to be dredged annually increased from less than 1 million m^3 in 1920 to more than 10 million m^3 in 1982. During the last 80 years, the cadmium concentrations in the river sediment increased by a factor of more than 50.

4.1 Assessment and Selection of Remediation Technology

Besides the cost of the remediation technique, major questions relate to contaminant loss pathways. Contaminant loss can occur through one or more pathways. An example of a confined disposal facility indicates that potential pathways for contaminant loss include surface runoff, effluent, seepage, leachate, dust, and uptake by plants and animals. Predictive techniques for estimating contaminant losses can be divided into categories of *a priori* techniques and techniques based on pathway-specific laboratory testing:

- *A priori* techniques are suitable for planning-level assessments.
- Techniques that use pathway-specific test data provide state-of-the-art loss estimates (generally more advanced)

The various types of sediment remediation methodologies can also be subdivided according to the mode of handling, i.e., "in place" or "excavation" (Table 2) or in relation to technology "containment" or "treatment". Important containment techniques include "in-situ capping" and "confined disposal facility". Regarding in-place treatment, biological processes may be applied. Apart from physical separation, excavated sediments can be treated to immobilize pollutants, mainly metals.

Table 2 Technology types for sediment remediation (US EPA, 1993)

	In place	Excavated
Containment	*In-situ* capping	Beneficial use
	Contain/fill	Capping/confined aquatic disposal
		Commercial landfills
		Confined disposal facility
Treatment	Bioremediation	Chemical
	Chemical	Biological
	Immobilization	Extraction
		Immobilization
		Physical separation
		Thermal treatment

A more general conceptual scheme related to excavated sediment material has been proposed by the TNO, the Netherlands' scientific technological organization (Van Gemert et al., 1988). The following two techniques are distinguished: "A-" technique is for large-scale concentration such as mechanical separation. These techniques are characterized by low cost per unit of residue, low sensitivity to variation, and they may be applied using mobile plants. "B-" techniques are decontamination procedures which are especially designed for relatively small operations. They involve higher operating costs per unit of residue. They are more complicated, need specific experience of the operator and are usually constructed as stationary plants. "B-"techniques include biological treatment, acid leaching, solvent extraction, etc. Characteristic applications and their limitations are included in an US EPA (1988) report.

Biological treatment has been used for decades to treat domestic and industrial wastewater and in recent years has been demonstrated as a technology for destroying some organic compounds in contaminated soils. Bioremediation or biorestoration may be used in certain cases for sediments with organic contaminants. However, since in large catchment sediments that would be contaminated only by organics are rare the expectations of the usefulness of these remediation techniques have been overestimated. The request for such procedures is often a simple indication of ignorance for the sediment pollution problems. Even in optimal cases there are many limiting factors to biodegradation processes, temperature, nutrients, oxygen being most important.

The primary application for *solvent extraction* is to remove organic contaminants such as halogenated compounds and petroleum hydrocarbons. Extraction processes may also be used to extract metals, but these applications which usually involve acid extraction, have not proven to be cost effective for contaminated sediments. Fine grained materials are more difficult to be extracted and the presence of detergents adversely impacts oil/water separation. The procedure is less effective for high-molecular compounds and very hydrophobic substances. In any case, careful selection of reagents and laboratory testing is required.

Solidification/stabilization is a commonly used term for immobilization techniques. Solidification refers to a change of physical properties, stabilization suggests chemical effects. Several factors negatively interfere with the objective to solidify or stabilize: organic compounds, oil and grease, inorganic salts such as nitrates, sulfates and chlorides, small particle sizes, volatile organic compounds, and low solids content.

In an attempt to rank individual technologies one can follow the proposals in the Draft Remediation Guidance Document prepared by the US EPA (1994). Three criteria have been used:

- *State of Development*. The immobilization method is mostly good except for non-removal and treatment alternatives. Such deficiencies should be overcome by research.

- *Potential Contaminant Loss*. It is particularly high for removal. This would favor in-situ techniques. In-situ capping could be the method of choice for areas where maintenance dredging is not essential.

- *Cost*. It is already clear at this point that the cost factor will mostly exclude treatment of large volume contaminated sediments. The only possibility would be the reduction of the volume by mechanical pretreatment then chemical/biological techniques could be potentially used.

It is quite obvious that technological options are more restricted for dredged sediments than for other waste materials in most cases. In particular, remediation techniques are often economically unacceptable because of the large volume of contaminated sediments.

Table 3 Ranking of remediation technologies (US EPA, 1994, also see text)

Technology	State of Development	Potential Contaminant Loss	Cost
No removal	Low/Medium	Low	Low/Medium
Removal	High	High	Low
Transport	High	Low	Low
Pretreatment	Medium	Medium	Low/Medium
Treatment	Low	Low/Medium	High
Disposal	High	Low/Medium	Low/Medium
Effluent/Leachate Treatment	High	Low	Low

4.2 Geochemical Engineering - Application to Contaminated Sediments

"Geochemical Engineering" (Salomons and Förstner, 1988) applies geochemical principles (such as concentration, stabilization, solidification, and other forms of long-term, self containing barriers) to determine mobilization and biological availability of critical pollutants (Figure 2).

Inclusion of the time factor moves beyond a traditional chemical approach. It also transcends the civil engineering approach of waste management, which usually devotes little attention to long-term emissions from the waste disposal site. "Because we have become accustomed to considering the filling period as the most important phase in landfill operation, we have forgotten that subsequent to the active working period there is an infinitesimally long time in which the site has to function as a depository for all materials unwanted in the biosphere " (Stief, 1987).

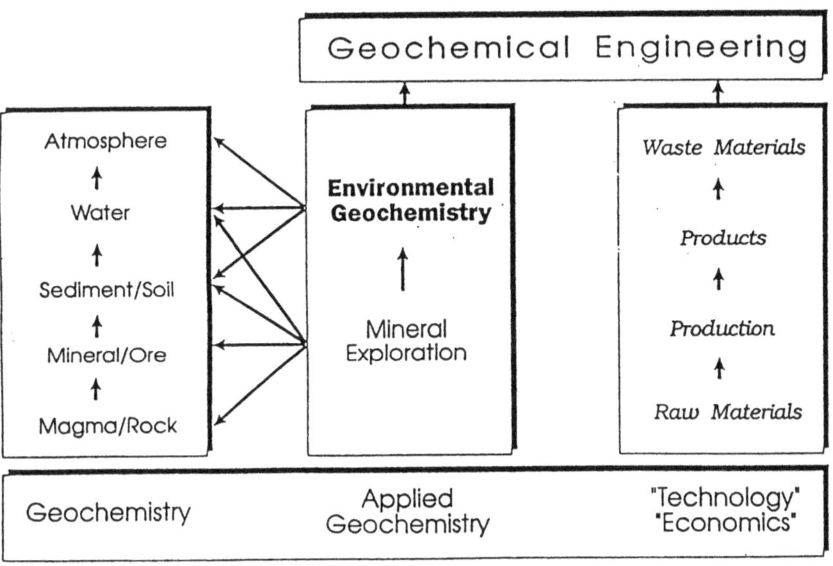

Figure 2 New developments in environmental geochemistry.

In modern waste management, the fields of geochemically oriented technology include:

- the study of material fluxes within and between the anthroposphere and "geospheres;"
- the optimization of elemental distribution at high-temperature processes;
- the selection of favorable milieu conditions for deposition of large-volume wastes;
- the selection of additives for the solidification and stabilization of waste materials;
- the development of test procedures for long-term prognoses of pollutant behavior in all kinds of waste depositories.

The conceptual approach of "storage capacity controlling properties" (see also the subsequent chapter by Salomons) forms the link between the geochemical cycles comprising driving forces such as organic matter/redox-sensitive elements and cycles of mobilizable pollutants. It plays a key role in the framework of the "chemical time bombs" (Stigliani, 1992) or "non-linear release" concept. In this respect, the potential of sediments, soils and waste materials to immobilize toxic chemicals is conceived as a first and in most cases preferential barrier against dispersion of these substances in both ground and surface waters or their transfer to terrestrial or aquatic biota. To make scientific objectives clear, it is useful to distinguish between two different mechanisms: the first one is direct saturation by which the capacity of sediment for toxic chemical becomes exhausted. The second way to "trigger" a time bomb is through a fundamental change in a chemical property of the solid matrix that reduces its capacity to adsorb (or keep adsorbed) toxic materials. Methodologies should be designed for assessing effects related to processes of "early diagenesis," i.e., mechanisms and effects by which solids are changed in their chemical form, involving a new equilibrium between solid and their dissolved species.

For dredged materials, mining residues and municipal solid waste, long-term immobilization of critical pollutants can be achieved by promoting less soluble chemical phases, i.e., by thermal and chemical treatment, or by providing respective milieu conditions. In extension of the initial conceptual model of coupled geochemical cycles, as designed by Salomons (1993). Figure 3 schematically indicates that selection of appropriate environmental conditions predominantly affect chemical gradients, whereas chemical additives are aimed to enhance capacity controlling properties in order to bind (or degrade) micropollutants. In general, micro-scale methods, e.g., formation of mineral precipitates in the pore space of a sediment waste body will be employed rather than using large-scale enclosure systems such as clay covers or retaining wall constructions. A common feature of geochemically-designed deposits, therefore, is their tendency to increase overall stability in time due to formation of more stable minerals and closure of pores, thereby reducing water penetration.

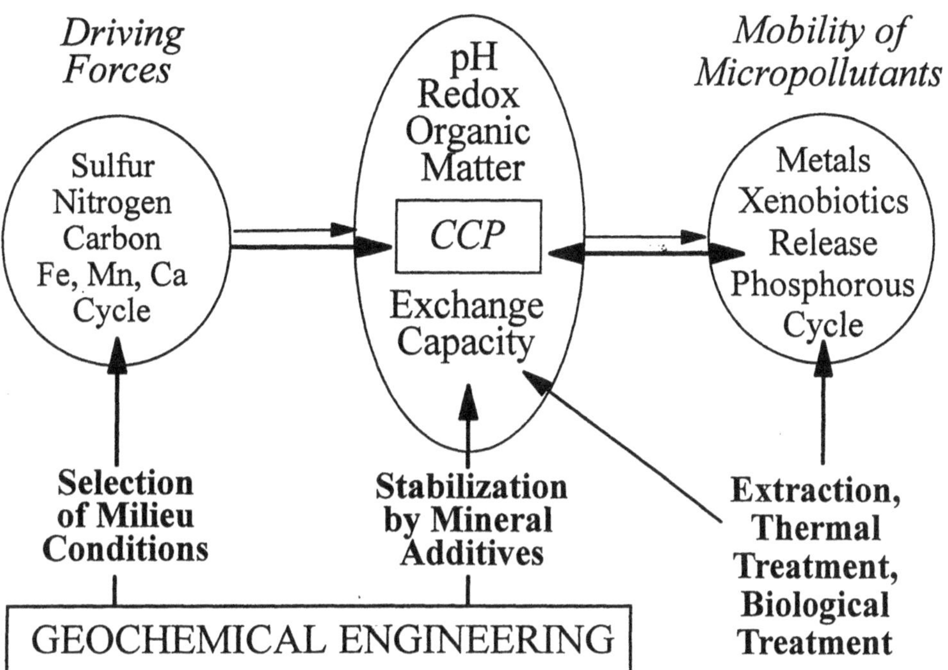

Figure 3 Interference of geochemical engineering methods with biochemical cycles.

4.3 Storage Under Permanent Anoxic Conditions

Regarding various containment strategies it has been argued that upland containment (e.g., in heap-like land fills) could provide more controlled management than containment in aquatic (marine) environment. However, contaminants released either gradually through a leaking barrier (including releases into groundwater) or catastrophically from failure of the barrier could produce substantial damage (Kester et al., 1983). On the other hand, near-shore marine containment (for example, in capped mound deposits), offers several advantages, particularly with respect to the protection of groundwater resources since the underlying water is saline and heavier than freshwater and inherent chemical processes are favorable for immobilization or degradation of priority pollutants.

In a review of various marine disposal options, Kester et al. (1983) suggested that the best strategy for disposal of contaminated sediments is to isolate them in a permanently reducing environment.

Disposal in capped mound deposits above the prevailing sea floor, disposal in subaqueous depressions, and capping deposits in depressions provide procedures for sediment containment (Bokuniewicz, 1982). In some instances it may be worthwhile to excavate a depression for the disposal site of contaminated sediments which can be capped with clean sediment. This type of waste deposition under stable anoxic conditions where large masses of polluted materials are covered with inter sediment is known as "sub-sediment deposit." The first example was planned for highly contaminated sludges from Stanford Harbor in the Central Long Island Sound following intensive discussions in the U.S. Congress (Morton, 1980).

Under sub-sediment conditions there is a particular low solubility of metal sulphides, compared to the respective carbonate, phosphate, and oxide compounds (Figure 4). One major prerequisite is the microbial reduction of sulphate. Thus this process is particularly important in the marine environment, whereas in an anoxic freshwater milieu there is a tendency for enhancing metal mobility due to formation of stable complexes with ligands from decomposing organic matter. Marine sulfitic conditions seem to repress the formation of mono-methyl mercury, one of the most toxic substances in an aquatic environment, by a process of dissociation into volatile dimethyl mercury and insoluble mercury sulphide (Craig and Moreton, 1984). There are indications that degradation of highly toxic chlorinated hydrocarbons is enhanced in the sulfitic environment relative to oxic conditions (Sahm et al., 1986; Kersten, 1988).

Laboratory studies on the evaluation and efficiency of stabilization processes were performed by Calmano et al. (1986). As an example, Figure 5 shows acid titration curves for Hamburg Harbor mud without and after addition of limestone and cement/fly ash stabilizers. Best results are attained with calcium carbonate since the pH-conditions are not changed significantly after the addition of $CaCO_3$. Generally, maintenance of a pH neutral or slightly basic environment favors adsorption and/or precipitation of soluble metals (Gambrell et al., 1983).

5. SUMMARY AND FUTURE OUTLOOK

Remediation techniques on contaminated sediments generally are far more limited than those for most other solid waste materials, except mine wastes. The widely diverse contamination sources in larger catchment areas usually produce a mixture of pollutants, which is more difficult to treat than industrial waste. Even if one has procedures at hand to reduce nine priority pollutants below the guideline values, the tenth, for example, mercury or PCB, may make the whole effort futile. There is a long retention time for sediments in larger catchments. Improvement at the source may need decades to become effective in sediments at the lower reaches and harbors close to the river mouth. For most sediments from maintenance dredging there are more arguments in favor of "disposal" than for "treatment." Considering world-wide dredging activities - more than 1 billion cubic meters per year - only a very small percentage of these materials can undergo "treatment", in more specific terms, solvent extraction, bioremediation, thermal desorption,, vitrification, etc. Mechanical separation of less strongly contaminated fractions, however, may become a useful step prior to final storage of the residues.

It is possible that economic considerations will increase the popularity of geochemically engineered solutions in the near future, in particular for large waste masses such as mine residues, dredged materials and fly ash. But even if traditional engineering continues to dominate practical solutions the specific potential of geochemistry to provide instruments for long-term assessment of processes should be used and expanded.

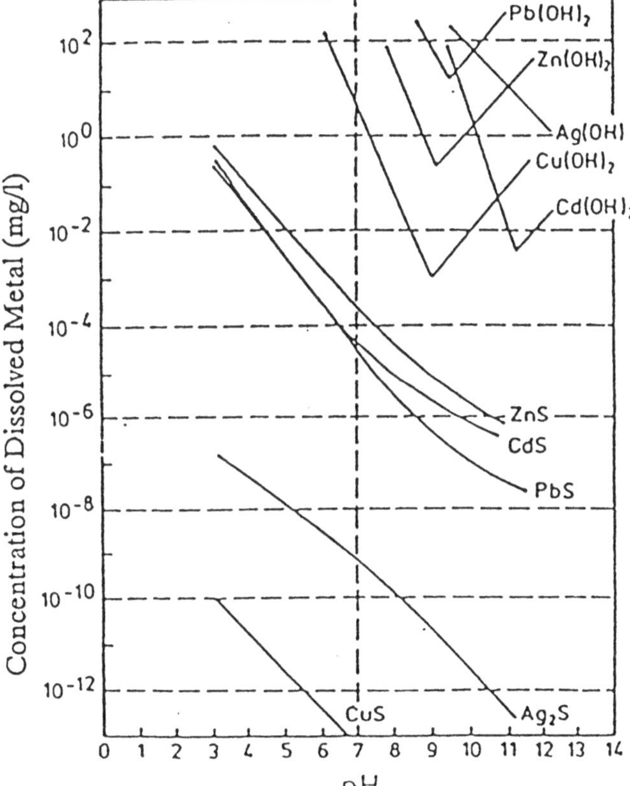

Figure 4.

Solubility of metal sulphides and metal oxides (from Ehrenfield and Bass, 1983)

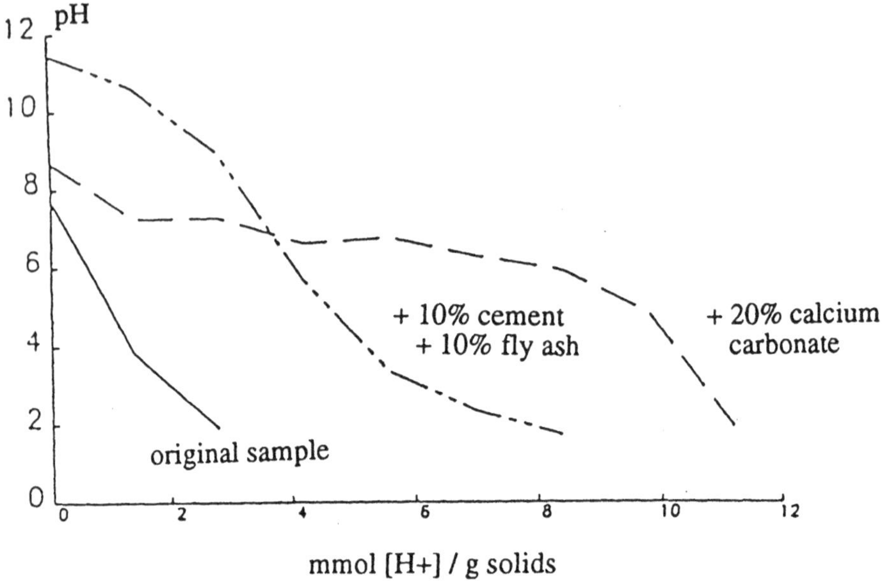

Figure 5 Effect of calcium carbonate and cement/fly ash additives on chemical stabilization of fine-grained sediments from Hamburg Harbor (Calmano et al., 1986).

REFERENCES

Ahlf, W. and M. Munawar (1988)"Biological assessment of environmental impact of dredged material," In Chemistry and Biology of Solid Waste - Dredged Material and Mine Tailings (W. Salomons and U. Förstner, eds.), pp. 127-142, Springer Verlag, Berlin

Ahlf, W., J. Gunkel, W. Liß, H. Neumann-Hense;, K. Rönnpagel, and U. Förstner (1992) " mikrobielle Biotests mit Sedimenten," In Biologische Testferfahren. Schriften-Reihe Verein WaBolu, No. 89, pp. 427-435, Gustav Fischer Verlag, Stuttgart

Alderton, D.H.M. (1985) In: Historical Monitoring, MARC Technical Rep. No 31, pp. 1-95, Monitoring and Assessment Research Centre, University of London

Anderson, J. et al. (1987) "Biological effects, bioaccumulation, and ecotoxicology of sediment associated chemicals," In: Fate and Effects of Sediment-bound Chemicals in Aquatic Systems (K.L. Dickson, A.W. Maki, and W.A. Brungs, eds.), pp.267-295, Pergamon Press, New York

Ankley, G.T., M.K. Schubauer-Berigan and J.R. Dierks (1991) "Predicting the toxicity of bulk sediments to aquatic organisms with aqueous test fractions: Pore water vs. elutriate," Environ. Toxicol. Chem. **10**:1359-1366

Anonymous (1990) Technische Verordnung über Abfälle (TVA). Der Scheweizerische Bundesrat (Swiss Federal Parliament), SR 814.015, December 10, 1990, Bern, Switzerland

Bokuniewicz, H.J. (1982) "Submarine borrow pits as containments for dredged sediments," In: Dredged Material Disposal in the Ocean (D.R. Kester et al., eds.), pp. 215-227, John Wiley &Sons, New York

Breemen, N. van (1987) " Effects of redox process on soil acidity," Neth. J. Agr. Sci., **35**:271-279

Calmano, W., U. Förstner, and J. Hong (1993) "Mobilization and scavenging of heavy metals following resuspension of anoxic sediments from the Elbe River," In: The Environmental Geochemistry of Sulfide Oxidation. (C.N. Alpers ansd D.W. BLowes, eds.), Proc. ACS Geochemistry Division Symposium, August 25-27,1992, Washington, DC., pp. 298-321. ASC Symposium Series 55

Calmano, W. et al. (1986) "Behavior of dredged mud after stabilization with different additives," In Contaminated Soil, (J.W. Assink and W.J. Van Den Brink, eds.), pp. 737-746, Martinus Nijhoff Publ. Dordrecht

Chapman, P.M. (1986) " Sediment quality criteria from sediment quality triad: Am example," Environ. Toxicol. Chem. **5**:957-964

Chapman, P.M., R.C. Barrick, J.M. Neff and R.C. Swartz (1987) " Four different approaches to developing sediment quality criteria yield similar values for model contaminants," Environ. Toxicol. Chem. **6**:723-725

Craig, P.J. and P.A. Moreton (1984) "The role of sulphide in the formation of dimethyl mercury in river and estuary sediments," Mar. Pollut. Bull. **15**: 406-408

DiToro, D.M. et al. (1990) "Toxicity of cadmium in sediments: The role of acid volatile sulfide," Environ. Toxicol. Chem. **9**:1487-1502

DiToro, D.M. et al. (1992) "Acid Volatile Sulfide predict the acute toxicity of cadmium and nickel in sediments," Environ. Sci. Technol. **26**:96-101

Ehrenfeld, J and J. Bass (1983) Handbook for Evaluating Remedial Action Technology Plans. EPA-600/2-83-076, Municpal Res. Lab., U.S. Environmental Protection Agency, Cincinnati, OH

Gambrell, R.P., C.N. Reddy and R.A. Khalid (1983) "Characteriazation of trace and toxic materials in sediments of lake being restored," J. WPCF, **55**:1271-1279

Golterman, H.L., P.G. Sly and R.L. Thomas (1983) "Study of relationship between water quality and sediment transport," Technical Papers in Hydrology No 26. UNESCO Press, Paris

Herms, U. and G. Brümmer (1978) "Löslichkeit von Schwermetallen in Siedlungsabfällen und Böden in Abhängigkeit von pH-Wert, Redoxbedingung und Stoffbestand," Mitt. Deutsche Boden-kundl.. Ges. **27**:23-43

Kersten, M. (1988) " Geochemistry of priority pollutants in oxic sludges: cadmium, arsenic, methyl mercury, and chlorinated organics," In: Chemistry and Biology of solid Waste - Dredged Materials and Mine Tailings (W. salomons and U. Förstner, eds.), pp. 170-213, Springer Verlag, Berlin

Kester, D.R. et al. (1983) Wastes in the Oceans, Vol.2 : Dredged-Material Disposal in the Oceans, New York, NY

Maaß, B., G. Miehlich and A. Gröngröft (1985) "Untersuchungen zur Grundwassergefährdung durch Hafenschlick-Spülfelder. II. Inhaltsstoffe in Spülfeldsedimenten und Porenwässern. Mitt. Dtsch. Bodenkundl. Ges. **43/I**:253-258

Morton, R.W. (1980) "'Capping' procedures as an alternative technique to isolate contaminated dredged material in the marine environment," In: Dredge Spoil Disposal and PCB Contamination: Hearing before the Committtee on Merchant Marine and Fisheries, House of Representatives, 96th Congress, USGPO Ser. No. 96-43, pp. 623-652, Washington, DC

Oberman, P. and S. Cremer (1992) " Mobilisierung von Schwermetalen in Porenwässern von belasteten Böden un Deponien: Entwicklung eines aussagerkräftigen Elutionverfahrens," Landesamt für Wasser und Abfall Nordrhein-Westfalen, Düsseldorf, Germany, Vol. 6, 127 p.

Patrick, W.H., B.G. Williams and J.T. Moraghan (1973) " A simple system for controlling redox potential and pH in soil suspensions," Soil Sci. Soc. Amer. Proc. **37**:331-332

Reynoldson, R.B. (1987) " Interactions between sediment contaminants and benthic organisms,: In: Ecological Effects of In Situ Sediment Contaminants, (R.L. Thomas et al., eds.), pp. 53-66, Hydrobiologia 149

Sahm, H., M. Brunner and S.M. Schobert (1986) "Anaerobic degradation of halogenated aromatic compounds," Microbial. Ecol. **12**:147-153

Salomons, W. (1993) "Non-linear reponses of toxic chemicals in the environment: A challenge for sustainable development," In: Chemical Time Bombs, (G.R.B ter Meulen et al., eds.), Proc. European State-of-the-Art Conf. on Delayed Effects of Chemicals in Soils and Sediments, pp. 31-43, Veldhoven/Netherlands

Salomons, W. and U. Förstner (eds.)(1988) Environmental Management of Solid Waste: Dredged Materials and Mine Tailings, Springer Verlag, Berlin

Schoer, J. and U. Förstner (1987) "Avschätzung von Grundwasser durch die Ablagerung meallhaltiger Feststoffe," Vom Wasser **69**:23-32

Sly, P.G. (1969) " Bottom sediment sampling," Proc. 12th Conf. Great Lakes Research, pp. 230-239

Stief, K. (1987) "Zukünftige Anforderungen an die Deponietechnik und Konsequenzen für die Sickerwasserbehandlung," In: Deponiesickerwasserbehandlung, UBA Materialien 1/87, pp.27-36, Erich Schmidt Verlag, Berlin

Stigliani, W.M. (1992) "Chemical time bombs: Predicting the unpredictable," In: Chemical Time Bombs (G.R.B ter Meulen et al., eds.), Proc. European State-of-the-Art Conf. on delayed Effects of Chemicals in Soils and Sediments, p. 12, Veldhoven/Netherlands

Tessier, A., P.G.C. Campbell and M. Bisson (1979) "Sequential extraction procedure for the speciation of particulate trace metals," Anal. Chem. **51**:844-851

U.S. Environmental Protection Agency (1988) Technology Screening Guide for Treatment of CERCLA Soil and Sludges, EPA/540/2-88/004, Office of Solid Waste and Emergency Response and Office of Remedial Response, Cinncinnati, OH

U.S. Environmental Protection Agency (1993) ARCS Assessment Guidancce Document, Great Lakes National Program Office, Chicago, IL

U.S. Environmental Protection Agency (1994) Remediation Guidance Document, (Draft) Oceans and Coastal Protection Division and Great Lakes National Program Office, Chicago, IL

Van Gemert, W.J.T., J. Quakernaat and H.J. Van Veen (1988) " Methods for the treatment of conaminated dredged sediments," In: Environmental Management of Solid Waste - Dredged Material and Mine Taillings (W. Salomons and U. Förstner, eds.), pp. 44-64, Springer Verlag, Berlin

CHAPTER 7

ASSESSMENT AND IMPACT OF LARGE SCALE METAL POLLUTED SITES

W. Salomons[1]

1. INTRODUCTION

In the US, Western, Central and Eastern Europe there are many large scale polluted sites that are simply too large to be cleaned up with available technologies. They are polluted by mining activities and smelters, and consist of large industrial sites, polluted sediments in water ways and impoundments and polluted soils. In addition new large scale polluted sites are being created under conditions of no or little environmental control. Since these large scale polluted areas are expected to be present for many years, it is necessary to take a long term view of the metal mobility and of the capacity of these areas to retain contaminants.

The capacity of soils and sediments to adsorb and retain contaminants depends on their composition. For organic micro-pollutants the organic matter is the component which has the strongest binding capacity. For inorganic contaminants, like heavy metals, the key property is the adsorbing capacity of both organic and inorganic particles. The adsorbing capacity is linked to the surface area and surface properties of particles, consequently, small particles like clay minerals have a high adsorbing capacity. Clay minerals as well as other components, like quartz and feldspar grains, are usually coated with hydrous manganese and iron oxides, and by organic substances. These coatings regulate, to a large extent, the adsorption process (Salomons and Förstner, 1984; Hart and Hines, 1994).

The individual particles and coatings determine the capacity of soils and sediments to retain contaminants. The actual mobility of contaminants is additionally determined by pH, redox conditions and the process of complexing agents like dissolved organic matter and inorganic anions. These parameters which control retention and mobility of contaminants are called *capacity controlling parameters or master variables.*

---

[1] GKSS, Geesthacht, Germany. On leave from: AB-DLO, AC Haren (GN). The Netherlands.

NATO ASI Series, Partnership Sub-Series, 2. Environment - Vol. 3
Remediation and Management of Degraded River Basins
Edited by V. Novotny and L. Somlyódy

For short term risk assessment (5 to 10 years), it is sufficient to understand how the master variables determine mobility (Figure 1). Much information is available in the literature. For long term risk assessment it is necessary to take the changes of the capacity controlling parameters into account, i.e., those caused by changes in major biogeochemical cycles of carbon, sulphur, iron, manganese and calcium. Human activities like civil engineering endeavors, drainage of land and changes in the use of land, disturb the geochemical cycles and hence the capacity controlling parameters. The result can be an increase in the rate of the release of contaminants into surface and groundwater and their higher bioavailability. The change in the rate of release is not necessarily linear but may exhibit non-linear characteristics. Due to these non-linear changes a sudden increase in pollutant levels may occur over short time periods in comparison with the time needed for the build up of concentrations of pollutants in the system (Chemical Time Bomb, Stigliani and Salomons, 1993). Long term aspects which must be taken into account for the target areas include ecotoxicological guidelines, sustainable agriculture, changing land use and term protection of groundwater resources.

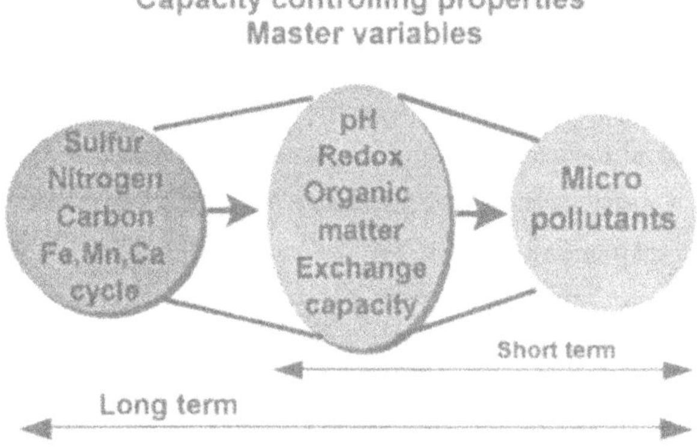

Figure 1. The link between the capacity controlling parameters and the major element cycles.

This chapter will present a framework for assessing how geochemical conditions at a polluted site affect metal mobility. Additional information is provided in chapters by Förstner and by DiToro and De Rosa. Emphasis will be on polluted soils and sediments. The large infrastructure changes in Eastern and Central Europe are expected to change geochemical conditions in soils and sediments and, hence, the mobility of stored pollutants. Past and present mining activities are a major cause of large polluted areas (Moore and Luoma, 1990). The processes involved with pollution from mines will be discussed in detail as well as methods to assess their impact.

2. CAPACITY AND INTENSITY CONTROLLING PARAMETERS

2.1 The Organic Matter as a Capacity Controlling Parameter

Organic matter in soil plays an important role for retention of pollutants. It has a high storage capacity for metals and organic micro-pollutants and, furthermore, it is concentrated at the interface between the soil and atmosphere as well as between soil and plants. It is directly exposed to atmospheric input of pollutants and through agricultural practice. High organic matter contents cause a high retention of micro-pollutants. This fact is recognized in the criteria for soils as they are used in the Netherlands (Vegter, 1994). Numerous models are available which describe the cycling of organic matter including living and dead material for the short term, however, long term models are still few. Nevertheless for agricultural soil, the change in organic matter content in topsoil can be dramatic. For example, a change of an agricultural land to pasture gives a threefold increase in organic matter content over a 100 year period, whereas conversion of pasture to potato cultivation gives a decrease of organic matter content from 2. 6% to 1.2% over a period of 7 years(Johnston et al, 1986). Figure 2 shows results of a model describing the decrease of organic matter content after the soil is left barren. A fast decrease of organic matter content takes place over the first three years. These examples show that organic matter is not a constant capacity controlling parameter in soils. Changes in the use of the soil will affect its capacity for retention of chemicals. In addition lower organic matter content will effect the distribution of chemicals over the soil and the soil solution.

Although organic matter is able to retain chemicals in the soil, its breakdown product is able to mobilize heavy metals and organic micro-pollutants in particular.

The organic matter cycle in soils changes due to acidification, changes in hydrology and alteration of land use. These changes are reflected in higher dissolved organic matter content and increased leaching of dissolved organic matter to surface waters.

The changes in the organic matter dynamics in soils in Scandinavian countries has caused a release of mercury through its association with dissolved organic matter (Meili, 1991). The mercury thus mobilized is transported to many lakes which are now in large numbers banned for fishing. Reservoirs in Canada have also shown elevated mercury levels due to the breakdown of organic matter causing mobilization of mercury and subsequently increased levels of mercury in fish (Verdon et al., 1991).

Dissolved organic matter not only mobilizes heavy metals but also effects its bio-availability. It can be generally stated that the ionic form of the metal is the most toxic. If this toxic form associates itself with either the suspended matter or dissolved ligands the toxicity of the water or ground water decreases (Kelly, 1988).

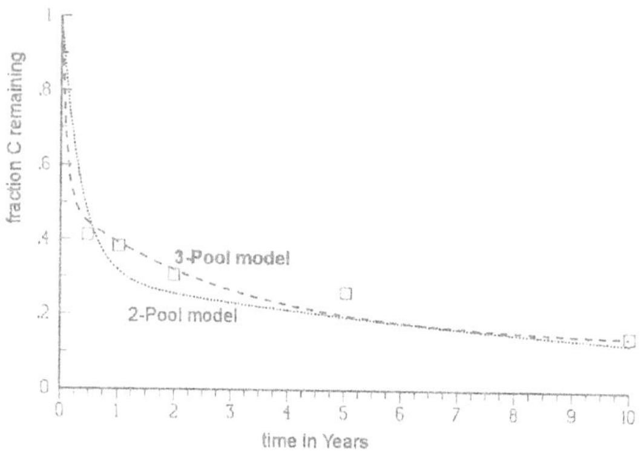

Figure 2. Results of a model describing organic matter in the soil with a model. Results show the decrease in organic carbon when a soil becomes barren. (Data point from Jenkinson and Rayner, 1977; Model by Bril, unpublished results).

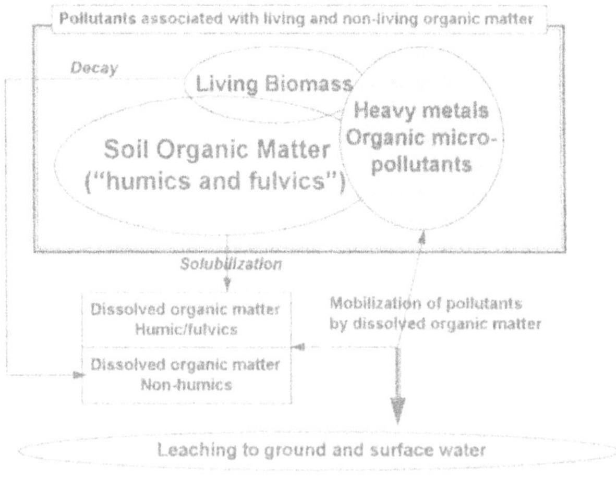

Figure 3. The role of living and dead organic matter in the mobilization of contaminants in the soil.

Usually, dissolved organic carbon levels in rivers vary in the range 1-10 mg/l, while solid or adsorbed organic carbon vary in the range 0.1-1 mg/l. Dissolved organic carbon is derived from the degradation of plant and animal tissues and has functional groups that can form chelates with metals. Humic and fulvic acids can form quite stable complexes (Tipping and Hurley, 1992). Copper, due to its chemical properties, is a metal which is strongly associated with organic matter in surface waters. Figure 4 shows the importance of this interaction. With an increase in dissolved organic carbon concentrations, the total dissolved copper concentration increases at the expense of the copper in the suspended (particulate) matter. The increase in dissolved copper is due to the formation of dissolved organic-copper complexes. Although the total dissolved copper concentrations increase, the concentrations of the toxic inorganic copper ion decreases. This figure shows that an increase in total dissolved copper due to complexing agents does not necessarily imply an increase in toxicity.

Although the discussion has been restricted to potentially toxic chemicals it should not be overlooked that nutrients are also associated with organic matter (Qualls and Haines, 1991) and an increased breakdown causes an elevated transport to the groundwater. For instance, organic phosphorous may account for more than 50% of the total phosphorous in the soil solution. Hydrophobic organic micropollutants have a strong tendency to associate themselves with dissolved organic matter and in this way enhanced leaching from soils and sediments is possible (colloid facilitated transport).

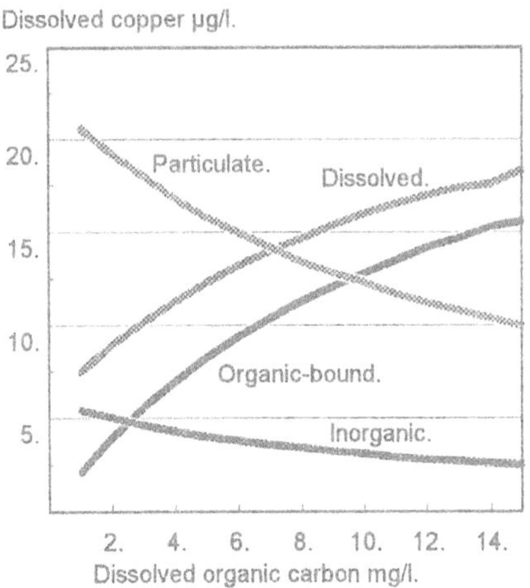

Figure 4. The influence of changing concentrations of natural dissolved organic matter on the distribution of copper over the adsorbed and soluble phase. For the soluble phase a distinction is made between copper-organic and copper-inorganic complexes.

2.2. Changes in Redox Conditions as Capacity and Intensity Controlling Factors

Strong changes in redox from oxic and anoxic conditions occur in areas with changing groundwater tables. This is extremely important for sediments and wetlands. The change from oxic to anoxic conditions takes place in a number of predefined steps controlled by the biochemistry of the systems. Examples of redox processes - involving the elements C, N, O, S, H, Fe, and Mn - are arranged in Figure 5 (Reddy et al., 1986) in the sequence of reactions observed in an aqueous system at various Eh-values.

Since the reactions considered (with the possible exception of the reduction of MnO_2 and FeOOH) are biologically mediated, the chemical reaction sequence is paralleled by an ecological succession of micro-organisms -aerobic heterotrophs, denitrifiers, fermenters, sulfate reducers, and methane bacteria (Stumm and Morgan, 1981). The driving force of reduction processes is the decomposition of organic matter by non-photosynthetic organisms, thereby obtaining a source of energy for their metabolic needs.

Water	Oxygen Reduction zone	
Soil	Oxygen Reduction zone Eh=>300 mV	*Aerobic Respiration*
Depth ↓	Nitrate Reduction zone Mn4+ Reduction zone Eh=100 to 300 Mv	*Facultative anaerobic Respiration*
	Fe3+ Reduction zone Eh=-100 to 100 Mv	
	Sulfate Reduction zone Eh=-200 to -100 Mv	*Anaerobic Respiration*
	Methane Formation zone Eh=<-200 Mv	

Figure 5. Various zones of redox in a waterlogged soil or sediment (Reddy et al., 1986).

Reduction processes involving oxidation of organic matter follow the sequence "aerobic transpiration", "denitrification", "nitrate reduction", "sulphate reduction" and "methane fermentation". In this succession the formation of soluble Mn(II) by reduction of Mn(III,IV)-oxides parallels or follows nitrate reduction, whereas the formation of soluble Fe(II) by reducing Fe(III)-oxides takes place near the conditions for sulphate reduction; the combination of the latter constituents for insoluble iron sulphides is characteristic for more strongly reducing environments. This sequence of redox changes, which in a sediment system comprises from few to ten's of centimeters, is also found in infiltrating river water in dune areas in the Netherlands (Stuyfzand, 1991). The scale however, is quite different and is in the order of ten to hundreds of meters. In plumes of landfill leachates in Denmark similar redox zones have been observed. Parts of these plumes are important sites for the degradation of organic micropollutants. The main degradation of organic micro-pollutants takes place in the zone of iron reduction. The process is microbially mediated with dissolved organic carbon in the leachate acting as a substrate (Lyngkilde and Christensen, 1992).

Apart from the redox conditions in soils and sediments, another important parameter is pH. Chemical conditions as they represent rivers, soils, sediment etc. in the environment can conveniently be depicted in so-called Eh-pH diagrams (Figure 6A). One of the impacts of mankind on the environment could be referred to as "interfering" with the Eh-pH diagram. Draining peat lands, flooding of soils and storage of dredged material on land are examples of man-induced changes in redox and pH conditions. These changes in conditions cause increased or decreased mobility of heavy metals. The changes in Eh and pH which cause an increase in the solubility of metals are shown as arrows in Figure 6B.

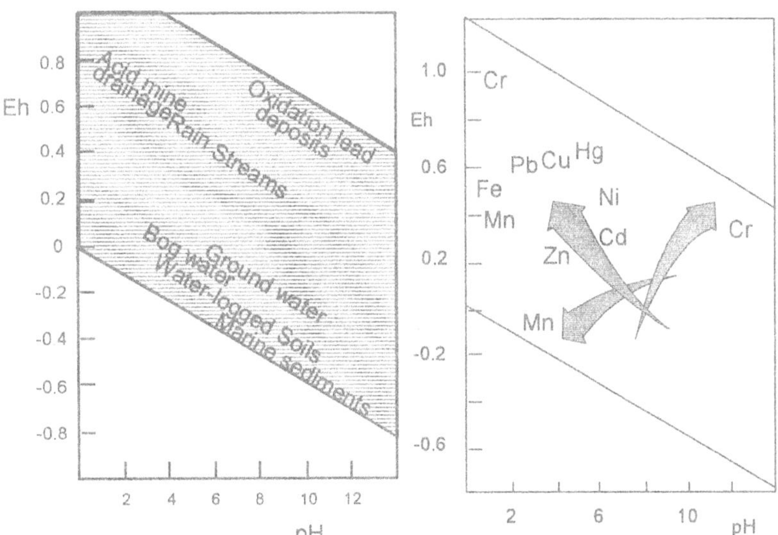

Figure 6. Major environments in terms of their Eh and pH range. (Based on Garrels and Christ (1965) and the shifts of increasing solubility with change in Eh and pH (Bourg, 1994).

Redox conditions influence the mobility of metals in two different ways. First, there are direct changes in the valences of certain metals. For example, under reducing conditions Fe^{3+} is transformed to Fe^{3+}. Similarly, manganese and arsenic are subject to direct changes in valence. Since the reduced ions are more soluble, increased concentrations of these metals have been observed in reducing environments such as ground waters and pore waters of sediments.

Indirect effects also occur when chemicals are associated with components which are subject to redox changes. This is the case for phosphorous and most metals which are (at least partly) associated or adsorbed onto iron and manganese hydroxides (Förstner, 1986, Figure 7). For long term predictions it is therefore not only necessary to look at variables like pH, competing ions, dissolved organic matter etc., which determine the initial adsorption, but even more so to look at the stability of the adsorbing phases. Iron in particular is an important capacity controlling parameter. Two stability diagrams for iron are presented in Figure 7. The first one (A) deals with the stability of iron in a freshwater environment, the second one (B) in a marine environment. The main difference is the presence of sulphate in the marine environment.

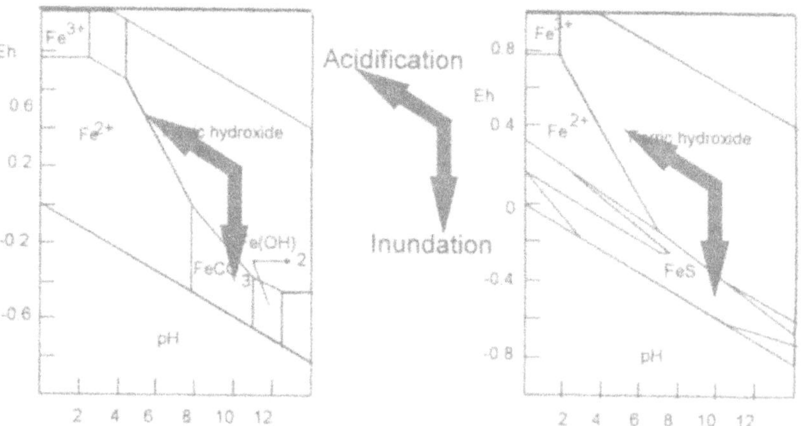

Figure 7. The Eh-pH diagrams for iron in a freshwater environment and in a marine environment.

Sulphate is used by bacteria and converted to sulphide (see the preceding section on sequence of bacterial processes in sediments). With the presence of sulphide, iron hydroxides become unstable and are converted to iron sulfides. This type of compound has a far less adsorbing capacity and the net result is a loss of the adsorbing capacity from the system. Inundation and acidification are examples of impacts which will cause iron hydroxides to become unstable.

When this occurs the metals associated with iron and manganese hydroxides are mobilized. This has been observed for instance for arsenic concentrations along infiltration routes of Rhine water or artificial recharge of water in dune areas. The area where iron hydroxides become unstable show a significant increase in arsenic concentrations (Stuyfzand, 1991).

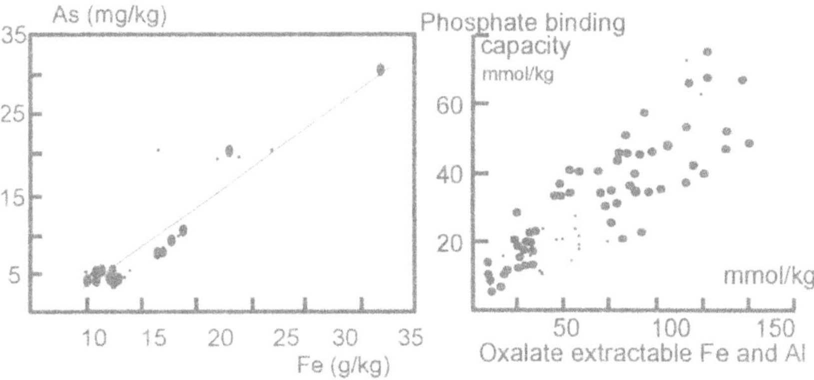

Figure 8. The correlation between iron in soils and arsenic and phosphorous contents.

Storage of contaminated dredged spoil on land causes an increase in the mobility of heavy metals in the surface layers of the dump site. Under reducing conditions most metals occur as sulphides with low solubility. Concentrations in the pore waters are independent of the total metal concentrations in the soil or sediments. Under oxygenated conditions sulphides become unstable and the metals become adsorbed onto the sediment/soil particles with an increase in overall solubility (see Figure 10). As a result an increased accumulation of metals in crops takes place (Driel van., and Nijssen, 1989). Reduced dredged material also contains pyrite. Oxidation results in the release of sulfuric acid. When not enough buffering capacity is present in the dredged material, the pH drop will result in an additional increase in dissolved concentrations (Figure 9).

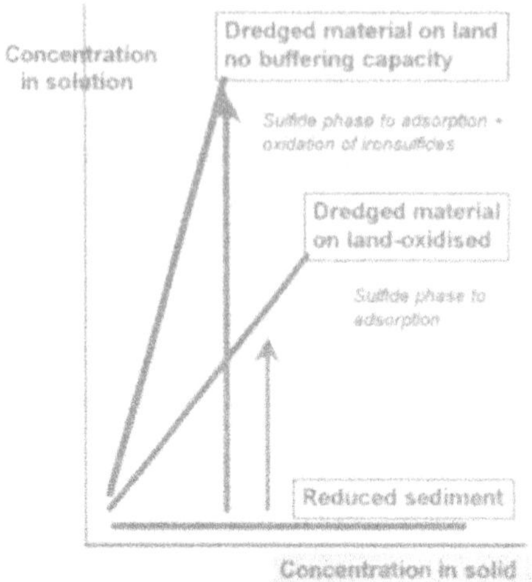

Figure 9.

The increase in concentrations when meta-sulfides become unstable in an oxygen rich environment.

Changes in redox conditions also take place when dredged material is removed from harbors and discharged into surface waters. Figure 10 shows the results of an experiment in which dredged material was suspended in river water and in seawater for extended periods of time to simulate dumping of dredged material in a highly turbulent environment. In both cases the dredged material contained cadmium sulfide which is unstable in surface water and remobilization of cadmium is expected. However, a great difference is observed between freshwater and marine environments. In both cases the cadmium sulfide decomposes. In the marine environment only a partial readsorption takes place because chloride ions form very strong cadmium chloride complexes and thus prevent readsorption. Chloride is hardly present in river water and a nearly complete readsorption and consequently low mobilization of cadmium takes place.

This sequence of sulphide formation to adsorption may be cyclic when sediments are transported on the coastal shelf. In the Dutch Wadden Sea concentrations of heavy metals decrease from West to East. This area is subject to frequent erosion and deposition of contaminated sediments. After each erosional episode the sulphides are unstable and heavy metals are partly released to the surface water and partly readsorbed on resuspended particles. This cyclic process results in a natural clean-up of contaminated sediments (Figure 11).

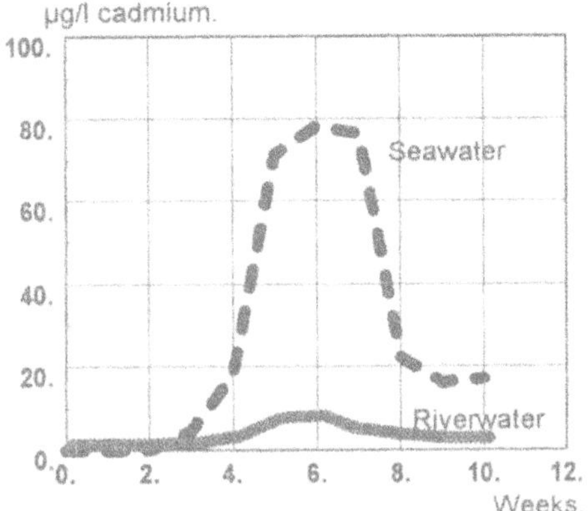

Figure 10

Release of cadmium from dredged material suspended in seawater and fresh (river) water.

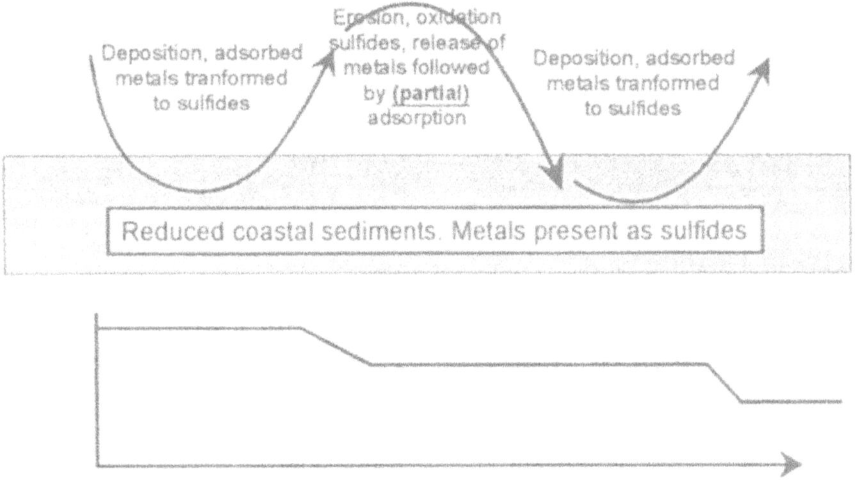

Figure 11. **Cyclic process of erosion and deposition of coastal sediments resulting in a decrease of metal levels in the sediments**

Redox conversions occur at many places in the hydrological cycle. In a number of cases this goes hand in hand also with strong changes in the pH as is the case when pyrite is present. Pyrite is an important constituent of near shore sediments, dredged material, wetland and mining wastes. The negative impact of both pH and redox changes involving pyrite is well-documented for mining wastes and for acid-sulphate soils (Dent, 1986).

Oxidation of pyrites, due to changes in redox conditions, plays an important role in the negative environmental impact caused by draining of wetlands (or acid-sulphate soils). It is well documented that wetlands can serve to protect adjacent water bodies from inputs of sulfuric and nitric acids (acid rain). As these acids pass through wetlands, sulfuric acid is reduced to insoluble sulphides and nitric acid is reduced to molecular nitrogen by denitrification. When wetlands dry out, the system may switch from one dominated by anaerobic chemical reactions to one controlled by aerobic reactions (Figure 12). Interestingly, Lake Blamissusjon in northern Sweden, known as "the most acidic lake in Sweden", was not acidified by acid deposition, but rather, as a result of drainage (for agricultural purposes) of the watershed soils surrounding the lake. Acidification occurred because the waterlogged soils were rich in pyrite, a compound containing sulphur in the reduced chemical form (S^{2-}). After drainage, oxidizing conditions prevailed, causing oxidation of the sulphur to sulfuric acid, which subsequently washed into the lake and acidified it.

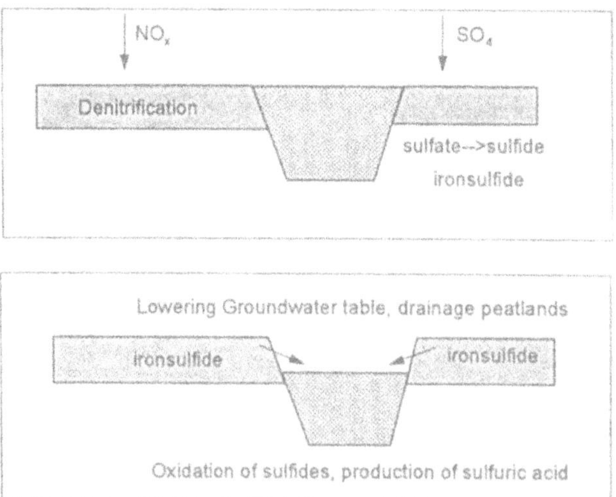

Figure 12. Changes when wetlands are being drained

2.3 Changes in pH and its Effect on Assimilative Capacity and Intensity Controlling Factors

Parameters which are relevant for adsorption in natural environments include pH (e.g., algal blooms), ionic strength (e.g., estuaries) and natural and man-made complexing agents. Natural complexing agents are organic matter and inorganic anions. An overriding factor, however, is the pH.

The adsorption of metals is highly dependent on the pH and is non-linear. It increases from near nil to near 100% as pH increases through a critical range 1-2 units wide (Figure 13) (Salomons and Förstner, 1984).

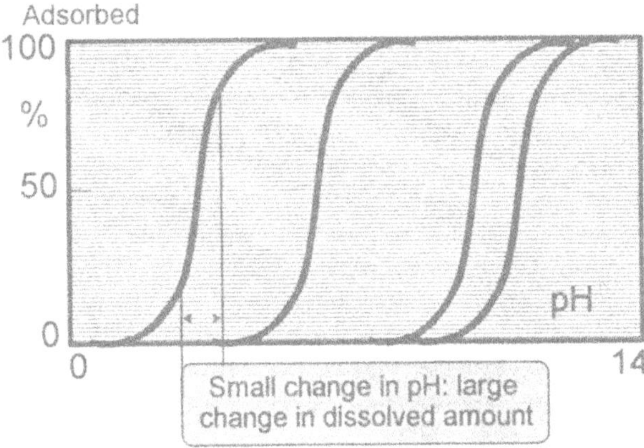

Figure 13. The adsorption of metals on solid surfaces. Metals show different adsorption edges at different pH values.

This means that a small shift in pH can cause a sharp increase or decrease in dissolved metal levels. pH depends, to a large extent, on the buffering capacity of the system. If the buffering capacity is slowly consumed, as it may happen through continuous input of acid from acid rain or acid mine drainage, a certain threshold will be exceeded and metals will start rather suddenly to desorb from the sediments or soils. The onset of adsorption or desorption is characteristic for each metal.

This non-linear response has also ecotoxicological significance. An illustrative example has been observed by the author in the River Rhine and its tributary, theRriver IJssel (Salomons, 1989). Travel time from the River Rhine to the monitoring point in the River IJssel is only a few days. However, over this time period the pH increases about 0.5 units. As a result the distribution of cadmium in suspended matter and water solution has changed due to increased adsorption (Figure 14). The decrease in dissolved cadmium causes a decrease in uptake by the bivalve Dreissena polymorpha. In this case there is a positive benefit (detoxification), however in a river or lake subject to acidic inputs the reverse would occur.

Figure 15 gives as an example of the relationship between pH and dissolved metal concentrations in a set of soils from Western Europe. These soils had different degrees of metal contamination, however, Figure 15 shows that pH is an overriding factor in determining concentrations in solution. Also quite clear, from these data, is the non-linear response of the metals to pH changes. If the buffering capacity in the form of calcium carbonate is present (e.g., from liming fields by farmers) the pH will remain above 6 and metals will have low concentrations in pore waters of the soil. Once the buffering capacity is exhausted, which in itself a non-linear phenomenon, pH decreases and metal concentrations in the soil solution increase rapidly.

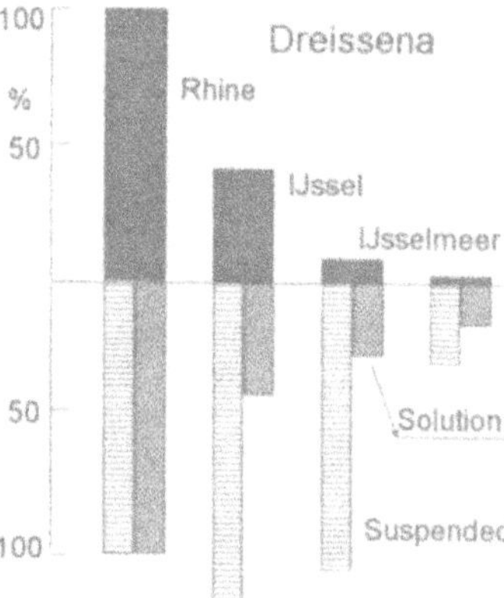

Figure 14. Relative concentrations of cadmium in Dreissena Polymorpha in the River Rhine, in its tributary the River IJssel and in the IJsselmeer (a lake is fed by the IJssel). The concentrations of cadmium in solution and in the suspended matter are shown below the X-axis. Concentrations in the River Rhine are given as 100% on the vertical axis.

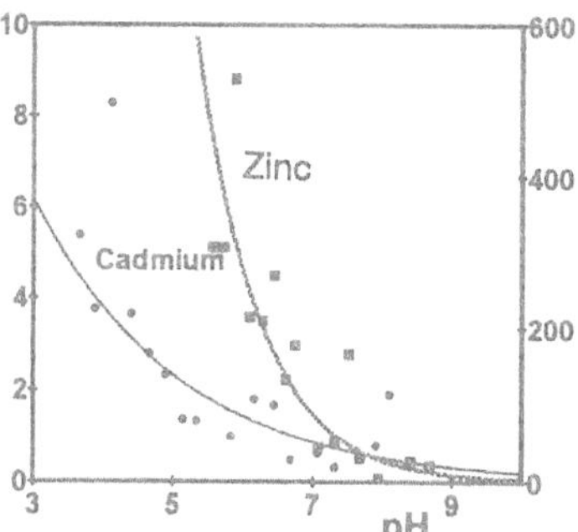

Figure 15. The relationship between pH and cadmium and zinc (in μg/l) in the soil solution for soils in the Netherlands.

3. CASE STUDIES

3.1 Heavy Metals, Acidification and Changing Land Use

Differences in the rate of transfer from soils to ground-and surface water exist and are caused by differences in properties of the chemicals and the soils. To illustrate these difference we calculated the time needed to remove different chemicals from one meter soil column by natural leaching (Figure 16). The soil chosen was a sandy soil because of its low retention properties.

For strongly hydrophobic compounds like PCB's and DDT or some heavy metals, the time needed is in the order of thousands to tens of thousands of years. For chemicals which are less strongly adsorbed like nitrate and potassium the removal time is less than 10 years. This long retention time of hydrophobic chemicals puts severe limits on the allowable inputs of contaminants from fertilizers and atmospheric deposition into the soil in order to prevent their further accumulation in the top soil. As an example consider the EC guidelines for organic micro-pollutants in groundwater. These guidelines are very restrictive, i.e., 0.1 μg/l of individual compounds regardless of their toxicological properties. If these guidelines are used for shallow groundwater then the amount which can be leached from agricultural soils is less than 0.3 g/ha of an individual compound. Prediction of these low leaching rates requires an extremely high accuracy of the sorption and transformation parameters of mathematical models (Boesten et al. 1993).

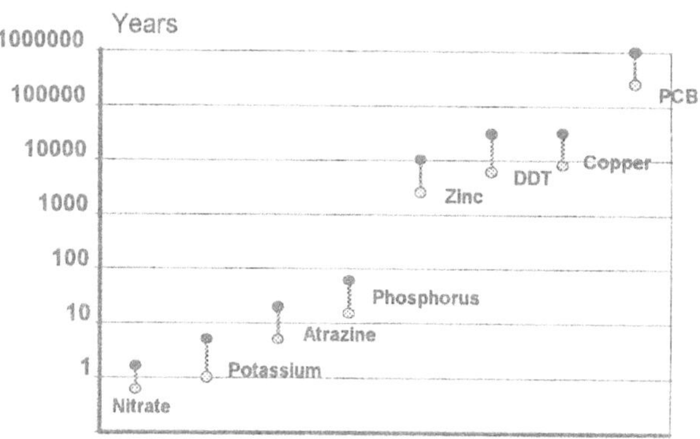

Figure 16. Time needed to remove by natural leaching chemicals from one meter of sandy soil (Salomons, 1993)

Heavy metal balances for agricultural soil in the Netherlands are presented in Figure 17. The situation refers to the 1980s and shows net accumulation of cadmium, copper and zinc in agricultural soils. Although the inputs have decreased due to lowering of copper content in animal manure, lower cadmium content in fertilizers and a decrease in atmospheric deposition of metals, the situation of net accumulation depicted herein had been going on for many years. A comparison of total heavy metals in arable soils in the Netherlands with those in forest soils shows indeed large differences (Figure 18).

The soil is subjected to atmospheric deposition. Although the quantities may be small it contributes to the continued accumulation of chemicals in the soil. Release to the ground water and removal by crops is rather low for heavy metals and for a number of organic chemicals. Therefore, even if the inputs decrease, most likely only the rate of accumulation will be reduced.

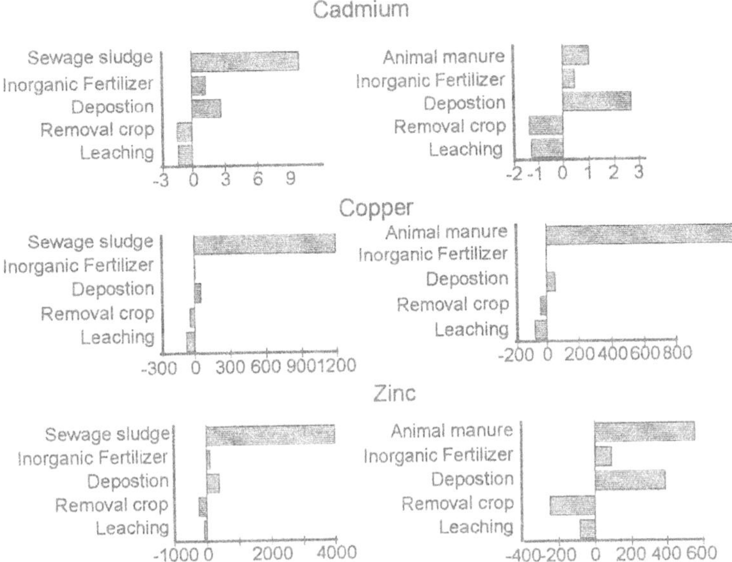

Figure 17. **Balances for cadmium, zinc and copper for arable soil in the Netherlands (Breimer and Smilde, 1986).**

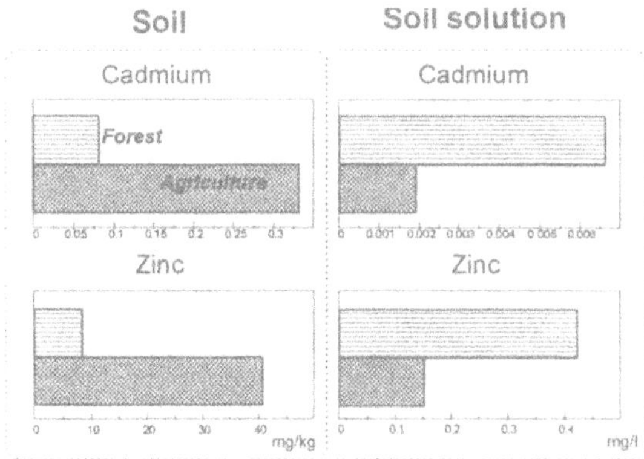

Figure 18. **The mean concentrations in the soil and in the soil solution of agricultural and forest soils in the Netherlands (Römkens, unpublished data)**

Calculations performed for cadmium accumulation in agricultural soil in Western Europe show a continuous increase with some leveling off for later years (Stigliani et al. 1993). Soils, at least in Western Europe, Canada and the U.S. are also subject to acid inputs from the atmosphere. In interpreting these kinds of predictions on metal accumulations and their impact on the environment, one should also consider the effects of acidification. Metal concentrations in the soil solution depend to a large extent on pH (see also Figure 15). The increase in dissolved cadmium and the change in speciation due to pH changes will both cause greater bioavailability of cadmium. The two effects of increased cadmium in the soil and greater uptake by crops due to pH change can be combined to determine their potential impact on human intake (Stigliani and Salomons, 1993). The results shown in Figure 19 show that cadmium concentrations, which at a certain time cause human uptake to stay below the WHO limits, will exceed the WHO limits when the acidity of the soil increases. These theoretical calculations show that it is necessary to take the whole system into account for predicting environmental impact. In this particular case, however, it is known that farmers apply lime to raise the soil-pH for maintaining crop productivity (Mahler and McDole, 1987), thus the situation of increased uptake may not occur under normal agricultural conditions.

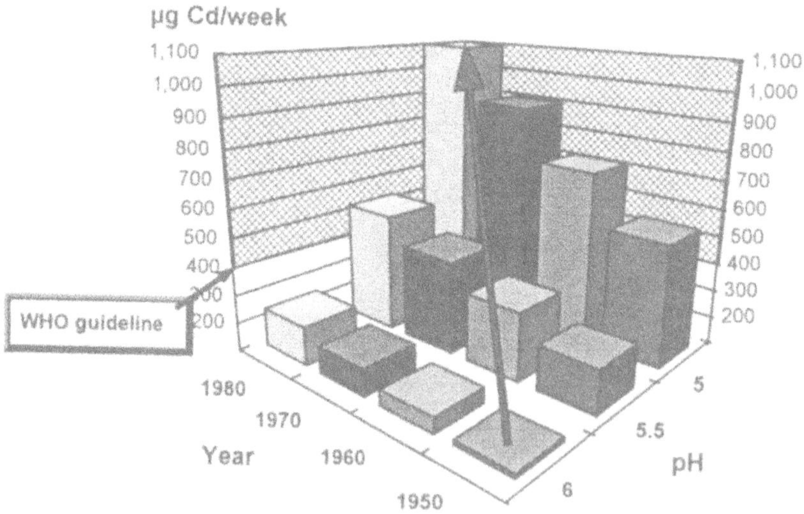

Figure 19. Estimated cadmium uptake by humans as a function of pH for the years 1960, 1970 and 1980 (Stigliani and Salomons, 1993)

In Eastern and Central Europe acidification of the soil as well as atmospheric deposition of heavy metals are continuing. The decreases in base cation deposition (Gorham, 1994; Hedin et al., 1994), partly due to removal of particles from stack effluents will reduce the buffering capacity of the atmospheric particles for SO_x and NO_x. Economic constraints in Eastern and Central Europe may reduce regular liming of soils by farmers and cause an increase in soil acidification and metal uptake.

The situation of increased metal accumulation and loading of agricultural soils combined with acidification is highly relevant for the present practice of turning agricultural land into forest. Present agricultural soils have higher concentrations of heavy metals compared with forest soils (Figure 19). These high concentrations are caused by agricultural practices and fertilizer residues or by application of sewage sludge (see also Figure 17). Despite higher concentrations in the soil, the concentrations in the soil solution are much lower.

On the other hand, forest soils have the highest concentrations of cadmium and zinc in the soil solution. These differences are simply due to differences in pH between agricultural (limed) and the more acidic forest soils. When agricultural land is converted to forest, the concentrations of metals in the soils are higher and, consequently, higher concentrations in the soil solution than in the present forest soils occur.

The mobility of contaminants in new forest soils will also depend strongly on the organic matter content. In soils with an initial low organic matter content, high concentrations of dissolved metals (e.g., cadmium in Figure 20) are expected to develop. This situation will occur when acidification proceeds faster than the development of a new equilibrium pool of organic matter in the soil. This new pool will be higher for the forest compared with the former agricultural soil. Figure 21 shows the counteracting effects of increase in organic matter content and decrease in pH on the solubility of cadmium.

The litter of natural forests is known for its high accumulation of metals and accumulation in detritus feeders, also Nuorteva (1990) has found high accumulation of metals in forest insects (beetles and ants). No information is available on metal cycling in artificial forests where the soil contains increased levels of metals. The expected higher concentrations in the most mobile and bio-available phase (soil solution) suggest increased accumulations.

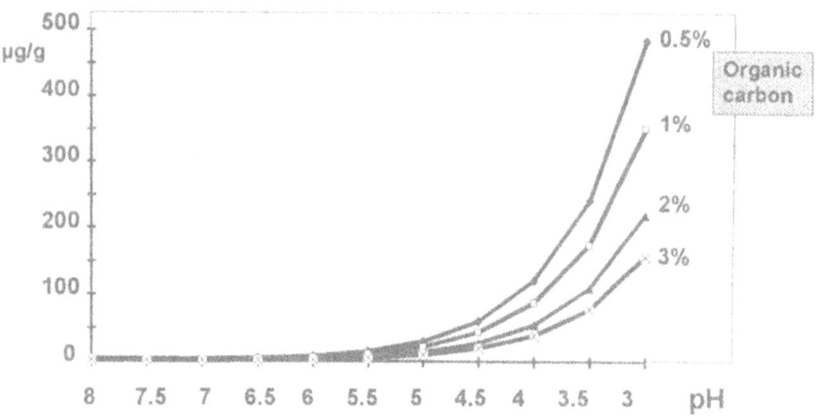

Figure 20. Concentrations of cadmium in the soil solution as a function of pH and four different organic carbon levels in the soil (theoretical calculations)

To illustrate these changes with actual data some preliminary results of a survey of forest soils of different ages are shown in Figure 21. The change in pH over a period of 70 years for forest in the Netherlands derived from former arable soil is shown in Figure 21. The first 20 years exhibited a large drop in pH of about 1.5 units. The dissolved organic carbon levels in the soil solution increased five times over a period of 70 years. The increased dissolved organic carbon levels together with the decrease in pH will increase the mobility of heavy metals (Kuiters, 1993) and organic micro-pollutants (Figure 21).

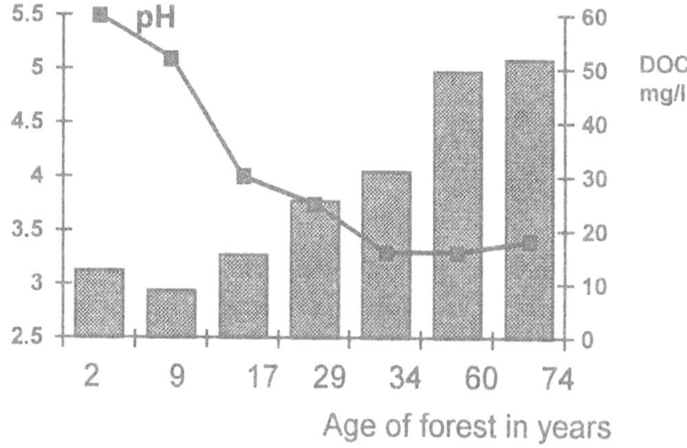

Figure 21. The decrease in pH and increase in dissolved organic carbon for forests of different ages in the Netherlands.

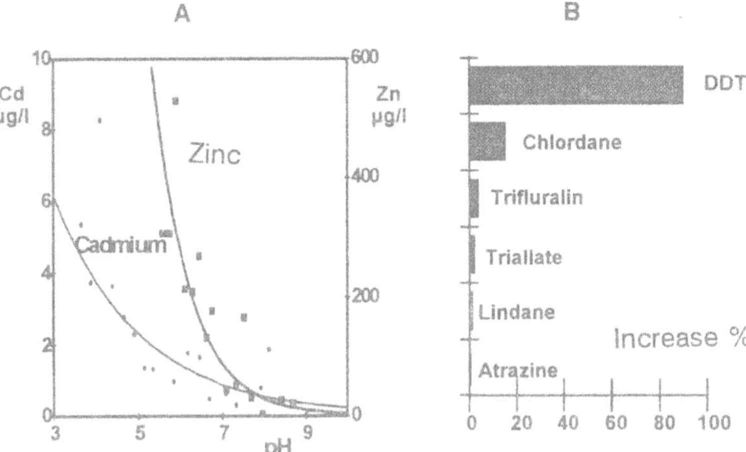

Figure 22. A. Relationship between pH and dissolved mental concentrations in the soil solution.
B. Increase of mobility of organic micro-pollutants due to the presence of dissolved organic carbon in the soil solution.

3.2 Environmental Impact of Large Scale Mining

3.2.1 Introduction

Mining is only one of the pathways by which metals enter the environment. Mining itself effects relatively small areas. Tailings deposits and waste rock deposits close to the mining area are the source of metals. When these deposits contain sulphides (pyrite) and there is access to oxygen, acid mine drainage will result. Depending on the nature of the waste rock and tailings deposits, the acid mine drainage (AMD) will contain elevated levels of metals. When these waste streams enter rivers a wider dispersion of metals both in solution and, after adsorption, in particulate form is possible. Erosion of waste rock deposits or direct discharge of tailings in rivers result in the introduction of metals in particulate form in rivers.

Because of the efficiency of modern commercial mining operations the loss of metals to the environment, as based on complete pathways of the metals, is low for the mining activity itself. However, localized impacts can be highly visible and if containment has not been practiced, as occurred often in the past (Moore and Luoma, 1990), a wide dispersion of pollutants over hundreds of kilometers is possible. This section will provide a short overview of the basic processes involved in the release of metals from mining activities and their fate after release. Tools which are available to determine potential impact are discussed.

3.2.2 Acid mine drainage (AMD)

Metals occur in different forms in ore bodies. For instance those metals which occur as associated with oxygen are in an inert form and a release from dump sites overburden with waste rock is not likely. On the other hand, metals which occur as sulphides are more likely to be released. Often they occur in association with pyrite. The oxidation of pyrite causes low pH values in the drainage waters from the dumps which contain high concentrations of toxic metals. Mining activities introduce particulate metals through erosion of dumps and through direct discharge of tailings and/or other mine waste (e.g., material removed during construction, overburden, etc.) in the river system. Release of metals by mining is through erosion of particulates and by dissolution as a result of AMD. Prediction of AMD is the key factor in predicting release of dissolved metals from active and past mining operations.

Within an active mine operation, AMD can be generated from a number of sources including waste rock dumps, ore stock piles, tailings deposits and the mine pit. A prerequisite for AMD is the generation of acid which is faster than its neutralization by alkaline materials in the waste, access of oxygen and water as well as the rate of precipitation which is higher than evaporation.

The most common mineral causing AMD is pyrite but also other metal sulphides will contribute. The oxidation of pyrite occurs in four steps (Kleinman et al. 1981):

$$2\,FeS_2 + 7O_2 + 2H_2O = 2Fe^{2+} + 4SO_4^{2-} + 4H^+ \qquad (a)$$

$$4Fe^{2+} + 10H_2O + O_2 = 4Fe(OH)_3 + 8H^+ \qquad (b)$$

$$2Fe^{2+} + O_2 + 2H^+ = 2Fe^{3+} + H_2O \qquad (c)$$

$$FeS_2 + 14\,Fe^{3+} + 8H_2O = 15\,Fe^{2+} + 2SO_4^{2-} + 16\,H^+ \qquad (d)$$

Stage I. Conditions above pH of 4.5, high sulphate and low iron concentrations with little or no acidity.

Reaction (a) proceeds both abiotically and by direct bacterial oxidation. Reaction (b) is abiotic and its rate decreases with decreasing pH.

Stage II. The pH in stage II is between 2.5 and 4.5; there are high sulphate levels, acidity and total iron are increasing. The Fe^{3+}/Fe^{2+} ratio is still low.

During this stage reaction (a) proceeds abiotically and by direct bacterial oxidation. Reaction (b) is determined predominantly by the activity of *T. Ferro-oxidans*.

Stage III. This stage occurs at pH values below 2.5; high sulphate as well as iron levels persist. The ratio of Fe^{3+}/Fe^{2+} is high.

Reaction (c) is totally determined by bacterial oxidation. Reaction (d) is determined by the rate of reaction (c).

The rate determining step in this whole sequence is the formation of Fe(III) (Singer and Stumm, 1970). These three steps are primary factors directly involved in the acid production process as recognized by (Ferguson and Erickson, 1988). All the above equations assume that the oxidized mineral is pyrite and the oxidant is oxygen. Other sulphide minerals such as pyrrotite and chalcocite are also amenable to oxidation and acid generation, however, research on these variations is limited. The intensity of acid generation by the primary factors is determined by chemical parameters like pH, temperature, oxygen concentrations in the gas and water phase, chemical activity of iron-III and surface area of exposed metal sulphides. The biological parameters include biological activation energy, population density of the bacteria, rate of growth and nutrient concentrations.

Secondary factors control the consumption or alteration of the products from the acid generation reactions. Neutralization of AMD can occur when carbonate minerals (calcite, dolomite or carbonates of Sr, Fe or manganese) are present. In tailings deposits often some CaO is left from the extraction process and neutralize AMD. The combined reaction of acid generation by pyrite oxidation and neutralization of the acid by calcium carbonate has been described by Williams (1982) as:

$$4FeS + 8CaCO_3 + 15\ O_2 + 6H_2O = 4Fe(OH)_3 + 8SO_4^{2-} + 8Ca^{2+} + 8CO_2$$

This reaction shows that two moles of calcium carbonate are necessary to neutralize the acid produced by one mole of pyrite. The total amount of carbonate is often not available because the precipitation of iron hydroxide will form coatings around the particles and a shield. Neutralization by carbonates is a relatively fast process and provides a short term buffering capacity. Acid is also consumed through reactions with silicate minerals. This long term buffering capacity is available when silicate minerals are weathered.

Tertiary factors are the physical aspects of waste materials or mine site that influence acid production, migration and consumption. The physical characteristics of mine waste and hydrological factors determine the intensity of acid generation. Oxygen advection and diffusion is promoted by coarse grained waste, which results in high production of acid. Fine grained material, on the other hand, limits oxygen diffusion and although the finer grained particles have a high surface area (promoting the reactions) the oxygen availability might be limiting. In fact, secondary and tertiary effects may be part of a "geochemical engineering" approach to control AMD.

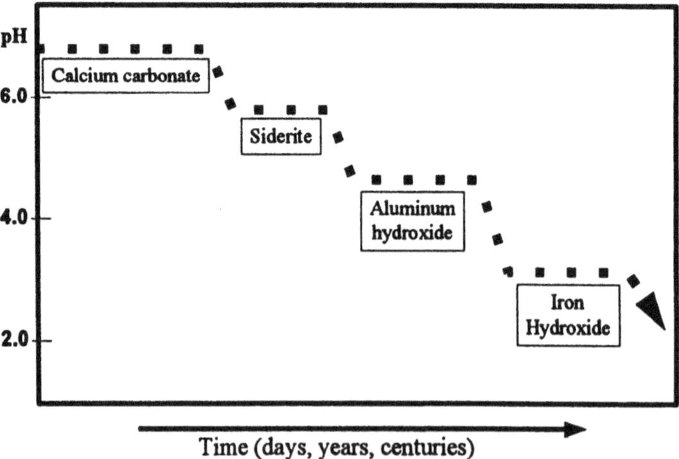

Figure 23. Stepwise consumption of buffering capacity in a hypothetical waste deposit and non-linear response of pH with time.

The development of the pH of AMD with time depends on the amount and nature of acid consuming minerals in the dump. The step-wise decrease of pH, with periods of constant pH, is shown in Figure 23. This step-wise degradation of pH and the kinetics involved have to be seriously taken into account in the interpretation of the various tests for AMD.

The onset of AMD can be fast, within one year dumps may start leaching AMD and even under high rainfall conditions, such as in the tropics, conditions may develop which cause AMD. However, there are also examples where it took years to develop significant AMD.

Several test procedures are available for predicting AMD. Ideally the test consists of three steps (Lapakko, 1992; Ferguson and Erickson, 1988; Robertson and Kirsten, 1989):

-static tests,

-kinetic tests, and

-mathematical modeling

Both static and kinetic geochemical tests are based on the assumption that geochemical reactions are the main factors which control AMD quality. The current static tests determine pyrite content or total sulphur content to calculate potential acidity. Titration procedures are used to determine the acid consuming ability. The net base (net neutralization) potential is calculated from the results by subtracting the acid potential from the base potential. In general, a negative net value indicates a potential acid producer. The static tests, by their nature, do not provide information on the time scales involved.

Geochemical kinetic tests involve weathering under laboratory or in-situ conditions in order to confirm the potential to generate net acidity, determine the rates of acid generation, sulphide oxidation, neutralization and metal depletion. Several tests are available which simulate some or a combination of the processes involved in AMD (Robertson and Kirsten, 1989; Hutchinson and Ellison, 1992). Testing involves leaching of representative samples and monitoring water quality over a period of months or years under laboratory or field conditions. Laboratory testing allows greatest control of factors but extrapolation to the field requires experience.

Although the basic chemical and biochemical processes involved in the production of AMD and heavy metal behavior are well known, application of this knowledge to the actual field conditions is still full of uncertainties. The combination of hydrological, biochemical, chemical and physical processes in actual waste dumps makes predictions during and after mine-life still a "moving target".

Oxidation of pyrite occurs almost immediately after waste dumps gain access to oxygen. Some results on the time frame involved in acidification of waste dumps and resulting drop in pH are shown in Figure 24. These results refer to the waste dump of a gold mine in Indonesia which contained several percent of pyrite. Over a period of less than one year the waste dump started to generate acid with an increase in metal levels. In this particular case, the waste rock contained abundant amounts of manganese and zinc.

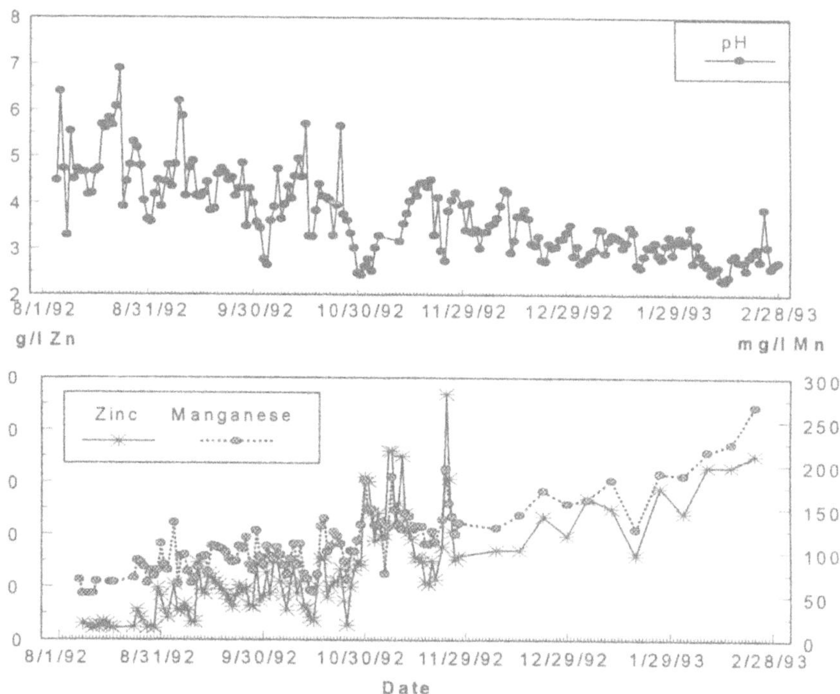

Figure 24. Change in the pH and resulting concentrations in manganese of a stream draining a waste rock dump (Kalimantan, Indonesia).

Prevention of AMD can be done by controlling the access of oxygen to the acid generating deposits. A very efficient measure, if local conditions allow, is flooding the waste deposits and, in this way, preventing access of oxygen. Other measures include capping with layers of material (clay, gravel, topsoil) and a revegetation program to prevent erosion. Bactericides have been investigated and are able to slow down the microbial processes, however, they don't last since they are subject to degradation.

Measurements taken after the end of mine-life are expensive as is illustrated by the well documented case of the Rum Jungle mine site in Australia (Harries and Ritchie, 1988) by the Clark Fork river system in Montana (Axtmann and Luoma, 1991; Moore and Luoma, 1990).

Well documented are also the landscape restoration efforts and, in particular, those of small lakes and depressions located in the brown coal areas of former Eastern Germany. During brown coal mining the groundwater level was lowered to a significant extent and the depressions in the landscape would be later on (years) gradually filled up with groundwater. Several methods have been developed to bio-geo-engineer the conditions in these artificial lakes (Glässer and Klapper, 1992) and make them suitable for recreation and/or natural habitats. In most cases, bio-geo-engineering efforts to remediate negative impacts of mining have been taken after the end of mine-life. These remedial measures should be taken preferably during the mine-life, taking local circumstances like available space, nature of the waste materials, hydrology and sensitivity of the surroundings (carrying capacity) into account.

3.2.3. Erosion of waste dumps and discharge of particulates in river systems

Erosion from active and inactive mining areas or direct discharge of tailings or waste rock introduces metals in particulate form to the aquatic environment. These materials will be partly weathered and contain metal forms or discrete minerals which are stable only under geochemical conditions of the mine site. In addition, the eroded material will still contain sulphides. The concentrations of the metals are not determined by adsorption reactions but primarily by release from the particulates (depending on circumstances) followed by adsorption. In the case of sulphides, a shift takes place from equilibrium concentrations of dissolved metals which are independent of the concentrations in the eroded material as determined by the low solubility of the products of sulphides, to concentrations which are determined by adsorption reactions (Figure 25A).

Since dissolution is a solid state reaction, kinetics play an important role. In addition, only part of the particulate metals will dissolve (Figure 25B). Part of the minerals will be locked up in a silicate matrix and protected against environmental changes. For modeling purposes it is important to make a distinction between the geochemically reactive part and inactive part (Salomons and Eagle, 1990). If modeling predictions have to be made for an active mine or mine-life, it is necessary to obtain representative samples of the ore body and carry out tests to determine the reactive metal part.

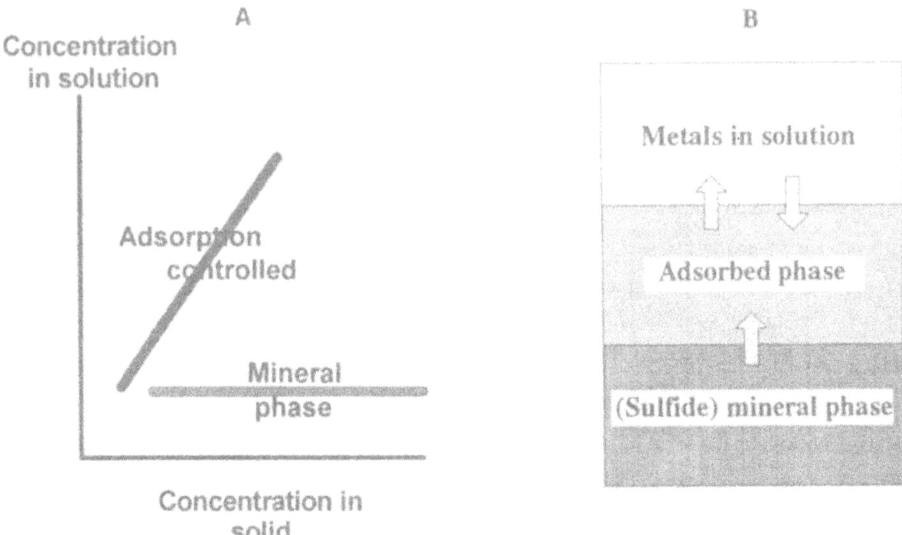

Figure 25. A. The correlation of the concentrations of metals in solution when discrete minerals are present (solubility product controlled) and when metals are adsorbed with the concentrations in the solid phase. B. Release of metals from sulphidic particles in eroded waste rock or tailings material.

It is a well known fact that adsorbed trace metals are concentrated in fine grained particles. The high surface area and the nature of fine grained particles (e.g., organic matter and/or clay) provide many sites for adsorption of dissolved metals or derived from dissolving discrete metal mineral particles. An example of the distribution of metals (in this case copper) in a river influenced by metal mining activity is shown in Figure 26. Copper is mainly present in the fractions <63 μm (silt +clay); this fraction is transported in suspension. Therefore, the sediment model has only to describe the transport of particulate copper in this particulate fraction and does not have to take into account transport of sand or pebbles.

Several constituents of sediment particles are involved in the adsorption and fixation of heavy metals. In connection with the problems arising from the disposal of contaminated dredge material, sequential extraction procedures have been developed which include the successive leaching of metals from sediments (Tessier et al.,1979; Förstner, 1993). These methods have been applied extensively for sediments and soils. Based on these operationally defined sequential extraction procedures, the metals present in soils and sediments can be classified in various categories (speciations).

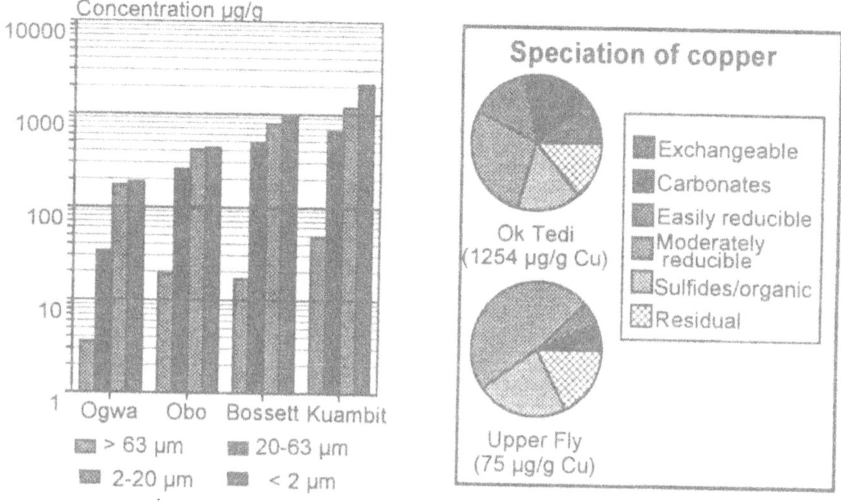

Figure 26. Concentration of metals in various grain size fractions along a river effected by mine inputs and the speciation of copper in a mine polluted river (Ok Tedi) and in an unpolluted tributary. (Salomons and Eagle, 1990).

The distribution of metals over various phases depends on the origin of the sediment or soil, but in general, copper is a metal which is more than others associated with organic matter. The amount present in the residual fraction, i.e., in the fraction remaining after all extraction's have been carried out, depends on the degree of contamination. In pristine river systems more metals will be associated with the inert fractions (Gibbs, 1973; Salomons and Förstner, 1980). In polluted rivers and soils more metals will be present in the chemically accessible forms. Such schemes give a first impression of the reactivity of metals on sediments and qualitative information on their reactivity under changing conditions. However, it is not yet possible to derive quantitative interpretation which could be used in mathematical modeling. A qualitative interpretation of mobility and biological availability for the various fractions is given in Table 1.

Hydrological circumstances (e.g., discharge) affect particulate metal concentrations in river systems. During periods of high discharge metal concentrations are low, whereas during low discharge the concentrations may increase by a factor of 5. The influence of the discharge is very strong (Salomons and Förstner, 1984) especially for metals which show high increases over base-line levels.

This phenomenon is probably caused by the following effects:

Table 1 Relative mobility and availability of trace metals

Metal species and association	Mobility
Exchangeable cations	High. Changes in the major cationic composition (e.g., in estuarine environments) may cause a release due to ion exchange
Metal compounds associated with iron and manganese hydroxides	Medium. Changes in redox conditions may cause a release but if sulphide is present insoluble Metal sulphides are formed
Metal bounds or fixed inside organic substances	Medium. After decomposition of organic matter
Metals associated with the sulphidic phase	Strongly dependent on environmental conditions. Under oxygen rich conditions, oxidation of sulphides occurs
Metals bound or fixed inside mineral particles	Low. After weathering and/or decomposition

- Proportional dilution as the discharge increases. Assuming a constant load of metals into the river system;
- Dilution as increased erosion during high discharge (surface runoff) causes a mixing of contaminated fluvial particulates with uncontaminated eroded soil particles;
- Difference in grain size composition. At low discharge suspended matter is relatively finer. The coarser particles are settled on the river floor. Whereas the pollutants are mainly associated with finer particles; hence the dilution caused by the little polluted coarse particles is absent.

On the other hand, if the river flood plain and other areas vulnerable to erosion are already contaminated, a high discharge will cause increased levels in the river and, as a result, the contaminants become more widely dispersed in the environment (Bradley 1984, Bradley and Cox 1986, 1987).

Physical mixing of sediments from different origins and with different metal loads causes changes in metal concentrations in sediments downstream. Natural tracers for sediment transport can assist in understanding the physical processes (Salomons and Mook, 1987). In the Fly river system in Papua New Guinea the differences in mineralogy of the middle Fly river and the Strickland were used to explain the decrease in metal concentrations in the lower Fly (Salomons and Eagle, 1990).

understanding the physical processes (Salomons and Mook, 1987). In the Fly river system in Papua New Guinea the differences in mineralogy of the middle Fly river and the Strickland were used to explain the decrease in metal concentrations in the lower Fly (Salomons and Eagle, 1990).

3.2.4. Modelling trace metal behavior in river systems

Models are available which describe adsorption processes in river systems and calculate the speciation of metals in surface waters (Honeyman and Santschi, 1988). For rivers, complex hydrology, sediment transport and biogeochemistry have to be coupled with models describing the behavior of metals. The few models actually applied to river systems have been for small streams and creeks. In a number of cases the fate of metals from AMD has been studied (Chapman et al. 1983, 1988). Pre-mine prediction or predictions based on mine plans are limited. In one case a geochemical model was used with simple sediment transport and hydrology routines (Salomons and Eagle, 1991) to predict the fate of copper containing tailings and waste rock in a major river system.

When large data series are available statistical models based on the K_d concept can be used to derive relationships which describe past behavior of the system. In this way the behavior of the past of the river system is considered as being a chemical experiment from which the necessary constants can be derived. Limited extrapolation is possible on the assumption that the properties of the tailings and waste rock do not change and also hydrological conditions remain the same. An example of the results of this kind of model for a location in the Fly river system is shown in Figure 27.

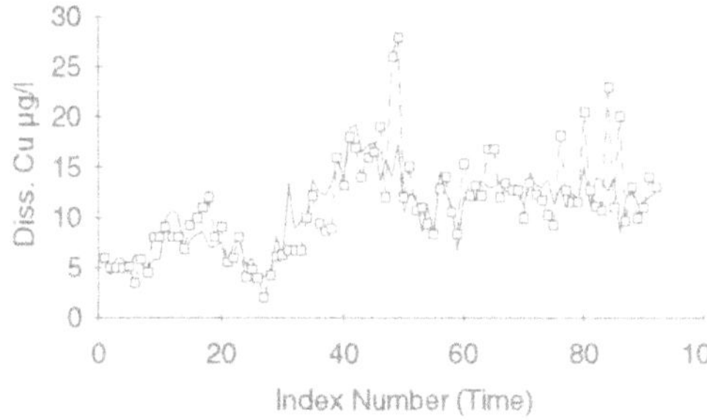

Figure 27.

Measured and predicted values for copper in the river system using the K_d approach. The index number refers to chronologically arranged samples for which a complete data set was available.

Results obtained with all coupled models are as reliable as the weakest link in the model. If data on hydrology, sediment transport, mine plan and predicted properties of tailings, waste rock and AMD are inaccurate then, consequently, predictions of metal concentrations will have large errors. Even relatively simple models, as those which predict pH of polishing ponds or rivers receiving acid mine drainage, are complex since. This is because processes like oxidation rates of Fe(II) and, if present,

REFERENCES

Axtmann E.V. and Luoma S.N. (1991) "Large scale distribution of metal contamination in fine-grained sediments of the Clark Fork river, Montana," U.S.A. Appl. Geochem. **6**:75-88.

Boesten, J.J.T.I, L.J.T. van der Pas, J.H. Smelt, M. Leistra and N.W. H. Houx (1993) " The behavior of pesticides in topsoil and subsoil in relation to possible contamination of ground water." Progress Report PCBB, April 1993.

Bourg, A.C.M. (1994) "Speciation of heavy metals in soils and groundwater and implications for their natural and provoked mobility." In: W. Salomons, P. Mader and U. Förstner (eds), Pathways, impact and engineering aspects of metal polluted sites. Springer Publ. Co., Heidelberg

Bradley, S.B. (1984) "Flood effects on the transport of heavy metals." Int. J. Environ. Studs. **22**:225-230.

Bradley, S.B. and J.J. Cox (1986) "Heavy metals in the Hamps and Manifold valleys, North Staffordshire, U.K.: Distribution in flood plain soils." Sci. Total Environ. **65**:105-128.

Bradley, S.B. and J.J. Cox (1986) "Heavy metals in the Hamps and Manifold valleys, North Staffordshire, U.K.: Partitioning of metals in flood plain soils." Sci. Total Environ. **65**:135-153.

Breimer, T. and K.W. Smilde (1986) " De effecten van organische mestdoseringen op de zware metaalgehalten in de bouwvoor van akkerbouwgronden," PAGV Themaboek Nr. 7 , Lelystad, The Netherlands, pp. 54-67

Chapman, B.M., Jones, D.R. and Jung, R.F. (1983) "Processes controlling metal attenuation in acid mine drainage streams." Geochim. Cosmochim. Acta **47**:1957-1973.

Chapman, B.M., Jung R.F. and Jones, D.R. (1988) "Heavy metal transport in streams-Field release experiments." In: Salomons and U. Förstner (Eds) Chemistry and Biology of Solid Waste: Dredged Material and Mine Tailings. Springer Publishing Co.

Dent, D. (1986) "Acid sulphate soils: a baseline for research and development." ILRI publication no. 39, Wageningen, The Netherlands.

Driel, W., Van and Nijssen, J.P.J. (1988) "Development of dredged material disposal sites: Implications for soil, flora and food quality." In: W. Salomons and U. Förstner (Eds) Chemistry and Biology of Solid Waste: Dredged Material and Mine Tailings. Springer Publishing Co., Heidelberg

Driel, W. van; and K.W. Smilde (1990) "Micronutrients in Dutch agriculture." Fertilizer Research **25**: 115-126.

Ferguson, K.D. and P.M. Erickson (1988) "Pre-mine prediction of acid mine drainage." In: W. Salomons and U. Förstner (Eds) Chemistry and Biology of Solid Waste: Dredged Material and Mine Tailings. Springer Publishing Co.

Förstner, U. (1986) " Changes in metal mobilities in aquatic and terrestrial cycles." In: J.W. Patterson (Ed) Speciation, Separation and Recovery of Metals. Lewis Publ. Chelsea, Michigan, pp 3-26.

Förstner, U. (1993) "Metal speciation - General concepts and applications." Intern. J. Environ. Anal. Chem. **51**:5-23.

Garrels, R.M., and C.L. Christ (1965) Minerals and Equilibria. Harper and Row, New York.

Gibbs, R.J. (1973) "Mechanism of metal transport in rivers." Science **180**:274-280.

Glässer, W., and H. Klapper (1992) "Stoffumsätze beim Füllprozess von Tagebauresteen." In: Bodem, Wasser und Luft. Umweltvorsorge in der AGF., Bonn, Germany

Gorham, E. (1994) "Neutralizing acid rain." Nature **367**:321.

Harries, J.R. and A.I.M. Ritchie (1988) "Rehabilitation measures at the Rum Jungle mine site." In: Environmental management of solid waste: Mine tailings and dredged material. (Eds. W. Salomons and U. Förstner). Springer Publishing Co., Heidelberg

Hart, B.T. and T. Hines (1994) "Trace elements in rivers." In: Salbu B., and E. Steinnes (Eds). Trace elements in natural waters. CRC Press. (In press).

Hedin, L.O., L. Geanat, G.E. Likens, T.A. Buishand, J.N. Galloway, T.J. Butler and H. Rohde. (1994) "Steep decline in atmospheric base cations in regions of Europe and North America." Nature **367**:351-354.

Higgins, R. (1988) "Environmental management of new mining operations." In: Environmental management of solid waste: Mine tailings and dredged material. (Eds. W. Salomons and U. Förstner). Springer Publishing Co., Heidelberg

Honeyman, B.D., and P.H. Santschi (1988) "Metals in aquatic systems: predicting their scavenging residence times from laboratory data remains a challenge." Env. Sci. Technol. **22**:862.

Hutchinson, I.P.G. and R.D. Ellison, (1992) Mine waste management. Lewis Publishers. 654 pp.

Jenkinson, D.S., and J.H. Raner (1977) "The turnover of soil organic matter in some of the Rothamsted classical experiments." Soil Science **123**:298-303.

Johnston, A.E., K.W.T. Goulding and P.R. Poulton (1986) "Soil acidification during more than 100 years under permanent grassland and woodland at Rothamsted." Soil Use and Management **2**:3-10.

Kelly, M. (1988) Mining and the freshwater environment. Elsevier Applied Science, Amsterdam

Kleinman, R., P. Crerar and R. Pacelli (1981) "Biogeochemistry of acid mine drainage and a method to control acid formation." Mining Engineering, March 1981.

Kuiters, A.T., (1993) "Dissolved organic matter in forest soils: sources, complexing properties and action on herbaceous plants." Chemistry & Ecology (In Press).

Lapakko, K. (1992). "Evaluation of tests for predicting mine waste drainage pH: Draft report to the Western Governors' Association." Minnesota Department of Natural Resources.

Lyngkilde, J. and T.H. Christensen (1992) "Fate of organic contaminants in redox zones of a landfill leachate pollute plume (Vejen, Denmark)." J. Contam. Hydrol. **10**:291-307.

Mahler, R.L. and R.E. McDole (1987) "Effect of pH on crop yield in northern Idaho." Soil Sci. Soc. Am. J. **79**:751-755.

Meili, M. (1991) "The coupling of mercury and organic matter in the biogeochemical cycle-Towards a mechanistic model for the Boreal forest zone." Water, Air and Soil Pollution **56**:333-347.

Meynink, B., J. Bril and W. Salomons (1994) A coupled hydrological and chemical with kinetics to predict acid mine drainage impact on river system. In preparation.

Moore, J.N. and Luoma, S.N. (1990) "Hazardous wastes from large-scale metal extraction." Env. Sci. Technol. **24**:1278-1285.

Nuorteva, P. (1990) Metal distribution pattern and forest decline: seeking Achilles' heel for metals in Finnish forest biocoenoses. Publ. Dept. Environ. Conserv., University of Helsinki No.11, Helsinki, Findland

Qualls, R.G. and B.L. Haines (1991) "Geochemistry of dissolved organic nutrients in water percolating through a forest ecosystem." Soil Sci. So. Am. J. **55**:1112-1123.

Reddy, K.R., T.C. Feijtel and W.H. Patrick Jr. (1986) "Effect of soil redox conditions on microbial oxidation of organic matter." In: Y. Chen and Y. Evnimelech (Eds) The role of organic matter in modern agriculture. Martinus Nijhoff.

Robertson, S., and Kirsten, B.C. (1989) Draft Acid rock drainage technical guide. Volume I. Prepared for the: British Columbia Acid Mine Drainage Task Force.

Salomons, W. (1994) "Non-linear and delayed responses of toxic chemicals in the environment." In. Arendt, F. Annokkée, G.J., Bosman, R., and van den Brink, W.J.: Contaminated Soil '93, 225-238. Kluwer Academic Publishers. The Netherlands.

Salomons, W. and U. Förstner (1980) "Trace mental analysis on polluted sediments. Part II. Evaluation of environmental impact." Environ. Technol. Lett. **1**:506-517.

Salomons W. and W.G. Mook (1987) "Natural tracers for sediment transport studies." Est. Coastal Shelf. Sci. **7**:1333-1343.

Salomons, W. (1989) "Fate and behavior of trace metals in a shallow eutrophic lake." In: Aquatic Ecotoxicology-Fundamental Concepts and Methodologies. (Eds A. Boudou and F. Ribeyre) CRC Press. Boca Raton, FL, pp 185-200.

Salomons, W. and Eagle, A.M. (1992) "Hydrology, sedimentology and the fate and distribution of copper in mine related discharges in the Fly river system-Papua New Guinea." Sci. Tot. Env. **97/98**: 316-334.

Salomons, W. and U. Förstner (1984) Metals in the Hydrocycle. Springer Publishing Co., Heidelberg

Singer, P.C. and W. Stumm (1970) "Acidic mine drainage: the rate determination step." Science. **167**:1121-1123.

Stigliani, W., and W. Salomons (1993) "Our fathers' toxic sins." New Scientist **140**:38-42.

Stigliani, W.M., P.R. Jaffe and S. Anderberg (1993) "Heavy metal pollution in the Rhine Basin." Env. Sci. Technol. **27**:786-793.

Stumm, W. and J.J. Morgan (1981) Aquatic Chemistry. Wiley, New York, London.

Stuyfzand, P.J. (1991) "Sporenelementen in duinwater en kunstmatig geinflitreerd opperlvaktewater in de kuststreek." In: G.E.M. van Beek en P.J. Stuyfzand (Eds) Sporenelementen in grondwater. Kiwa Mededeling nummer 118.

Tessier, A., P.G.C. Campbell and M. Bisson (1979) "Sequential extraction procedure for the speciation of particulate metals." Anal. Chem. **51**:844-851.

Tipping, E. and M.A. Hurley (1992) "A unifying model of cation binding by humic substances." Geochim. Cosmochim. Acta **56**: 3627-3641.

Vegter, J. (1994) "Soil Protection in the Netherlands." In: W. Salomons, P. Mader and U. Förstner (Eds): Pathways, impact and engineering aspects of metal polluted sites. Springer Publ. Co., Heidelberg

Verdon, R., D. Brouard, R. Lalumiere, M. Laperle and R. Schetagne (1991) " Mercury evolution (1978-1988) in fishes of the La Grande hydroelectric complex, Quebec, Canada." Water, Air and Soil Pollution **56**:405-417.

Williams, E.G. (1982) Factors controlling the generation of acid mine drainage. Report to the United States Bureau of Mines.

CHAPTER 8

GROUNDWATER REMEDIATION AND MODELING

Peter Shanahan[1]

1. INTRODUCTION

Because of the author's vantage point, this chapter is necessarily based on experience in groundwater remediation in the United States. Much of that experience has been gained over the last fifteen years in responding to the requirements of the Comprehensive Environmental Response, Compensation and Liability Act or CERCLA, a law passed by the U.S. federal government in 1980. CERCLA is more commonly known as "Superfund," although the prefix "Super" is an anachronism dating back to the time when $1.6 billion was thought to be a lot of money to complete a nationwide cleanup of the effects of hazardous waste disposal.

The Superfund law has been roundly criticized because it has proven so expensive and has created a highly adversarial setting in which industry is held responsible for site cleanup. A particularly good critique is that written by Freeze and Cherry (1989). The expense of Superfund is real: gathering the information to design an intelligent and effective cleanup and then carry it out is expensive. This is particularly so because the Superfund law requires that the groundwater resource itself be cleaned up instead of, for example, treating contaminated water after it has been withdrawn and is to be used. The adversarial setting of Superfund however is created by the enabling legislation. Much of the expense of Superfund has been in legal proceedings, which could be avoided or reduced with a less adversarial setting.

The United States Superfund program does not offer a good model for other countries looking to address groundwater contamination and hazardous waste. Nevertheless, many elements of the program are worth incorporating and the lessons from the program are essential. Freeze and Cherry (1989) emphasize the great value and savings in programs which prevent groundwater pollution in the first place,

[1] HydroAnalysis, Inc., Acton, Massachusetts, USA

NATO ASI Series, Partnership Sub-Series, 2. Environment - Vol. 3
Remediation and Management of Degraded River Basins
Edited by V. Novotny and L. Somlyódy

as opposed to programs which attempt to undo past pollution. They also stress that the need to remediate contaminated groundwater should be a site-specific decision, with some contaminated aquifers which are unused for drinking-water supply being left unremediated or minimally remediated. Finally, they recommend that a flexible and unbureaucratic process is needed to oversee groundwater cleanups, and that the culpability placed by Superfund on industry for past—and then legal—practices is counterproductive.

2. GROUNDWATER CONTAMINATION

Groundwater contamination can take as many forms as there are sources of contamination and geological environments. In the U.S., as the result of the Superfund law, considerable emphasis has been placed during the last fifteen years on contamination by synthetic organic chemicals. Nevertheless, "traditional" and even natural pollutants are capable of rendering groundwater unfit for consumption as convincingly as industrial chemicals.

The potential importance of non-industrial sources may be seen in a comprehensive review of groundwater contamination in the United States (OTA, 1984). The OTA study lists 33 different sources of groundwater contamination organized under six different categories (Table 1). Sources that the OTA designates as most important are shown in boldface in Table 1 and include diffuse sources like on-site wastewater disposal (septic systems and cesspools) and pesticide and fertilizer application. It is beyond the scope of this chapter to address remediation of all types of groundwater problems and therefore the chapter excludes diffuse sources which might best be addressed by management (for example, land-use controls) rather than engineering remedies. The focus is instead on localized sources such as waste disposal sites, landfill, and industrial contamination sites.

2.1 Nature of Groundwater Contamination

The great attention paid to groundwater contamination over the last fifteen years has significantly changed accepted concepts of the nature of groundwater contamination. Most significant are issues involving contaminants that are denser than water and the character of groundwater "plumes."

Table 1 Sources of Contamination

Designed discharges	Activities with incidental
On-site wastewater disposal	Irrigation
Injection wells	Pesticide application
Land application	Fertilizer application
Storage, treatment and disposal facilities	Animal feeding operations
Landfills	De-icing salt application
Open dumps	Urban runoff
Residential disposal	Atmospheric deposition
Surface impoundments	Mining and mine drainage
Waste tailings	Activities altering flow patterns
Waste piles	Oil and gas production wells
Material stockpiles	Other wells
Graveyards	Excavation
Animal burial	Natural sources
Above ground storage tanks	Surface-water interaction
Underground storage tanks	Natural leaching
Containers	Salt-water intrusion
Open burning and detonations	
Radioactive disposal	
Transport and transmission	From OTA (1984)
Pipelines	
Materials transport	

2.1.1 Dense nonaqueous phase liquids

Compounds which are denser than water create a special class of groundwater contaminants often called dense nonaqueous-phase liquid or DNAPL. Most common chlorinated organic solvents, including chlorinated methanes (carbon tetrachloride and chloroform, for example), ethanes (trichloroethane or TCA), and ethenes (tetrachloroethylene or PCE, trichloroethylene or TCE) are heavier than water, as is coal tar. The chlorinated solvents have particular significance because of their very common usage in industry and elsewhere as a degreasing compound. They are used in many industrial operations—including virtually every manufacturing operation that requires metal working—and are also used for dry cleaning and in many household products. The solvent TCE is the groundwater contaminant found most commonly at U.S. Superfund sites, occurring at one third of the sites in an early Superfund program survey (U.S. EPA, 1984).

The behavior of DNAPL in the subsurface was demonstrated in a series of laboratory experiments by Schwille (1988). A dense, poorly miscible liquid tends to travel vertically in both unsaturated and saturated soil. As it travelsthrough the soil, DNAPL finds its way into the pore space between grains where it is held by capillary forces. A residual of tightly held DNAPL, which is resistant to gravity flow, thus remains in the soil. A small spill of DNAPL may be fully taken up as soil residual, and thus is limited in its vertical migration. However, larger spills, or repeated spills, lead to greater vertical penetration of the DNAPL. With enough material released, DNAPL will eventually penetrate both the unsaturated zone and underlying saturated aquifer, only coming to rest when its reaches a geologic material of sufficiently low permeability to stop DNAPL. DNAPL will collect at such strata, forming pools of separate-phase liquid within the aquifer.

Subsurface pools of DNAPL act as extremely long-lived sources of contaminant to the ground water. A pool of DNAPL is depleted by dissolution into the groundwater which flows past the pool. Given the low solubility and dissolution rates of these compounds, a DNAPL pool is an effectively infinite source. Although the rate of contaminant dissolution is slow, the compounds are toxic at very low concentrations, thus the low rate of dissolution is sufficient to create a significant groundwater problem. The persistence of pools of DNAPL creates special problems of remediation, a topic discussed further in later parts of this chapter.

2.1.2 Geometry of groundwater plumes

Localized groundwater contamination sources, such as DNAPL pools, typically result in a plume of groundwater contamination. Experience from hazardous waste site investigations and studies at experimental aquifer sites (Mackay *et al.*, 1986; Garabedian *et al.*, 1991) have shown that plumes will very often form into long narrow zones of contamination with surprisingly little spreading. Exceptions are sites of complex and heterogeneous hydrogeology, situations of extremely slow groundwater movement, locations where intermittently pumped wells cause groundwater flow patterns to change, or other situations in which groundwater mixing is promoted. At the many locations where such effects are absent however, a long narrow plume will form, starting at the contaminant source and extending many kilometers in the direction of groundwater flow.

The character of these observed plumes shows that dispersion, which was thought just ten years ago to be a very significant factor in attenuating plumes, is unimportant. Field experience and detailed experimental aquifer studies have shown that vertical and transverse dispersion have little effect in diluting or spreading groundwater plumes. Rather, a ntaminant will typically limit itself to selected strata of soils in which hydraulic conductivity is higher. Even slight differences in hydraulic conductivity have been found to make great differences in preferential contaminant and groundwater flow.

The evolving perception of how contaminants behave in the ground both complicates and simplifies groundwater remediation. The potential to create pools of DNAPL that act as limitless pollutant sources is clearly detrimental. On the other hand, the tendency of plumes to low a narrow path may in some cases limit the extent of the aquifer requiring remediation. These and other aspects of groundwater contaminant transport are discussed below in the context of aquifer remediation alternatives and case-study examples.

3. GROUNDWATER REMEDIATION

3.1 Remediation Objectives

Design and implementation of a groundwater remedy will depend in large measure on the remediation objectives. A U.S. Environmental Protection Agency guidance document on groundwater cleanup for Superfund (U.S. EPA, 1988) defines the following possible objectives:

1. **Preventing exposure to contaminated groundwater** - This entails treating or providing an alternative to contaminated groundwater used as drinking water. Alternatively, use of the contaminated groundwater could be restricted through legal or other controls.

2. **Protecting uncontaminated groundwater** - This entails containing contaminated ground water so that the extent of contamination cannot increased.

3. **Restoring contaminated groundwater** - This entails reducing contaminant levels in the aquifer to below concentrations required for drinking water use.

4. **Protecting environmental receptors** - This entails reducing contaminant levels in the aquifer to below concentrations safe for biological receptors at points of groundwater discharge.

The Superfund program has essentially required that the more stringent of Objectives 3 or 4 be met in groundwater remediation. However, this may be technically impossible and financially ruinous. In many situations, Objectives 1 and 2 can be far more practical and realistic. Considerations in the selection of remedial objectives are numerous and transcend simply technical aspects. Factors to be weighed include the technical feasibility of extracting contaminants, the feasibility of providing an alternative water supply, the potential uses of groundwater in the area, the ability to impose restrictions on groundwater use, and the ability to monitor and control contaminant movement (U.S. EPA, 1988).

3.2 Pump-and-Treat Systems

By far, the dominant groundwater treatment alternative at hazardous waste sites is the "pump-and-treat" system. Pump-and-treat entails installing one or more pumping wells downgradient of the contaminant sources, pumping continuously to remove contaminated groundwater, and treating the removed groundwater prior to discharge to surface water, reinjection to groundwater, or disposal to a sewer. When first implemented, it was anticipated (and in many cases calculated) that the system would need only a few years or less to remove sufficient contamination to restore the aquifer to drinking-water standards. What was found was that soon after an initial period of decrease, contaminant concentrations in the pump-out well leveled off at a decreased but finite level which was very often above the target clean-up level. In situations where multiple wells were installed along the axis of a plume, downgradient wells were found to achieve desired concentration levels but not wells nearest the source. If the pumping well near the source area was shut off, concentrations began to rise, often achieving their former, pre-remediation levels.

This unforeseen behavior in early pump-and-treat systems, and the further investigations which that behavior engendered, led to the interest in DNAPL and eventually the findings discussed above. Groundwater contamination was persistent because subsurface accumulations of DNAPL were the source of continuing contaminants in the groundwater. In many cases, a single nearby pump-and-treat well eventually captured all contaminant emanating from the source area, thus allowing other wells eventually to clean-up. But, the one or few wells nearest the source were required to operate indefinitely in order to control migration from an effectively limitless source. In a significant departure from past regulatory practice, the U.S. EPA recently issued guidelines that recognize that some sites can never be fully cleaned up and establishes procedures for those cases (U.S. EPA, 1993).

The concept of long-term containment by pump-and-treat wells is now routinely incorporated in hazardous waste site cleanups. At some sites, source control wells, which are located near the source and expected to operate indefinitely, are distinguished from gradient control wells, which are located in the downgradient plume and expected to remove contamination within a period of time. Despite the prospects of indefinite operation, pump-and-treat systems remain a viable and cost-effective option for many sites and continue to be used frequently.

The treatment component of pump-and-treat systems generally draws upon established treatment technologies for industrial wastewater. Most common are air stripping for volatile organic compounds and activated carbon treatment for nonvolatile organic contaminants. Pretreatment to remove iron and other precipitates is often required to prevent fouling of the treatment media. Biological, ultraviolet, and other treatment alternatives have been used with some organic contaminants. Detailed information on groundwater treatment alternatives is available from Nyer (1992).

3.3 Groundwater Remediation Alternatives

The continued reliance on pump-and-treat systems does not so much reflect the attractiveness of the technology as the paucity of good alternatives. A variety of techniques is available, but perhaps the most attractive alternatives are those still in active development.

Table 2 summarizes available groundwater remediation technologies. The remediation of groundwater contamination is inevitably linked to the removal of the contaminant source, and thus source control technologies such as source removal, soil flushing, *in situ* stream stripping, and soil vapor extraction are included in the table. Some of the bioremediation alternatives may also be applicable to contaminated soils, although this technology is most often considered as a groundwater remedy.

Table 2. GroundWater Remediation Alternatives

Technology	Operating Mechanism	Effectiveness
Source removal	Eliminates the source of groundwater contamination such as leaking storage tank, surface impoundment, contaminated soil, etc.	Effective and straightforward.
Soil flushing	A flushing agent (water, solvent, or surfactant) is passed through contaminant source in unsaturated soils to physically and/or chemically remove the source.	Few if any successful applications. May cause contaminant to spread further and thus worsen problem.
In situ stream stripping	Similar to soil flushing except steam is used to enhance volatilization of organic compounds.	Same as for soil flushing.
Soil vapor extraction	A vacuum is applied to unsaturated soils to withdraw volatile organic compounds.	Effective and frequently used for volatile organics in sandy soils. Does not address DNAPL below the water table.
Air sparging	Introduces air into aquifer via an injection well. Volatile organic compounds removed by volatilization into injected air.	Suitable for sites with uniform, sandy soils. Relatively new technology which appears to have been effective at some sites.
Hydraulic containment (pump-and-treat)	A pumping well is used to capture contaminated groundwater and prevent its further migration.	Most frequently used groundwater remediation alternative. See text for further discussion.
Physical containment	The source area is enclosed by low-permeability barriers (slurry walls or sheet piles).	Requires presence of relatively shallow underlying low-permeability layer. Used successfully under these conditions.
Natural *in situ* bioremediation	Relies on naturally occurring microorganisms to degrade organic compounds.	Frequently used successfully for petroleum hydrocarbons in aerobic (shallow) groundwater after floating product or source removal.
Active *in situ* bioremediation	Introduces nutrients and/or microorganisms into the subsurface to stimulate and enhance biodegradation of organic compounds	Successful applications are few. Requires favorable geologic media and contaminants amenable to biodegradation.
Natural attenuation	Contamination is monitored while plume dissipates by biodegradation, chemical reactions, volatilization, and dilution.	Unlikely to be successful if a continuing source is present (such as DNAPL).

Technologies to treat contaminated groundwater can be seen to be few. In-situ biodegradation (whether as a natural attenuation or enhanced process) is a proven technique at sites contaminated by petroleum hydrocarbons but not at chlorinated solvent sites.

4. GROUNDWATER MODELING

Groundwater modeling can be a valuable and useful tool in investigating and cleaning up contaminated sites. Models can be effective in answering such questions as:

1. **What is the character of the existing situation?** Here a groundwater model is used to interpret and interpolate between field observations. While a model is never a substitute for actual field data, modeling can be helpful in interpreting field measurements and in directing the collection of additional field data. This is possible because models can consider more points in space and time than can be measured by a finite field program. Models can be particularly powerful in posing "What if?" questions—in other words, proposing and testing hypotheses about the subsurface geology and hydrology. If model tests show certain subsurface conditions are possible or likely, then additional field studies can be conducted to verify the hypothesis.

2. **How might the existing situation change over time?** Here the model is used to predict potential future effects of waste disposal or water-supply development. This use of the model helps to evaluate the potential for groundwater changes and whether future adverse effects on the environment are a realistic concern.

3. **How effective are remediation measures and groundwater monitoring?** This is an area in which groundwater modeling can be particularly powerful. Models can be used at av relatively low cost to screen and evaluate a large number of alternative engineering designs. Models have been used in this way to determine the engineering approach or monitoring system design that is most protective of the environment and most cost-effective.

In addition to answering these specific questions, modeling is useful in a more general sense.

Modeling is a rigorous methodology requiring detailed and comprehensive attention to the processes affecting environmental behavior. If the model ignores an important process, it is unlikely that the model can be calibrated or verified against field data. Thus, the model serves as a structured and rigorous framework to gather and interpret knowledge of the environmental system. It is a framework that frequently forces an analyst to confront limitations in his conception of the system and to develop new understanding of how the groundwater system behaves.

Most groundwater models fall into two classes: flow models and contaminant transport models. Flow models simulate the distribution of force potential (measured as the elevation of groundwater in wells) and the groundwater flow field (the direction and rate of groundwater flow). The output from flow models shows the direction of groundwater flow and thus the direction in which contaminant is most likely to travel. The groundwater flow field is required input to contaminant transport models. Contaminant transport models consider the movement of contaminants as the result of groundwater flow (advection), the spread of contaminant due to mixing (dispersion), and the removal or transformation of contaminant due to radioactive decay and physical, chemical and biological reactions. Both flow and transport models are discussed in turn below.

4.1 Groundwater Flow Models

4.1.1 Groundwater flow

The basic physical unit in groundwater flow is the aquifer. An aquifer is defined as a geologic formation or group of formations that contains sufficient water and is sufficiently permeable to yield appreciable quantities of water to wells and springs. There are two main types of aquifers. The free-surface aquifer is one in which a water table forms the aquifer's upper boundary. The elevation of the water table changes over time in response to inflows and outflows of water. A free-surface aquifer is also called a water-table, phreatic, or unconfined aquifer. The confined aquifer is an aquifer that is bounded from above and below by less permeable formations. Water in a confined aquifer is under pressure. Confined aquifers are also called artesian or pressure aquifers.

Flow of groundwater is driven by gravity and pressure forces. The combined effect of these forces is captured in the force potential. The force potential is also referred to as the piezometric head, the fluid potential, the hydraulic head, the potential, or simply the head, the term we use here. Whatever the terminology, the head measures the potential energy of the groundwater at a point, and thus its propensity to flow. Flow occurs from areas of high head to areas of low head.

Darcy's law defines the basic concept of groundwater flow. The law states that the rate of flow of groundwater per unit area is proportional to the gradient in the hydraulic head. The coefficient of proportionality is defined as the hydraulic conductivity of the soil or rock in which the groundwater flows. Hydraulic conductivity is an inverse measure of the resistance to flow in the soil. Tight soils such as clays have a high resistance to flow and thus a low hydraulic conductivity. Hydraulic conductivity is also called permeability. Darcy's Law is given as the following equation:

$$Q = K\, i\, A \tag{1}$$

where Q is the rate of groundwater flow $[L^3/T]$;

K is the hydraulic conductivity [L/T]; and

i is the head gradient [L/L], equal to the change in head, Δh, over a certain horizontal distance, $\Delta \ell$; and

A is the cross-sectional area through which flow occurs $[L^2]$.

4.1.2 Basic principles of flow models

The modeling of groundwater flow is basically solving mathematically for the head over time and space. The solution considers the influence of boundary conditions where flow enters or leaves the groundwater system. Typical boundary conditions are wells, rivers, springs and seeps, and impermeable barriers to flow. The ground surface is a boundary through which rainfall recharge travels to become inflow to the groundwater system. The solution for the head also considers the change in the quantity of water stored in the soil. Under high pressure, both the water and soil will compress and thus effectively store water. If the pressure increases or decreases, water enters or leaves storage.

A complete model of groundwater flow considers the balance between the change in storage with time and the flow of groundwater in the vertical and the two horizontal directions. This model may be effectively simplified in many problems. Typical simplifications are:

Steady-state flow - Changes in flow and storage over time are insignificant and the system is modeled as steady, or unchanging with respect to time.

Quasi-three-dimensional - Vertical flow is negligible except between geologic formations. The system is modeled as a series of layers with horizontal flow within each layer and vertical flow between layers.

Two-dimensional - Vertical flow is insignificant throughout the system. Groundwater is modeled as flowing horizontally within a single layer.

Homogeneous aquifer - The hydraulic conductivity and storage properties of the soil are assumed to be uniform throughout the formation.

Isotropic aquifer - The hydraulic conductivity is assumed to be the same in all directions. Usually, aquifers are assumed to be isotropic horizontally, but with a lower hydraulic conductivity vertically.

Most models include one or more of the assumptions outlined above.

The discussion above addresses only the problem of saturated flow. Unsaturated flow, such as percolation through the soil column, is a related phenomenon, but one that is more difficult to model. The major difficulty is the highly nonlinear behavior of unsaturated soil. The parameters used to describe moisture flow in soils may vary over several orders of magnitude as the soil changes from very dry to saturated. This nonlinearity makes the equations for unsaturated flow difficult to solve numerically and creates the practical problem of collecting sufficient data to characterize a particular soil. As a consequence, unsaturated flow modeling is usually highly simplified when applied to groundwater contamination site problems.

4.1.3 Application to groundwater assessment and remediation

Groundwater flow models are frequently used in site investigations in the initial (and perhaps final) analysis of contaminant movement. Flow models do not address contaminant movement per se. Nevertheless, by knowing the speed and direction of water movement, one can make an initial assessment of contaminant movement. The fate and transport of contaminants in groundwater are determined in large part by the hydraulics of the groundwater system: contaminants are carried along by the flowing groundwater. The fate and transport of contaminants is also determined by dispersion and reactions, although these are often secondary to advection in many groundwater systems.

The use of flow models to approximate contaminant transport may at first seem misdirected. After all, contaminant transport models construct a more detailed representation of contaminant fate. Unfortunately, as discussed in more detail below, contaminant transport models require input data that are considerably more difficult to obtain and far less certain than those needed for flow models. The advantages of using a flow model are the greater certainty of its results, and the reduced requirements of its input data. Generally, a flow model gives a conservative transport prediction in the sense that it will predict a faster rate of contaminant migration than would a transport model. This is because dispersion and especially reactions tend to retard plume movement. These retarding effects are ignored by flow models which thus predict more rapid transport of contaminants by groundwater flow.

There are numerous examples in the literature of site assessments employing groundwater flow models to estimate pollutant transport. Yazicigil and Sendlein (1981) use a flow model to estimate pollutant transport and compare remediation alternatives for a manufactured gas plant site in Ames, Iowa, USA contaminated by coal-tar compounds. Mercer *et al.* (1983) use a flow model to evaluate contaminant transport from the Love Canal in upper New York State, USA and Anderson *et al.* (1984) use flow models to complete a thorough comparative analysis of remediation alternatives for a landfill in New Jersey, USA. These studies are all recommended as examples of well conceived applications of flow models to groundwater contamination problems.

The calculation of groundwater flow is based upon a single set of principles, thus there is relatively little gradation in the complexity of modeling techniques. The simplest techniques are analytical solutions to Darcy's Law. More complicated approaches are computerized versions of analytical solutions, followed by more complex two-dimensional and three-dimensional finite-difference or finite-element models.

An important distinction in any of these techniques is dependence upon what Bear (1979) defines as the hydraulic and hydrodynamic approaches to groundwater flow. The hydraulic approach treats the aquifer as a unit and neglects vertical flow and vertical differences in the flow rate within the aquifer. Hydraulic conductivity is characterized by an aquifer-averaged property, the transmissivity. Vertical flow is presumed to be negligible within the aquifer. The approach works very well for most problems. It fails in aquifers where there is significant vertical flow within an aquifer (although it will work where the flow is primarily between aquifers).

The hydrodynamic approach considers vertical variations within an aquifer. Hydraulic properties are not averaged over the aquifer thickness, but rather the full complexity of the vertical soil column is accounted for. This approach is appropriate when there are significant vertical variations in the aquifer materials. It is also used in analyses of water table aquifers in which the saturated thickness varies significantly.

The simplest method of groundwater flow analysis is a direct evaluation of field-measured heads. For these the rate of flow per unit width of aquifer, Q′, is determined by an aquifer-averaged version of Darcy's Law:

$$Q' = T\, i\, b \qquad (2)$$

where Q' is the flow per unit width of aquifer [L^2/T];
T is the aquifer transmissivity [L^2/T] = Kb; and
b is the aquifer thickness [L].

The head gradient must be measured in the direction perpendicular to lines of constant head. The speed of groundwater flow is determined as:

$$V = \frac{Q'}{bn} = \frac{K}{n}\frac{\Delta h}{\Delta t} \qquad (3)$$

where V is the groundwater flow velocity [L/T]; and
n is the effective porosity [-].

The groundwater flow velocity is seen to depend upon the porosity, which is a measure of the fraction of the aquifer that is open to flow (i.e., the pore space or void space). By dividing by porosity, one determines the groundwater flow velocity as actually traveled water in the soil pore spaces. This is roughly the speed at which a nonreactive contaminant will move.

The next step in complexity is to compute the head gradient rather than simply measure it. Such computations are essential if one anticipates installing remedial measures such as pumping wells that will change the gradient. Groundwater hydrology texts such as Freeze and Cherry (1979) or Bear (1979) give a variety of analytical solutions for particular problems. For example, there are one-dimensional solutions for the problems of flow to wells, flow to a stream, or flow to drains.

More complex analytical solutions consider two-dimensional flow in an assumed uniform flow field. These solutions are often used to design the pumping well component of pump-and-treat systems. Javandel and Tsang (1986) and Lundy and Mahan (1982) describe the use of "capture-zone" models which determine the area of the aquifer which flows to a pumping well. Pertinent equations define the envelope of the capture zone, given by the distance from the capture zone centerline, y, as an implicit function of the distance from the pumping well, x:

$$y = -x \tan\left(\frac{2\pi T i}{Q_p} y\right) \tag{4}$$

where Q_p is the pumping rate of the well [L^3/T].

Far upgradient of the pumping well, the distance from the capture zone centerline to the edge of the capture zone is:

$$y = -\frac{Q_p}{2 T i} \tag{5}$$

Downgradient from the pumping well there exists a stagnation point, at which the groundwater velocity in the x direction is zero. The location of the stagnation point downgradient of the pumping well is given by:

$$x = \frac{Q_p}{2 \pi T i} \tag{6}$$

Figure 1 illustrates a capture-zone solution. The capture-zone solution assumes a uniform aquifer of constant transmissivity, a single pumping well, and a uniform hydraulic gradient. Despite these assumptions, the model is a useful tool, applicable to many situations.

Uniform Flow Analytical Model

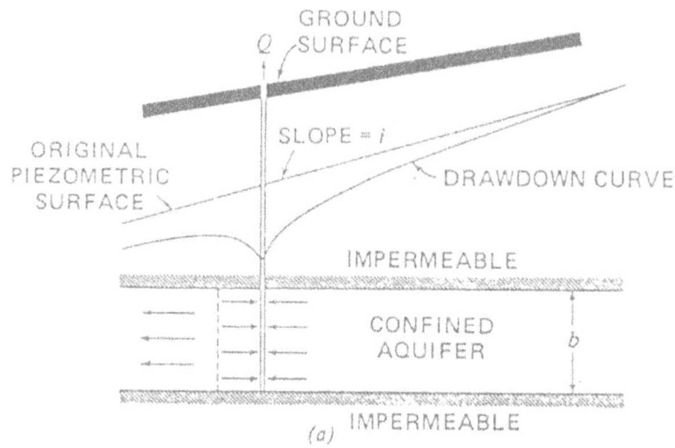

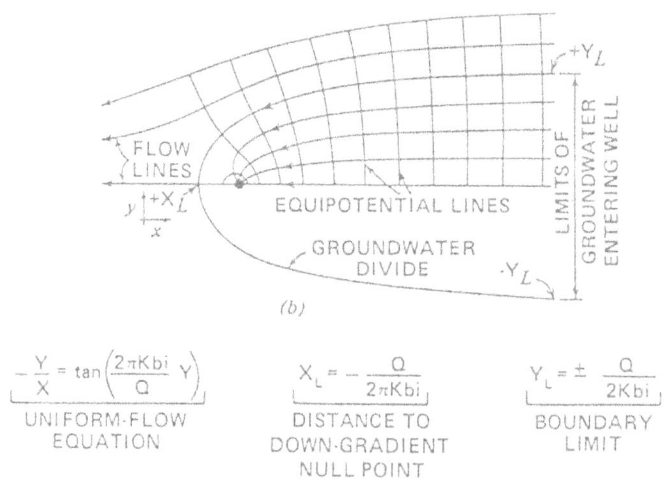

LEGEND:

● Pumping Well

Where:
Q = Well Pumping Rate
K = Hydraulic Conductivity
b = Saturated Thickness
i = Hydraulic Gradient
π = 3.1416

Figure 1. **Analytical capture - zone analysis (from US EPA, 1987)**

Javandel and Tsang (1986) provide solutions for two- and three-well capture-zone systems. However, if many wells affect the flow field, hand calculation of the analytical solutions becomes impractical. Bonn and Rounds (1990), Nelson (1978), and Cosgrave *et al.* (1989) describe computerized solutions of these problems.

4.1.4 Numerical models

A system influenced by complex boundary condition or exhibiting nonuniform flow patterns is best analyzed using groundwater flow models that solve the flow equations by finite-difference or finite-element techniques. There are a variety of readily available computer programs to employ in such studies. Both two-dimensional and three-dimensional models are available. Two-dimensional models suffice for a great many problems. Three-dimensional models are required when multiple aquifers must be considered or when vertical flow is significant.

Care is required in selecting computer programs for evaluating groundwater contamination problems. Several criteria should be met:

- The program should be widely known and recognized by experts in the field of groundwater modeling.
- The program should be well documented.
- The program should be verified through test cases or against analytical solutions to prove that it correctly solves the governing equations.
- The program should be easy to use and free of computational instabilities.
- Increasingly, compatibility with personal computers is desirable.

These requirements narrow the field of potentially useful computer programs.

A general principle in selecting a computer model for any application is that it be the simplest model adequate for the problem to be solved. Modeling is an exercise that is complicated and difficult. But the job of modeling can be simplified if the modeling tool itself is kept as simple as possible. Although models with many capabilities and features are often attractive, their data requirements can be onerous and their results unsatisfactory without complete data. Most modelers agree that the best model for any job is the model with the least complexity and fewest data requirements that still incorporate the primary features of the system being modeled.

4.1.5 Available flow models

A catalog of available computer models of both groundwater flow and transport has been published by the American Geophysical Union (van der Heijde et al., 1985) but is now somewhat dated. An up-to-date and comprehensive collection of groundwater models and information is offered by the International Ground Water Modeling Center (IGWMC) located in Golden, Colorado, USA and Delft, The Netherlands (addresses are provided in the appendix). The IGWMC provides computer programs and user support at nominal prices. The following is a brief description of some of the more widely used groundwater flow models.

Probably the most widely used groundwater flow model is the United States Geological Survey Modular Flow Model or MODFLOW (McDonald and Harbaugh, 1988). This program is able to simulate a wide variety of conditions, including:

- two-dimensional or three-dimensional flow in confined or unconfined aquifers,
- steady state or transient conditions,
- anisotropic, heterogeneous aquifers,
- areal recharge and evapotranspiration,
- flow to wells and drains, and
- flow through riverbeds.

MODFLOW is a finite-difference model which was originally developed with two alternative solution algorithms. Development work on the model continues, with new versions and additional solution algorithms (Hill, 1990; McDonald *et al.*, 1991), water-balance modules for stream infiltration (Prudic, 1989), and automatic parameter estimation (Hill, 1992). Because of its wide use, a cottage hydrogeological industry has sprung up to simplify the preparation of MODFLOW program input and to produce graphical output. ModelCad, a program written and distributed by Geraghty and Miller, Inc. is an example. MODFLOW itself is available directly from the U.S. Geological Survey or from IGWMC and many private vendors.

One of the earliest groundwater flow models, but still in wide use, is the Prickett-Lonnquist Aquifer Simulation Model (PLASM) (Prickett and Lonnquist, 1971). This finite difference model was developed with a different philosophy than MODFLOW. Rather than one comprehensive and complex program, the model consists of a set of several short programs which can be easily modified and combined to generate a digital simulation of a particular aquifer situation. Prickett and Lonnquist present FORTRAN programs capable of simulating:

- two-dimensional and quasi-three-dimensional non-steady state flow problems in heterogeneous aquifers,
- unconfined and confined flow,
- time varying pumping and recharge,
- evapotranspiration,
- water exchange between surface and groundwater, and
- transition from confined to unconfined conditions.

As with MODFLOW, PLASM is a finite-difference model.

For irregularly shaped aquifers, the finite-element method offers much greater flexibility than the finite-difference method used in MODFLOW and PLASM. A well known finite-element groundwater flow model is AQUIFEM, a two-dimensional (single-layer) model developed at the Massachusetts Institute of Technology (Townley and Wilson, 1980). This code is oriented toward the most common situation encountered by groundwater modelers: a two-dimensional, roughly horizontal aquifer of large areal extent. The finite-element formulation requires fewer node points to accurately represent the discretized aquifer and allows deformation of the grid to simulate observed boundary conditions. AQUIFEM is a complex program incorporating several options:

- anisotropic, heterogeneous, phreatic or confined aquifers,
- transient or steady-state conditions,
- time varying boundary conditions (including specified head or specified flow conditions),
- pumping and recharge wells,
- drains and springs,
- excavation dewatering,
- areal recharge and evapotranspiration, and
- exchange of water between aquifer and surface-water bodies.

Aside from the user's manuals and other original reports on these programs, an excellent reference is a recent textbook by Anderson and Woessner (1992). The book has a very practical orientation, and includes numerous case studies and examples. It provides guidance on modeling in general as well as on some of the less used features of the three codes discussed above.

4.1.6 Particle tracking models

Groundwater flow models are described above as effective tools to approximate contaminant transport. Several computer codes have been developed to use the results of numerical flow models in just this fashion. The underlying concept is known as particle tracking: a "particle" of water is tracked from a selected origination point along the predicted groundwater flow paths, to various destinations through time. Alternatively, a particular end point may be selected, and the flow paths leading to that point may be backtracked through time in the past. In either mode, particle tracking is a powerful tool for visualizing and understanding the groundwater system and predicting likely past or future paths of groundwater contaminants.

Pollack (1988, 1989) developed the particle tracking model MODPATH for use as a post-processor to MODFLOW. The computer program is able to compute and plot travel paths for multiple particles either forward or backward through time. Cumulative travel time along each path is recorded and may also be plotted. MODPATH is available from the IGWMC.

Shafer (1987a and b) developed the model GWPATH employing generally similar principles as Pollack, but with an interactive user interface. Unlike MODPATH, GWPATH was not developed specifically to interface with MODFLOW. It accepts an input head distribution in a generic format designed to accommodate information generated from either field data or flow model outputs. The program has similar capabilities as GWPATH but can also produce plots of groundwater velocity vectors and capture zones defined by travel time.

4.2 Groundwater Transport Models

4.2.1 Basic principles of groundwater contaminant transport

Contaminants respond to many factors in the subsurface environment. As discussed above, contaminants are advected by flowing groundwater. However, the nature of porous media is such that contaminants do not travel as contiguous masses. In traveling between and around the solid particles that make up the porous medium, contaminants tend to spread, or disperse. Contaminants may also undergo physical transformations. Adsorption of dissolved contaminants to aquifer solids is an important physical reaction in most aquifers; volatilization (evaporation) may be important in shallow free-surface aquifers. Chemical reactions may transform contaminants to different chemical species. Radioactive compounds are removed by radioactive decay. Finally, bacteria in the soil may biodegrade some contaminants.

The various influences on contaminant in groundwater are represented in the one-dimensional form of the advection-dispersion equation:

$$\frac{\partial C}{\partial t} = - V \frac{\partial C}{\partial x} + D_l \frac{\partial^2 C}{\partial x^2} + \frac{\rho_b}{n} \frac{\partial S}{\partial t} + R \qquad (7)$$

where C is the concentration of dissolved contaminant [M/L^3];

t is time [T];

V is the groundwater velocity [L/T];

D_l is the longitudinal dispersion coefficient [L^2/T];

x is a curvilinear distance coordinate aligned with the groundwater flow direction [L];

ρ_b is the bulk density of the porous medium [M/L^3];

S is the mass of the contaminant adsorbed on the solid part of the porous medium per unit mass of solids [M/M];

R is a generalized reaction term [M/L^3/T].

This form of the equation neglects vertical and transverse dispersion but otherwise illustrates the major influences on contaminants in the subsurface.

Inclusion of dispersion in transport modeling requires the specification of the dispersion coefficient as a model parameter. The dispersion coefficient is a somewhat artificial parameter in that it attempts to capture the tendency of nonuniform aquifer materials to cause contaminant spreading. The dispersion coefficient is difficult and time-consuming to measure in the field, and is then highly uncertain. Often, the most expedient way to estimate the dispersion coefficient is by calibrating the transport model to observed contaminant patterns. Calibration must be based on conservative (nonreactive) chemicals such as salt. Literature estimates may also be used, but it is important to recognize that the dispersion coefficient is highly dependent on the local spatial distribution of hydraulic conductivity. Therefore, it is inadvisable to estimate the dispersion coefficient of a particular aquifer by using data developed for unrelated systems. Anderson (1979) gives a good review of field measurements of dispersion and the problems of the dispersion coefficient. More recently, Gelhar *et al.* (1992) critically reviewed available field measurements. Other recent studies at experimental aquifers, cited above, have shown the lesser importance of vertical and transverse dispersion.

Adsorption and desorption may be major influences on many organic chemicals in groundwater. In modeling sorption reactions, it is typically assumed that an equilibrium, reversible, linear adsorption isotherm describes the distribution between dissolved and adsorbed fractions of the chemical (Freeze and Cherry, 1979). Under this model, the dissolved fraction and the sorbed fraction are related as:

$$S = K_d\, C \tag{8}$$

where K_d is the partition coefficient or distribution coefficient [L^3/M].

If this relation is assumed and substituted into the equation for contaminant transport, it manifests itself as an apparent reduction in contaminant travel velocity, known as retardation. The degree to which contaminant travel is slowed is represented by the retardation factor, R_f :

$$R_f = \frac{V}{V_c} = 1 + \frac{\rho_b}{n} K_d \tag{9}$$

where, V_c is the retarded contaminant velocity.

The retardation factor is a measure of how much slower the contaminant moves than the groundwater. For example, a typical retardation factor for benzene is about 6, indicating benzene will travel about six times slower than the flowing groundwater. Some compounds such as PCBs and higher molecular weight PAH have a great tendency to absorb to aquifer solids (high partition coefficients) and effectively do not travel in groundwater. The limitations of the retardation factor approach are that adsorption is often nonlinear, is rarely perfectly reversible, and may not be an equilibrium process. Nevertheless, the linear partition model has become the standard assumption made by groundwater modelers. Until considerably more basic laboratory data are developed, it is an assumption consistent with the available data base and a reasonable representation of observed behavior in the laboratory and field.

Several papers in the literature furnish useful overviews of methods to determine retardation factors for various contaminants. The best approach is one that is often impractical: laboratory experiments with the contaminants and aquifer materials in question. (See Goerlitz, 1984 for a discussion of appropriate methodology.) In a more practical approach, several researchers have proposed essentially similar empirical equations to determine the retardation factor from basic contaminant and aquifer material properties. Fetter (1992) provides a good summary of empirical equations.

Many models include a second reaction term to approximate reactions other than adsorption. Usually, a first-order decay or degradation process is assumed. This may be either biologically-mediated degradation, volatilization, radioactive decay (for radioactive substances) or chemical reaction. Since these processes are lumped in a single parameter, and are assumed to all be first-order processes, the model representation is necessarily approximate and uncertain. Thus, this reaction term is less widely employed than the retardation-factor term. Nevertheless, it is important to recognize that reactions other than adsorption and desorption may occur in aquifers. A very subtle manifestation of such reactions is the conversion of a contaminant to other contaminants. An investigator may be misled by the appearance of "new" compounds into the conclusion that other contamination sources exist, when in fact the known contaminants have reacted to form new compounds. This is an area of emerging research and information continues to be developed on the degradation of specific compounds.

4.2.2 Basic principles of transport models

Approaches to contaminant transport modeling include both analytical models and more complicated numerical models. For many problems, the simpler and more easily applied analytical approaches are very effective. At some sites, particularly those where there is extensive field information, numerical models offer the ability to simulate contaminant transport in greater detail.

As discussed above, contaminant transport may be assessed in a preliminary fashion by using flow models rather than contaminant transport models. In this case, groundwater flow models are applied as an approximate approach to determine contaminant movement. Flow models do not address contaminant movement per se. Nevertheless, by predicting the speed and direction of water movement one can make a good assessment of contaminant movement because contaminants are carried by the flowing groundwater.

The advantage in using flow models to determine contaminant transport is the greater certainty of the results and the lesser burden in input data. Generally, a flow model will give a transport prediction that is conservative in the sense that it will predict a faster rate of contaminant migration than would a transport model. This conservatism is introduced by ignoring the retarding effects of reaction on contaminant movement. However, there may be errors when dispersion is an important phenomenon that causes low concentrations of contaminant to move in advance of the main body of advected contaminant. Nonetheless, as indicated above, recent results from experimental field sites have shown that dispersion is a less important phenomenon than previously believed and may be neglected in many cases (Molz *et al.*, 1986; Lehr, 1988).

4.2.3 Application to groundwater assessment and remediation

The difficulties and limitations discussed above are compounded further when performing an actual site assessment. The critical problem in site assessment is definition of the source term. Contaminant transport models require as input the quantification of a source of contaminant. Usually the composition and the quantity of material deposited at a waste site and its release rate over time are very poorly known. Further, contaminant often travels through an unsaturated soil zone before it enters the groundwater, a process that is difficult to model. Thus, transport model application to site assessment is an approximate and uncertain exercise. A recent paper explores the problems of uncertain transport parameters and contaminant source history using contaminant transport modeling (Rogers, 1992).

Despite its problems, transport modeling can be effective in site assessments. One approach is to complete comprehensive field and laboratory investigations to collect the data necessary for accurate modeling of transport. This comprehensive approach is expensive and justified only for those very few sites that represent a serious imminent threat to sizable human populations.

More commonly appropriate is a second approach, in which transport modeling serves as a heuristic tool. In this approach, the modeler makes reasonable assumptions based on available field data concerning the aquifer materials and chemicals of interest. The modeler also estimates the contaminant source in a conservative manner (that is, tending to overestimate source strength). If it can be demonstrated that the assumptions are uniformly conservative, then it can be confidently assumed that the transport model represents a situation that is more adverse than the actual. Thus, the modeling study simulates an upper bound on the severity of environmental impacts.

Another heuristic approach to transport modeling can be used to evaluate the relative effectiveness of remedial action alternatives. This approach is appropriate when the modeler is able to determine reasonable aquifer parameter values, but has little information on contaminant source strength. For the purposes of comparing different remedial actions, the modeler simply assumes an arbitrary unit source strength. This very approximate approach cannot determine the absolute effectiveness of any remedial action. Nevertheless, the model will preserve the relative effectiveness of remedial alternatives. For example, if the model predicts that remedial alternative A performs better than alternative B, which in turn is better than C, then the same is likely to be true in the field situation.

The heuristic approach may be used to further evaluate remediation schemes. If the modeler has been conservative in his estimation of source strength and other model parameters, then it may be possible to determine the sufficiency of a remedial action. Consider the situation in which a transport model has been used conservatively to predict the existing conditions and it can be verified that the model-predicted existing conditions are worse than the actual existing conditions. The model is then used to predict future conditions following a remedial action, again being conservative in assumptions affecting the severity of the predicted conditions. If the model prediction indicates acceptable remediation, then the remedial action can be confidently recommended. The degree of confidence in this recommendation is directly related to the degree of conservatism in the modeling approach.

The heuristic use of transport models in site assessments is widespread, however it is rarely reported in the scientific literature. Examples may be found in some of the reports filed as a part of the U.S. Environmental Protection Agency Superfund program—site investigation and remedial action plan reports for many Superfund sites include such examples. The scientific literature contains several reports on comprehensive transport studies. Anderson (1979) reports on a number of these studies and cites references.

4.2.4 Modeling approaches

Transport models cover a wide range of sophistication and numerical complexity. The user of these models should bear in mind that applications of transport models are more limited by the availability of accurate data describing the groundwater system than by computing power. A useful overview of models for groundwater transport is given by Javandel *et al.* (1984).

The simplest transport modeling techniques are those based on analytical solutions to the advection-dispersion-reaction equation. One widely used model is that proposed by Wilson and Miller

(1978, 1979). The model determines the concentration distribution due to a continuous source of contaminant introduced uniformly throughout the thickness of a two-dimensional aquifer. One may construct similar models for instantaneous and continuous sources in one-, two- and three-dimensional flow fields. A recent U.S. Geological Survey report by Wexler (1992a) is a good catalog of analytical solutions for which a diskette of IBM-PC computer programs is available (Wexler, 1992b). More complicated applications of these analytical solutions entail accounting for multiple sources and the influence of boundaries. Bear (1979) discusses the mathematics of solving these more complicated situations.

Further complexity in representing mass transport requires digital computer programs to solve the transport equations numerically. As with flow models, a wide variety of approaches is available and includes two-dimensional and three-dimensional solutions by such techniques as the finite-difference and finite-element methods. The choice of the dimensionality of the solution is critical and is dictated by the circumstances of the site being modeled.

4.2.5 Available transport models

As with groundwater flow models, groundwater transport models should be selected with care. The criteria given earlier for selecting flow models fully apply to transport models as well. Van der Heijde *et al.* (1985) catalog available computer programs for numerical models of both groundwater flow and contaminant transport. Javendel *et al.* (1984) give a broader overview of transport modeling and include listings of several computer program codes.

Probably the most widely used computer program for simulation of mass transport is the U.S. Geological Survey method of characteristics model (MOC) by Konikow and Bredehoeft (1978). Unfortunately, this model treats transport in two dimensions only and in its original version neglects chemical reactions. A revision to the model, by Goode and Konikow (1989) incorporates first-order decay and contaminant adsorption. The MOC model simulates both groundwater flow and transport on a finite difference grid. The flow model component considers a variety of boundary conditions including leakage from an adjacent aquifer, recharge, and no-flow and fixed-head boundaries. MOC is a finite-difference model, and requires a fixed-size grid in either dimension (i.e., constant Δx and Δy). Preprocessors are available to simplify input and include the ModelCad program mentioned above. The MOC model is available from IGWMC.

The MOC model was modified by Rifai *et al.* (1987) for simulating dissolved oxygen and biodegradation kinetics in the subsurface. This model has been successfully applied to simulations of biodegradation at several contamination sites (Rifai *et al.*, 1988; Rifai and Bedient, 1990). The model is distributed by the Center for Subsurface Modeling Support (CSMoS) operated by the U.S. EPA (see appendix for address).

The Random Walk model, developed by Prickett *et al.* (1981), uses an entirely different simulation and solution approach. Random Walk is based on the concept that dispersion in porous media is a random process. Particles move with two types of motion:

1 along streamlines, in the direction of mean flow (advection), and
2 with a random motion governed by scaled probability curves related to flow length and longitudinal and transverse dispersion coefficients (dispersion).

The Random Walk model simulates rather than solves the dispersion equation, thereby drastically reducing the computer processing time required. Enough particles (up to several thousand) must be included in the contaminant transport solution so that the locations and density or particles as they move through a flow model are adequate to describe the contaminant of interest.

The Random Walk program incorporates a finite-difference algorithm to solve the two-dimensional steady or transient flow problem in a heterogeneous aquifer under confined or unconfined conditions. This flow component of the model is essentially the same as that used in the PLASM model. A modified version of the program includes a three-dimensional Random Walk transport component and accepts the flow field computed by MODFLOW in lieu of its own simulation of the flow field (ETA, 1989). The original, two-dimensional version of the Random Walk code is available from IGWMC.

A relatively new transport model, MT3D, employs a modular construction similar to that used in MODFLOW for flow simulation. MT3D uses a hybrid method of characteristics technique to solve the transport equations in three dimensions. It does not include a flow component, and thus the ground-water flow field must be computed in another model and imported into MT3D. An interface with MODFLOW is available. The MT3D code has capabilities to model linear and nonlinear sorption, and first-order decay or biodegradation. It is available from the U.S. EPA CSMoS.

5. MODELING CASE STUDIES

This section of the chapter provides selected examples of the use of models for groundwater contamination investigations and remediation. All are sites in the United States addressed by the author's consulting practice.

5.1 Analytical Capture-Zone Model

The site is a paint manufacturing facility located in the state of Virginia (Figure 2). Monitoring wells installed on the site found contamination by a variety of organic solvents including PCE, toluene, xylene, and ethyl benzene. The most contaminated monitoring well had approximately 75 mg/l of total organic solvents; nevertheless, contamination was fairly widely distributed on the site although it was found only in the shallow ground water (upper 9 meters). Based on regional information, the aquifer is about 30 meters thick, consisting of fine sand and silt. On-site testing indicated a hydraulic conductivity of approximately 1.7 x 10^{-3} cm/sec. Water levels measured in wells showed a hydraulic gradient of 0.018. A pump-and-treat system was recommended for the facility, with a single pumping well discharging to an air stripper to remove volatile organics. Because contamination was widespread on the facility, it was necessary to control flow from an area approximately 150 meters across. The half-width of the capture zone is thus 75 meters, which can be used in Equation 5 to compute a pumping rate. Based on the hydraulic conductivity and aquifer thickness given above, the transmissivity is 13 m^2/day. From Equation 5, the necessary pumping rate is calculated as 25 l/min. For conservatism, a rate of 40 l/min was used in the actual design.

Figure 3 shows the analytically computed capture zone superimposed on the site map. The capture zone encompasses the desired area of contamination, but the analysis says little about the amount of water withdrawn from the uncontaminated depth of the aquifer. For that analysis, a simple MODFLOW simulation was performed, looking at a vertical slice of the aquifer down the capture zone centerline. The groundwater flow equations do not include a gravity term and thus are as applicable to a vertically two-dimensional domain as a horizontal domain. The vertical head distribution predicted by MODFLOW is shown in Figure 4 and shows that the pumping well, if screened only 7.5 meters into the aquifer, will withdraw most of its water from upper 9 meters of the aquifer.

Figure 2. Analytical capture-zone model example: Site map showing concentrations of total volatile organic compounds

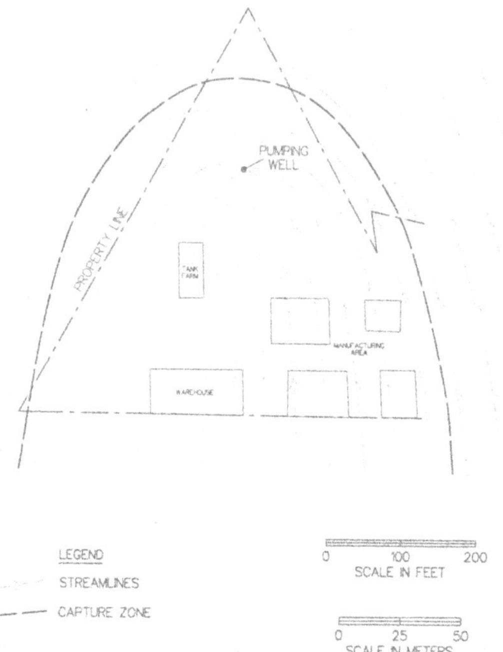

Figure 3 Analytical capture model example: Capture zone and groundwater streamlines

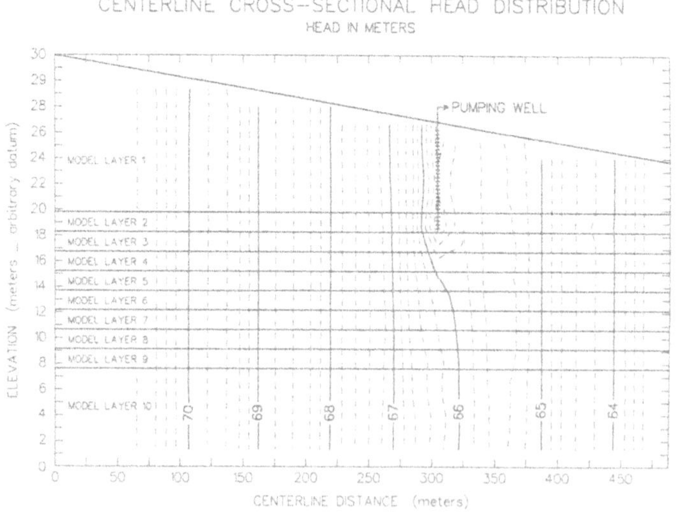

Figure 4. Analytical capture zone model example: Analysis of vertical flow using MODFLOW

5.2 Numerical Capture-Zone Model

This example is based on the case study published by Cosgrave *et al.* (1989). The site is a former coal-tar refinery located in the alluvial aquifer along the Mississippi River east of St. Louis, Missouri. The aquifer is 25 to 30 meters thick and consists of sand and gravel at depth grading into fine sand in the shallow part of the aquifer. Contaminants at the site include volatile organic solvents which were used in manufacturing coal-tar enamel paint as well as polynuclear aromatic hydrocarbons from the coal tar. Contamination was found in an area of former surface impoundments and solid waste disposal areas (Figure 5).

Figure 5. Numerical capture zone example: Site map

The transmissivity at this site was measured as 800 m²/day, much greater than at the site described above. The gradient however is nearly flat: 0.0012. A system consisting of two pumping wells (10A and 10B in Figure 5), each pumping at 4 liters per minute, was initially installed. The pumping wells were installed near the downgradient property line to prevent contaminants from migrating off-site. Effluent from the pumping wells was directed to an existing process wastewater treatment system.

An extensive monitoring well network had been installed on the site and water quality was monitored during the pump-and-treat system operation. Not surprisingly, it was found that contaminant migration from the source areas was increasing. The system design, though effective for controlling off-site migration, was exacerbating the extent of on-site contamination. After this discovery an array of seven source control wells (SCW1 through 7 in Figure 6) was designed and installed. The design employed a computer program based on the uniform flow equations able to account for the effects of multiple wells. The program, which is described more fully by Cosgrave *et al.* (1989), calculates the groundwater flow streamfunction and plots streamlines. The predicted flow field is shown in Figure 6.

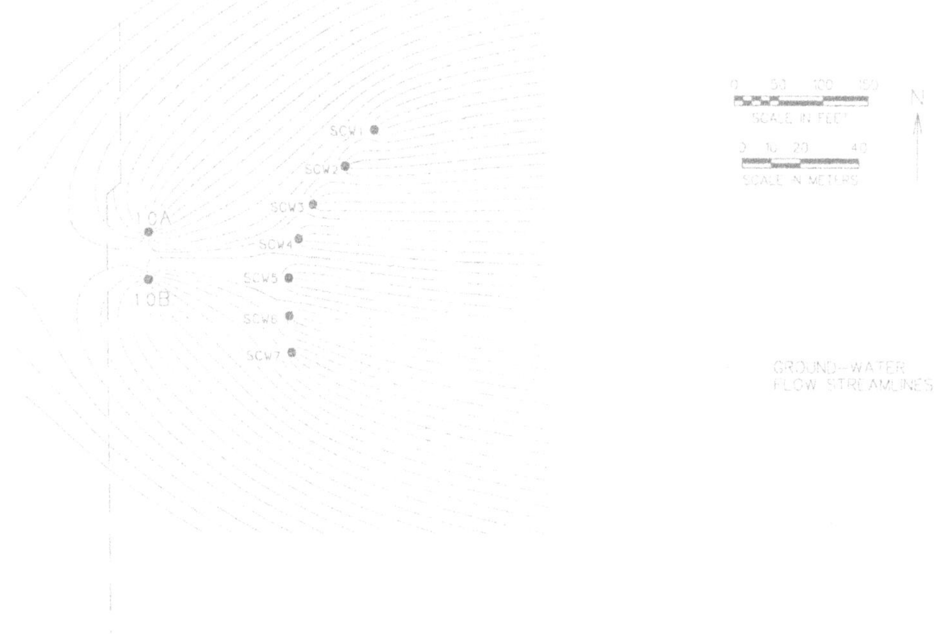

Figure 6. Numerical capture zone example: Predicted stream lines for gradient and source control wells

The dense monitoring well network at this site allowed the predictions to be tested against field data. The streamfunction and the potential functions are orthogonal for uniform ground-water flow; thus, hydraulic head maps determined from field measurements should be perpendicular to the predicted streamlines. Figure 7 shows that this is generally the case at this installation, providing confirmation of the utility of the capture-zone method.

Figure 7. Numerical capture-zone model example: Comparison of model predictions with field data

5.2.1 Numerical groundwater flow model (MODFLOW)

There are numerous examples in the literature of applications of the MODFLOW program; see for instance the many examples in Anderson and Woessner (1992). Literature examples are typically of complicated site studies, in which a calibrated and verified site model is developed and used. However, the MODFLOW model can be very useful for constructing simple design studies. The example shown in Figure 4 is just such a use of MODFLOW. The following provides a similarly simple application of MODFLOW for site remedial design.

The site is a long active wood treating facility at which railroad ties are pressure treated with creosote and in the past pentachlorophenol (Figure 8). Drippage from treated ties and at the pressure cylinder as well as former wastewater lagoons and spray treatment fields have caused widespread soil contamination by creosote DNAPL and groundwater contamination by polynuclear aromatic hydrocarbons.

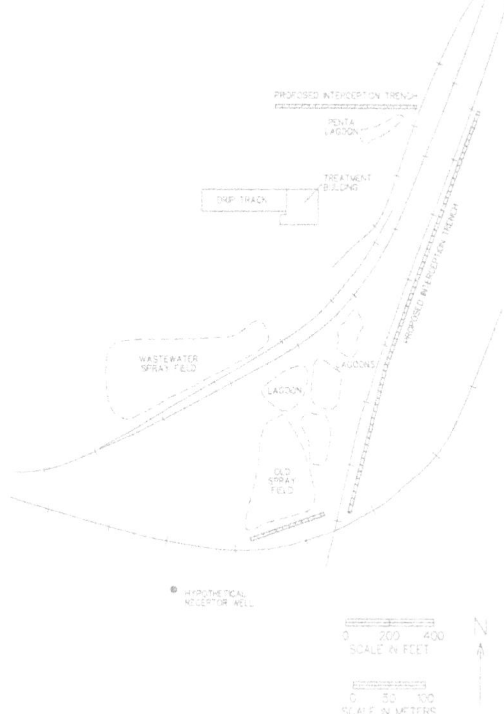

Figure 8. Numerical groundwater flow model example: Site map

The site is located in Illinois on extensive plains of glacial till. The shallow soils have a high clay content, but are somewhat more conductive near the land surface. The till, which is about 7.5 meters thick, is underlain by a confined sand aquifer approximately 5 meters thick. Most contamination is confined to the shallow till which, though it has a low transmissivity, acts as a shallow aquifer. Groundwater in the shallow till flows radially from the treatment works, which is located more or less on a surface- and groundwater divide, to surrounding creeks and intermittent streams.

A Superfund feasibility study specified a system of excavated and pumped trenches as the means to intercept contaminated groundwater from the shallow till aquifer. Altogether, 1.3 kilometers of interception trenches ringing the contaminated areas of the site were proposed. The plan also called for a pilot study of a short section of trench to test the constructibility of the trench and judge its effectiveness. The success of the pilot trench test would depend upon the ability to collect meaningful hydraulic head data and to conduct the test for a sufficient period of time. To address these questions, a simple MODFLOW model of the test trench was developed.

Figure 9 shows the MODFLOW finite difference grid. The model was constructed of three layers, a 6-meter thick upper layer of moderately transmissive till; a 1-meter thick layer of low permeability till; and a lower 5-meter thick layer to represent the confined sand aquifer. Because only a small section of the site was modeled, constant-head boundaries were used to mimic the wider pattern of groundwater flow. The trench was modeled using the MODFLOW drain capability. It was presumed that a level-sensing sump pump would operate in the trench, maintaining the water level in the trench at a roughly constant elevation. The model was run in transient mode, examining the predicted hydraulic head distribution in each model layer during a presumed 6-day pumping test. The hydraulic head predicted in the upper till is shown in Figure 10. The information from the model was used to recommend locations for monitoring wells in the shallow and intermediate till for observations during the pilot test.

5.3 Analytical Groundwater Transport Model

The wood-treating site discussed in the previous example was also the subject of a transport modeling study. Although most site contamination occurs in the shallow aquifer, there is limited contamination of the underlying sand aquifer by polynuclear aromatic hydrocarbons (PAH). Contamination is found in the sand aquifer only under the most contaminated portions of the site. Nonetheless, private drinking water wells are located downgradient of the site and migration of contaminants to private wells was a concern. The contaminant transport study was undertaken to examine this problem.

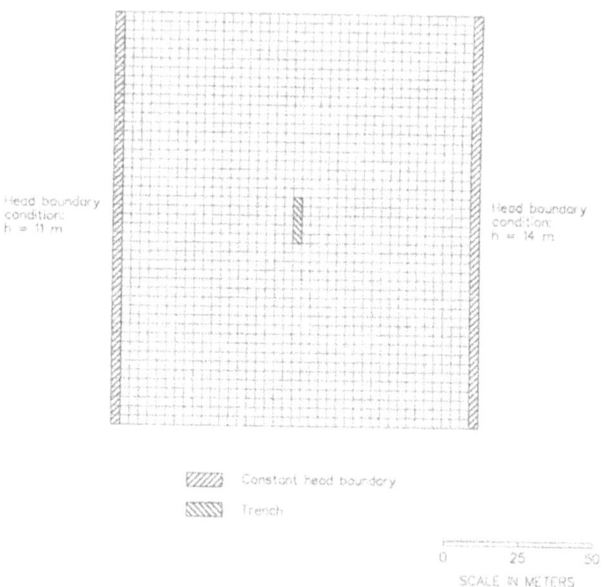

Figure 9. Numerical groundwater flow model example:MODFLOW model of pilot trench study

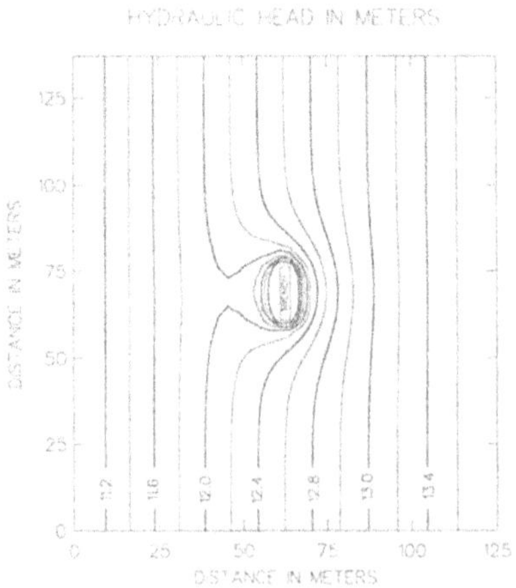

Figure 10. Numerical groundwater flow model example: Predicted head after six days of pumping from trench

The flow in the confined sand aquifer is highly uniform and thus amenable to an analytical modeling approach. Because contributions from several sources need to be superimposed in the solution, the computerized analytical model HPS (Horizontal Plane Source) developed by Galya (1987) was used. Other calculations were performed to estimate the long-term rate of contaminant release from the contaminated shallow soils to the underlying sand aquifer, a needed input to the model. Groundwater flow at the sand aquifer is essentially due south with a gradient 0.001. An aquifer test determined the hydraulic conductivity to be 1.2 x 10^{-2} cm/sec; aquifer porosity was estimated to be 45%.

In order to be conservative in the analysis, a hypothetical well was assumed to be located at the site property line (Figure 8), about 600 meters from the nearest source area. From Equation 2, the groundwater velocity is computed as 8 meters per year, indicating it would take about 70 years for groundwater from the contaminated area to reach the receptor well. This is comparable to the time the facility had operated, thus raising a potential concern. However, PAH are strongly adsorbed to soil and thus substantially retarded compared to groundwater flow. The PAH are a family of compounds consisting of two or more fused benzene rings. The modeling exercise considered a range of representative PAH compounds: naphthalene as a two-ring, anthracene as a three-ring, benzo(a)anthracene as a four-ring, and benzo(a)pyrene as a five-ring PAH. The propensity of PAH compounds to adsorb to soils increases with their molecular size. Retardation factors computed for these four PAH using measured or assumed site properties were 8.7, 83, 8,100, and 32,000, respectively. Thus, the predicted arrival times change from 70 years for groundwater, to 600 years for naphthalene, to 5,800, 570,000, and over 2 million years for anthracene, benzo(a)anthracene, and benzo(a)pyrene. Consideration of retardation changed an imminent threat to a very distant threat.

The HPS model was used to compute concentrations as a function of time at the receptor well. Figure 11 shows the results for naphthalene, comparing results when contaminated soils are left in place or are partially excavated and removed. The curve reflects the reductions in concentration as the contaminants from different sources are leached out over time. The high retardation of these compounds implies very slow transport to receptors, but also very slow removal from soil and remediation in the groundwater.

Numerical Groundwater Transport Model (MOC). The U.S. Geological Survey MOC model was used in this study to evaluate the effects of a former manufactured gas plant site located in an industrial area of Baltimore, Maryland. The facility operated from the mid-1800s to the 1960s, gasifying coal and producing coal tar as a byproduct. Coal tar contaminates shallow soil on the site and has caused groundwater contamination by PAH and monoaromatic hydrocarbons (benzene, toluene, xylene, and ethylbenzene).

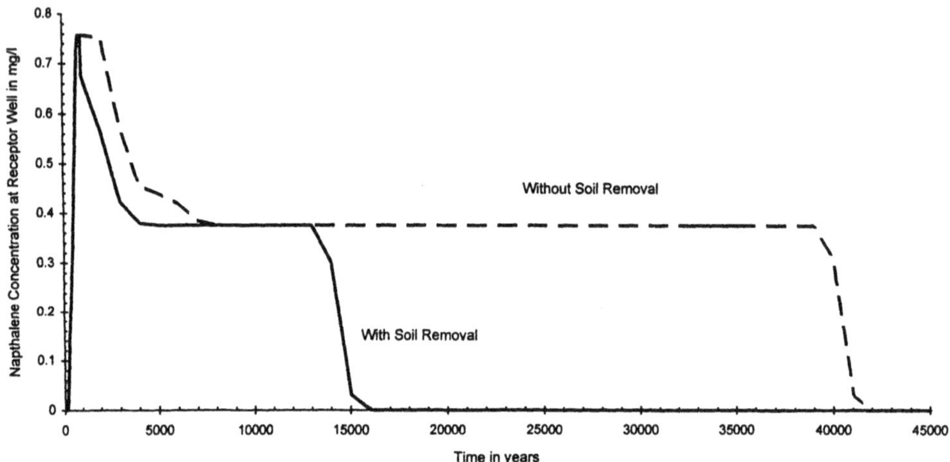

Figure 11. Analytical groundwater transport model example: Predicted concentrations of napthalene versus time at hypothetical downgradient receptor well

The site is located in the western part of Baltimore in the outcrop area of the Patuxent aquifer, a sand and gravel aquifer that supplies much of the water supply for the city. The aquifer dips steeply to the east and is a confined aquifer under much of the city. Historically, industry in Baltimore has depended upon groundwater for industrial water supply, with the result that the amount and location of groundwater withdrawal has varied over time and space. Changing rates of industrial pumping in the past have caused groundwater flow patterns to vary. The site may have been a source of contamination to the Patuxent since 1900 or before, and thus any contaminant plume would have been affected by the historically changing groundwater flow patterns.

Although the site itself has been very extensively investigated, there is very limited off-site data. A concern is the possible extent of the contaminant plume in the Patuxent aquifer and its fate over time. However, laboratory studies and field data suggest that biodegradation may have significant influence on the plume, although the extent and rate of biodegradation in the deep aquifer is not known.

A groundwater transport model was developed to provide a tool for assessing potential off-site contaminant migration, studying the possible effects of biodegradation, and helping design future groundwater monitoring. The model grid is shown in Figure 12. Only a portion of this area was significant for transport model predictions, but the widespread pumping centers required a larger area be simulated for groundwater flow. The model hydraulic component was calibrated against extensive water-level surveys

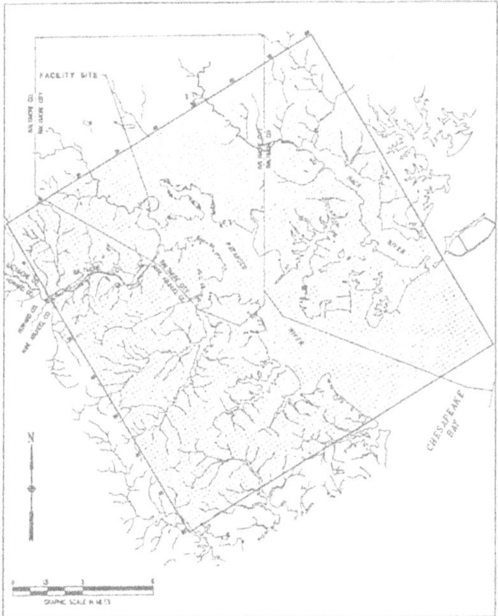

Figure 12. Numerical groundwater transport model example: Site area and model grid

completed in 1945 and 1982 as well as against more recent near-site water-level measurements. The contaminant transport component was calibrated against the distribution of chloride which had entered the aquifer via intrusion from the Chesapeake Bay estuary. Extensive sensitivity simulations were also performed.

Figure 13 illustrates the contaminant transport predictions of the model. This figure shows the effect of different assumptions regarding biodegradation (expressed as the biodegradation half-life). Without biodegradation (Figure 13a), the plume is predicted to extend 3000 meters. With a relatively modest rate of biodegradation (half-life of two years), the plume is attenuated in a very short distance off-site. The transport model proved useful in focusing attention on biodegradation as a critical aspect for further study. Model results also showed that no sources of drinking water were threatened by the plume now or in the future, as the predicted extent of the plume under worst-case assumptions was within the area of the aquifer already contaminated by salt-water intrusion.

This chapter has given an overview of groundwater modeling with an emphasis on practical application to assessment and remediation of hazardous waste sites. Although there are a wide variety of modeling approaches and computer programs available to model groundwater, the keys to a successful model investigation are:

- careful prior definition of the questions to be answered by the model;

- development of adequate data to use the model; and

- selection of the simplest model adequate for the investigation.

Two major types of groundwater models are available: groundwater flow models and groundwater contaminant transport models. Groundwater flow models generally require less data and more easily measured properties than groundwater transport models, and therefore can be used with more confidence. In many cases, a flow model can be used to make a rough assessment of contaminant transport inasmuch as contaminant movement is primarily determined by groundwater movement. Flow models usually give a conservative assessment of contaminant transport because they neglect dispersion and chemical reactions which tend to slow plume migration.

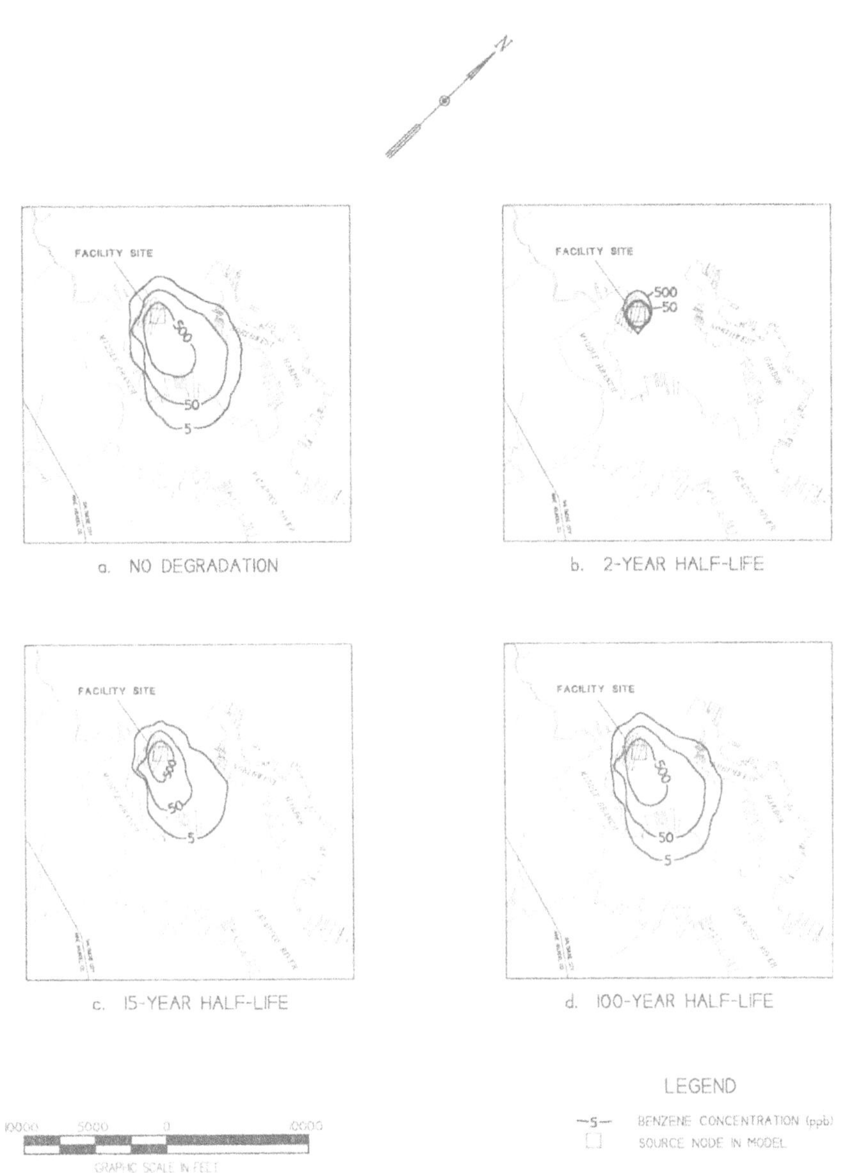

Figure 13. Numerical groundwater transport example: Effect of biodegradation on predicted benzene plume

Groundwater transport models have greater data demands and are usually less certain than flow models. Nonetheless, they can be extremely useful in many site investigations and remedial designs. Transport models can often be used very effectively in a heuristic mode. For example, contaminant transport can be simulated with very conservative assumptions, thus giving confidence that the model outcome is a conservative assessment of migration potential. Another form of heuristic use of transport models is to assume a unit contaminant load and compare remediation alternatives. Although these predictions have no value as absolute predictors of performance they can supply very useful and accurate information on relative performance of different remedies.

The value of groundwater modeling for site investigation and remediation is illustrated through selected case studies. Case studies illustrate the use of models in both a conventional mode and a simplified mode. In a conventional mode, the model is calibrated and verified against field data in what is generally a fairly rigorous and extensive effort. In a simplified mode, models can be constructed quickly without field data calibration for use as a design and investigation tool.

REFERENCES

Anderson, P.F., C.R. Faust, and J.W. Mercer (1984) " Analysis of Conceptual Designs for Remedial Measures at Lipon Landfill, New Jersey," Ground Water. 22(2):176-190

Anderson, M.P. (1979) "Using Models to Simulate the Movement of Contaminants through Groundwater Flow Systems," Critical Reviews in Environmental Control 9(2):97-156.

Anderson, M.P., and W.W. Woessner (1992) Applied Groundwater Modeling: Simulation of Flow and Advective Transport. Academic Press, Inc., New York.

Bear, J., (1979) Hydraulics of Groundwater. McGraw-Hill Inc., New York.

Bonn, B. and S. Rounds (1990) DREAM: Analytical Ground-water Flow Programs. Lewis Publishers, Boca Raton, Florida, USA.

Cosgrave, T., P. Shanahan, J.C. Craun and M. Haney (1989) "Gradient Control Wells for Aquifer Remediation: A Modeling and Field Case Study." In: Solving Ground Water Problems with Models Conference, February 7-9, 1989, Indianapolis, Indiana. National Water Well Association, Dublin, Ohio.

ETA (1989) Three Dimensional Solute Transport using the Random-Walk Algorithm and MODFLOW: User's Manual for PREMOD3 Preprocessor and User's Manual for RAND3D Three Dimensional Random Walk. Engineering Technologies Associates, Inc., Ellicott City, Maryland, USA and Thomas A. Prickett and Associates, Urbana, Illinois, USA. September 1989.

Fetter, C.W. (1992) Contaminant Hydrogeology. Macmillan Publishing Company, New York.

Freeze, R.A. and J.A. Cherry (1979) Groundwater. Prentice-Hall, Inc., Englewood Cliffs, New Jersey.

Freeze, R.A. and J.A. Cherry (1989) "Guest Editorial: What has gone wrong," Ground Water **27**(4):458

Galya, D.P. (1987) "A Horizontal Plane Source Model for Ground-Water Transport," Ground Water **25**(6):733-739

Garabedian, S.P., D.R. LeBlanc, L.W. Gelhar, and M.A. Celia (1991) " Large-Scale Natural Gradient Tracer Test in Sand and Gravel, Cape Cod, Massachusetts. 2. Analysis of Spatial Moments for a Nonreactive Tracer," Water Resources Research **27**(5):911-924

Gelhar, L.W., C. Welty, and K.R. Rehfeldt (1992) "A critical review of data on field-scale dispersion in aquifers," Water Resources Research **28**(7):1955-1974

Goerlitz, D.F. (1984) "A column technique for determining sorption of organic solutes on the lithological structure of aquifers," Bulletin of Environmental Contamination and Toxicology **32**(1):37-44.

Goode, D.J., and L.F. Konikow (1989) "Modification of a Method-of-Characteristics Solute-Transport Model to Incorporate Decay and Equilibrium-Controlled Sorption or Ion Exchange," Water-Resources Investigations Report 89-4030. U.S. Geological Survey, Reston, Virginia, USA.

Hill, M.C. (1990) "Preconditioned conjugate-gradient 2 (PCG2), a computer program for solving ground-water flow equations," Water-Resources Investigations Report 90-4048. U.S. Geological Survey, Reston, Virginia, USA.

Hill, M.C. (1992) A computer program (MODFLOWP) for estimating parameters of a transient, three-dimensional, ground-water flow model using nonlinear regression. Open-File Report 91-0484. U.S. Geological Survey, Reston, Virginia, USA.

Javandel, I., C. Doughty and C.F. Tsang (1984) Groundwater Transport: Handbook of Mathematical Models. Water Resources Monograph 10. American Geophysical Union, Washington, D.C.

Javandel, I. and C.-F. Tsang (1986) "Capture-Zone Type Curves: A Tool for Aquifer Cleanup," Ground Water **24**(5):616-625

Konikow, L.F. and J.D. Bredehoeft (1978) Computer Model of Two-Dimensional Solute Transport and Dispersion in Ground Water. Techniques of Water-Resources Investigations, Book 7, Chapter C2. U.S. Geological Survey, Reston, Virginia.

Lehr, J.H. (1988) "Editorial: An Irreverent View of Contaminant Dispersion," Ground Water Monitoring Review **8**(4):4-6.

Lundy, D.A. and J.S. Mahan (1982) "Conceptual Designs and Cost Sensitivities of Fluid Recovery Systems for Containment of Plumes of Contaminated Groundwater." National Conference on Management of Uncontrolled Hazardous Waste Sites, Washington, D.C., November 29-December 1, 1982. Pg. 136-140.

Mackay, D.M., D.L. Freyberg, P.V. Roberts, and J.A. Cherry (1986) "A Natural Gradient Experiment on Solute Transport in a Sand Aquifer 1. Approach and Overview of Plume Movement," Water Resources Research **22**(12):2017-2029

McDonald, M.G. and A.W. Harbaugh (1988) A Modular Three-dimensional Finite-difference Ground-water Flow Model. Techniques of Water-Resources Investigations, Book 6, Chapter A1. U.S. Geological Survey, Reston, Virginia.

McDonald, M.G., A.W. Harbaugh, B.R. Orr, and D.J. Ackerman (1991) A method of converting no-flow cells to variable-head cells for the U.S. Geological Survey modular finite-difference ground-water flow model. Open-File Report 91-536. U.S. Geological Survey, Reston, Virginia.

Mercer, J.W., L.R. Silka, and C.R. Faust (1983) "Modeling Ground Water Flow at Love Canal, New York," Journal of Environmental Engineering, ASCE **109**(4):924-942

Molz, F.G., O. Güven, J.G. Melville and J.F. Keely (1986) Performance and Analysis of Aquifer Tracer Tests with Implications for Contaminant Transport Modeling. EPA/600/2-86/062. Robert S. Kerr Environmental Research Laboratory, U.S. Environmental Protection Agency, Ada, Oklahoma, USA. July 1986.

Nelson, R.W. (1978) "Evaluating the Environmental Consequences of Groundwater Contamination. Obtaining Location/Arrival Time and Location/Outflow Quantity Distributions for Steady Flow Systems," Water Resources Research **14**(3):416-450. March 1978.

Nyer, E.K. (1992) Groundwater Treatment Technology, 2nd Edition. Lewis Publishers, Boca Raton, Florida, USA.

OTA (1984) Protecting the Nation's Groundwater from Contamination. Report No. OTA-O-233 and OTA-O-276. Office of Technology Assessment, U.S. Congress, Washington, D.C. October 1984.

Pollack, D.W. (1988) "Semianalytical computation of path lines for finite-difference models," Ground Water **26**(6):743-750

Pollack, D.W. (1989) Documentation of computer programs to compute and display pathlines using results from the U.S. Geological Survey modular three-dimensional finite-difference ground-water flow model. Open-File Report 89-381. U.S. Geological Survey, Reston, Virginia, USA.

Prickett, T.A., and C. G. Lonnquist (1971) "Selected Digital Computer Techniques for Groundwater Resource Evaluation," Bulletin 55. Illinois State Water Survey, Champaign, Illinois, USA.

Prickett, T.A., T.G. Naymik, and C. G. Lonnquist, 1981. " A 'Random-Walk' Solute Transport Model for Selected Groundwater Quality Evaluations." Bulletin 65. Illinois State Water Survey, Champaign, Illinois, USA.

Prudic, D.A. (1989) Documentation of a computer program to simulate stream-aquifer relations using a modular, finite-difference, ground-water flow model. Open-File Report 88-0729. U.S. Geological Survey, Carson City, Nevada, USA.

Rifai, H.S., P.B. Bedient, R.C. Border, and J.F. Haasbeek (1987) BIOPLUME II--Computer Model of Two-Dimensional Transport under the Influence of Oxygen Limited Biodegradation in Ground Water; User's Manual, Version 1.0. Department of Environmental Science and Engineering, Rice University, Houston, Texas, USA.

Rifai, H.S., and P.B. Bedient (1990) "Comparison of Biodegradation Kinetics with an Instantaneous Reaction Model for Groundwater," Water Resources Research **26**(4):637-645

Rifai, H.S., P.B. Bedient, J.T. Wilson, K.M. Miller, and J.M. Armstrong (1988) "Biodegradation Modeling at an Aviation Fuel Spill Site," Journal of Environmental Engineering ASCE **114**(5):1007-1029

Rogers, L. (1992) "History Matching to Determine the Retardation of PCE in Ground Water," Ground Water **30**(1):50-60

Schwille, F. (1988) Dense Chlorinated Solvents in Porous and Fractured Media. Lewis Publishers, Chelsea, Michigan, USA. (Translation of *Leichtflüchtige Chlorkohlenwasserstoffe in porösen und klüftigen Medien.*)

Shafer, J.M. (1987a) "Reverse pathline calculation of time-related capture zones in nonuniform flow," Ground Water **25**(3):283-289

Shafer, J.M. (1987b) GWPATH: Interactive Ground-Water Flow Path Analysis. Bulletin 69. Illinois State Water Survey, Champaign, Illinois, USA.

Townley, L.R., and J.L. Wilson (1980) Description of and user's manual for a finite element aquifer flow model AQUIFEM-1. Technology Adaptation Program Report 79-3. Ralph M. Parsons Laboratory, Department of Civil Engineering, Massachusetts Institute of Technology, Cambridge, Massachusetts, USA.

U.S. Environmental Protection Agency (1984) Extent of the Hazardous Release Problem and Future Funding Needs, CERCLA Section 301(a)(1)(C) Study. Office of Solid Waste and Emergency Response, U.S. Environmental Protection Agency, Washington, D.C. December 1984.

U.S. Environmental Protection Agency (1987) Guidelines for Delineation of Wellhead Protection Areas. Report No. EPA/440/6-87-010. Office of Ground-Water Protection, U.S. Environmental Protection Agency, Washington, D.C. June 1987.

U.S. Environmental Protection Agency (1988) Guidance for Remedial Actions for Contaminated Ground Water at Superfund Sites. OSWER Directive 9283.1-2. Report EPA/540/G-88/003. Office of Emergency and Remedial Response, U.S. Environmental Protection Agency, Washington, D.C. December 1988.

U.S. Environmental Protection Agency (1993) Guidance for Evaluating the Technical Impracticability of Ground-Water Restoration. OSWER Directive 9234.2-25. Office of Solid Waste and Emergency Response, U.S. Environmental Protection Agency, Washington, D.C. September 1993.

van der Heijde, P., Y. Bachmat, J. Bredehoeft, B. Andrews, D. Holtz and S. Sebastian (1985) Groundwater Management: The Use of Numerical Models. Second Edition. Water Resources Monograph 5. American Geophysical Union, Washington, D.C.

Wexler, E.J. (1992a) Analytical Solutions for One-, Two-, and Three-Dimensional Solute Transport in Ground-Water Systems with Uniform Flow. Techniques of Water-Resources Investigations, Book 3, Chapter B7. U.S. Geological Survey, Denver, Colorado.

Wexler, E.J. (1992b) Analytical Solutions for One-, Two-, and Three-Dimensional Solute Transport in Ground-Water Systems with Uniform Flow—Supplemental Report: Source Codes for Computer Programs and Sample Data Sets. Open-File Report 92-78. U.S. Geological Survey, Denver, Colorado.

Wilson, J.L., and P.J. Miller (1978) "Two Dimensional Plume in Uniform Ground-Water Flow," Journal of the Hydraulics Division, ASCE **104**(HY4):503-514

Wilson, J.L., and P.J. Miller (1979) "Closure, Two Dimensional Plume in Uniform Ground-Water Flow," Journal of the Hydraulics Division, ASCE **105**(HY12):1567-1570

Yazicigil, H. and L.V.A. Sendlein (1981) " Management of Ground Water Contaminated by Aromatic Hydrocarbons in the Aquifer Supplying Ames, Iowa," Ground Water **19**(6):648-665

APPENDIX: Sources for Computer Programs

The International Ground Water Modeling Center distributes most of the codes discussed in this chapter for nominal prices:

International Ground Water Modeling Center
Colorado School of Mines
Institute for Ground-Water Research & Education
Golden, CO 80401-1887 USA
1 (303) 273-3103
Fax: 1 (303) 273-3278

International Ground Water Modeling Center
TNO Institute of Applied Geoscience
P.O. Box 6012
2600 JA Delft
The Netherlands
31.15.697214
Fax: 31.15.564800

The U.S. EPA Center for Subsurface Modeling Support distributes the BIOPLUME II, MT3D and other codes at no charge:

Center for Subsurface Modeling Support
U.S. Environmental Protection Agency
Robert S. Kerr Environmental Research Laboratory
Post Office Box 1198
Ada, OK 74820 USA
1 (405) 436-8500

The U.S. Geological Survey distributes USGS computer programs from the National Center in Reston, Virginia and publications from the Book Sales office in Denver, Colorado for nominal costs:

U.S. Geological Survey
National Water Information Center
National Center WGS-Mail Stop 437
12201 Sunrise Valley Drive
Reston, VA 22092 USA
1 (703) 860-7000

U.S. Geological Survey
Books and Open-File Report Sales
Box 25286
Denver, CO 80225 USA
1 (303) 236-7477

Thomas A. Prickett sells versions of Random Walk and PLASM:

Thomas A. Prickett & Associates
6 G.H. Baker Drive
Urbana, Illinois 61801 USA
1 (271) 384-0615

Engineering Technologies Associates, Inc. sells RAND3D, the three-dimensional Random Walk code:

Engineering Technologies Associates, Inc.
3458 Ellicott Center Drive
Ellicott City, MD 21043 USA
1 (410) 461-9920

John M. Shafer sells the GWPATH program:

Dr. John M. Shafer
University of South Carolina
Earth Science Research Institute
901 Sumter
Columbia, SC 29208
1 (803) 777-4421

Geraghty & Miller, Inc. sells the ModelCad preprocessor for MODLFOW, MOC, and other models. They also sell versions of MODFLOW and other codes.

Geraghty & Miller, Inc.
1895 Preston White Drive, Suite 301
Reston, BA 22091 USA
1 (703) 476-0335
1 (703) 476-6372

GEOCOMP Corp. sells AQUIFEM:

GEOCOMP Corp.
342 Sudbury Road
Concord, MA 01742 USA
1 (508) 369-8304

CHAPTER 9

WASTEWATER TREATMENT TECHNOLOGY, ECONOMY AND POLITICS

Petr Grau[1]

1. INTRODUCTION

Numerous system conditions are necessary for successful abatement of water pollution. Some are generally valid, others are country and/or region specific. The specific conditions include climate, hydromorphology, the level of socio-economic and technological development, political factors, the state and efficiency of local administration, traditions, etc.

Some of the specific conditions are quite stable in the long run, such as climate and hydromorphology. Others may slowly change and several may undergo rapid transformation and development in particular periods of time. In Central and Eastern Europe, many of the country and region specific conditions have been rapidly changing since the end of 1989. The changes are reflected in every aspect of life and influence all individuals and institutions. Water pollution control and abatement policies are no exception.

2. REGULATORY ISSUES

Recently, trends in national legislation have been widely discussed. In the early stage of democratic transformation (1991-1992), the former Czechoslovakia (similarly as other former communist countries) introduced an enormous number of new laws and regulations. Many of them have had to be modified and updated several times since.

[1] AquaNova International, a.s., Prague, Czech Republic

NATO ASI Series, Partnership Sub-Series, 2. Environment - Vol. 3
Remediation and Management of Degraded River Basins
Edited by V. Novotny and L. Somlyódy

2.1 Legislation

It is a well-known fact that appropriate environmental legislation is a compromise between wishes (mostly of environmental groups and activists) and the economic reality faced by the producers and population in general. Many pressure groups formulate wishes, frequently disregarding the technological and economic reality, while legislators attempt to define achievable goals. Excessively stringent standards, which are far from economic reality, can be discouraging and counter productive. Polluters use all legal and sometimes (if under heavy pressure) even less legal means to get softer and achievable standards, waivers, extensions, etc. If standards are far from reality, too many dischargers develop political pressure that turns at least the professional opinion against environmental legislators. On the contrary, standards which are too soft are easily achievable, however, water quality continues to be unacceptable and below the minimum acceptable quality limits. Legislators and environmental authorities are then criticized for achieving too little and both the public and professional opinions turn against them.

Much depends on the political strength of the industry in a respective country. In the new democracies, after a complete overhaul of industries and their privatization, the power of industries is generally weak. The political system is unbalanced and the environmental fundamentalism (typically naive and radical) may prevail. The most dangerous period, when substantial errors can be made, is thus the initial stage of a freshly born democracy. The second dangerous period is that of economic recovery and acceleration. Several observers are of the opinion that in the Czech Republic such a period just started. Conflicts between wishes and reality seem to emerge again.

Thus the most difficult task for legislators is to correctly anticipate the future political, economic and technological development of the country as well as the development of environmental science and technology world-wide. Equally important is to foresee the ability of the country to absorb new technologies and apply them within local conditions. The least risky way is to set up short- and long-term goals and standards. An inspiring example can be found in the European Union Council Directive 91/271/EEC, concerning urban wastewater treatment. A stepwise process was adopted, for instance, requiring wastewater treatment for urban areas of more than 15 000 population equivalent (PE) at the latest by December 2000 and for urban areas between 10 000 and 15 000 PE cities at the latest by 31 December 2005, etc.

2.2 Emission (Effluent) and Ambient (Water Quality) Standards

In the former Soviet bloc (CEE) countries ambient receiving water quality standards had been used for decades as a principal regulative tool in water pollution control. Discharge permits were based on evaluation of three sets of data:

- water quality upstream of the discharge;
- load intended to be discharged into the stream; and
- resulting calculated concentration downstream.

Ideal mixing of wastewater with the receiving water was presumed and the discharge permits were based on the calculated resulting receiving water body quality.

A permit system based on the ambient standards is, no doubt, based on correct environmental logic. i.e., efficiency. Theoretically, each source of pollution would receive as much treatment as necessary for the receiving body. In practice, however, application of ambient standards causes enormous administrative and political problems in existing socio-economic systems (Helmer, 1987). Also it results in the most expensive administration of water quality imaginable.

2.3 Czech Water Quality Standards

After extensive evaluation of permit systems applied in various countries and considering the regulation 91/271/EEC (EC Council Directive of 21 May 1991), a concept of combined standards, based primarily on the emission standards but considering also ambient (receiving water quality), has been approved by the Czech Government (decree 171 of 26 February 1992).

Emission standards (effluent limitations) include two different values for each indicator; a value to be achieved prior to the end of the year 2005 and a more stringent value for the year 2005 and after. The decree states that (local) water authorities are obliged to consider also ambient standards (related to the low flow Q and published in the same document), and make them gradually obligatory according to local conditions. Obviously, the application of ambient (water quality) standards is not meant as downgrading of emission standards but rather making them more stringent wherever justified by existing or desired water quality of the receiving water body. Thus, this system *de facto* divides receiving water bodies into those which are effluent and those which are water quality limited (see Chapter 1 by Novotny and Somlyódy). The set of emission standards for municipal wastewater is abstracted in Appendix 1.

The concept behind the new regulation is to achieve general and basic clean-up of the country rivers before the end of the year 2004. This should be achieved by mostly applying the emission standards. Starting from the year 2005, a set of more stringent standards will be applied. It is also anticipated that ambient standards will play a more important role. Thus two distinctly different periods are presumed, the basic clean-up and the final tuning. Wastewater treatment plants planners know in advance what will be required within a foreseeable period of time and can adjust the design accordingly.

Top priority is to protect the quality of water resources which are used as a source of potable water. Drinking water supplies for cities with more than 50 000 inhabitants relies on 75% of the surface water bodies. Even for many small urban centers the proportion of surface water resources is very significant because many of them are supplied from regional drinking water supply systems. Watersheds of drinking water reservoirs are in most cases located in tributary headwater valleys and in the upper, mountainous, watersheds. Also, these watersheds have been controlled by much more stringent restrictions than other water resources. In spite of that, nitrogen, originating mainly from agricultural lands, and phosphorus from treated wastewater and soil erosion, contribute to eutrophication which is serious at the majority of drinking water reservoirs. Because of the described characteristics of the water resources and water supply systems, two different categories of ambient standards have been traditionally instituted:

a) ambient (water quality) standards for officially declared river basins used as drinking water resources;
b) ambient standards for all other surface waters.

Therefore, the set of ambient standards, abstracted also in Appendix 1 (Indicators III), recognizes at least two major uses of surface water bodies in the Czech Republic.

Contrary to some other countries in the region, categorization of receiving streams into more than two use categories, for instance water quality for irrigation, for industrial use, for recreation, etc., has never found enough support to be instituted.

2.4 Pollution Source Size Scale-up Factor

A prevailing opinion among regulators is that the emission (allegedly more efficient) standards are too soft on small pollution sources. Such opinion is superficial because it does not take into account the flow variations as related to the pollution source size.

In Figure 1, the effluent standards for BOD are shown together with the expected peak hour dry weather flow. Simple kinetic calculations as well as practical experience indicate that the flow variation introduces doubling of the effluent concentration for source size 500 and 100 000 PE, respectively. In other words, in an 8 h composite sample, BOD concentration would be twice as high for the source of 500 PE compared to the source of 100 000 PE (except at night) just because of the larger flow variation.

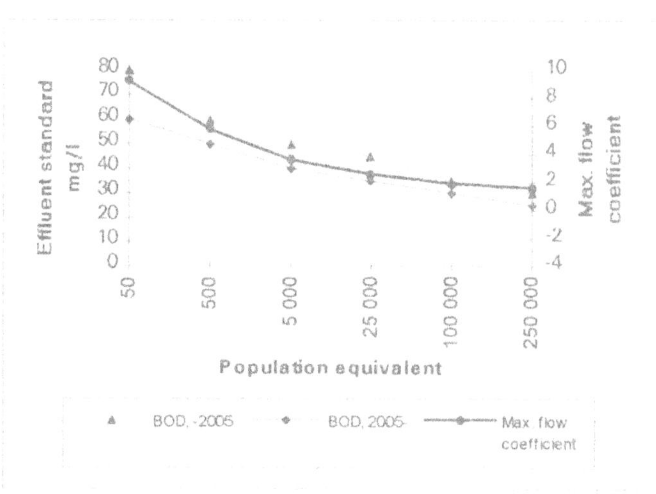

Figure 1: Effluent standards in relation to the source size

Approximate investment costs per PE for plants of different size are shown in Figure 2 (1 US$ ≈ 28 CZK).

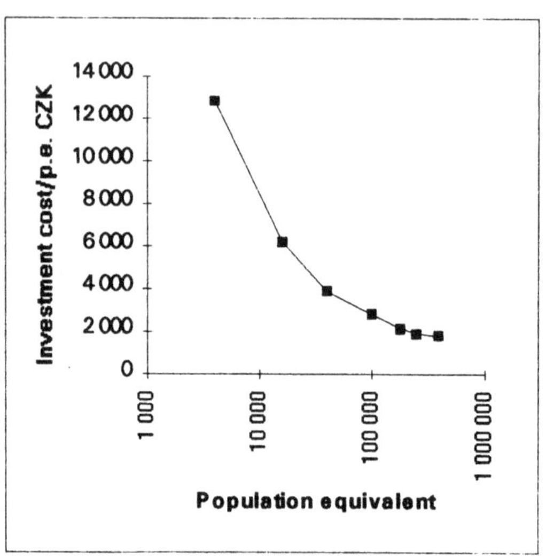

Figure 2: Investment costs in relation to the source size

Size of urban centers and population distribution according to the size of an urban area is shown in Table 1. Total population of the Czech Republic is 10.5 million, residing in 5 768 urban centers.

As can be seen from the table, 78% of all urban centers are smaller than 1000 inhabitants, with cumulative frequency of inhabitants of 16%. 95% of agglomerations are smaller than 5 000 inhabitants, with cumulative frequency of inhabitants of 35%. Providing appropriate wastewater treatment to small communities represents an important and costly program.

Operation costs depend on the plant size, type of process technology and on the extent of the plant exploitation. This is shown in Table 2 for one fraction of the operation cost, energy consumption (Kos, 1993).

Table 1. Urban areas and population distribution in the Czech Republic

Size category of agglomerations (inhabitants)	Frequency of agglomeration S	Cumulative frequency of agglomeration S	Frequency of inhabitants	Cumulative Frequency of inhabitants
thousands	%	%	%	%
>100	0.10	100.00	22.41	100.00
50 - 100	0.31	99.90	12.37	77.59
25 - 50	0.47	99.58	8.56	65.22
10 - 25	1.47	99.12	12.77	56.66
5 - 10	2.27	97.64	8.84	43.90
2.5 - 5	3.97	95.37	7.69	35.05
1 - 2.5	13.16	91.40	11.15	27.36
0.5 - 1	21.24	78.24	8.34	16.21
0.25 - 0.5	25.66	57.00	5.20	7.87
0.1 -.0.25	24.13	31.35	2.37	2.67
0.01 - 0.1	7.21	7.21	0.30	0.30

Table 2. Specific energy consumption in small treatment plants (5 000 PE)

Plant exploitation (% of nominal capacity)	20	40	60	80	100
Type of process technology	Specific electric energy consumption (KWh/kg BOD removed)				
Activated sludge process	6,5	3,75	2,5	2,0	1,4
Trickling filters	1,15	0,6	0,55	0,5	0,4
Rotating bio-contractors	3,25	1,75	1,25	1,0	1,0

If plant loading is far below the design capacity, especially for the activated sludge plants, the specific energy consumption increases significantly. The data applies to the best technology in the country. Older or less advanced activated sludge systems may need up to 4 kWh/kg BOD removed at full loading.

It can be concluded that equal or similar effluent standards for all size sources of pollution would result in enormous investment and operation costs at small plants. Also, using the activated sludge process for small plants is questionable. Less expensive technologies are urgently needed.

3. LESS EXPENSIVE TECHNOLOGIES

The terms "low-cost", "alternative", "appropriate", "adequate", "competitive", etc. wastewater treatment technologies are not well defined and their use is arbitrary. Research on less expensive technologies is rudimentary, conferences are infrequent and interest among the academics is sparse. In many countries, there is an urgent need to solve water quality problems at affordable costs. The direct link between money spent on research and development of less expensive technologies and the speed of pollution abatement seems to be unrecognized by national and international funding institutions. Mainstream, i.e., the northern hemisphere, developed countries technologies are frequently applied with little knowledge and recognition of local conditions.

There are certainly strong natural laws which have to be obeyed everywhere. In contrast, the idea of an optimum practical wastewater treatment technology *per se* and its general applicability is merely a fiction. Many experienced, independent international consultants indicate that reviews of unsuccessful projects reveal underestimation of country specific conditions and of their potential development and transformation during the project life-time.

3.1 Low-cost Treatment Plants

The term "low-cost treatment plants" is well known in the CEE countries. Since essentially everything was state owned, financed, approved and operated, there were numerous institutional means and steps in which every project was thoroughly reviewed and "economized" (down scaled).

The first concern was to eliminate equipment which would have to be imported from the hard currency regions. That frequently included instrumentation, automatic control, computers, special mechanical equipment, polymeric flocculants and a number of other commodities.

Second, all construction materials had to be local (i.e., from national production) or imported from the former COMECON countries (for instance, aluminum alloys and stainless steel can be hardly found at treatment plants).

Third, investment costs were cut, sometimes up to absurdity. The single exception was the requirement to design treatment plants for capacity expected 20 - 25 years after their commissioning. Unfortunately, much of the pre-investment could not be utilized for various reasons.

Many of the elements of low cost technology, as described above, can be found in serious recommendations of international consulting and financing institutions. Indeed, if implemented wisely and with appropriate construction quality control (which was not the case in the CEE countries), they can lead to inexpensive and functioning installations. It is also extremely important that the financial review and cost cutting ought not to be limited only to investment (capital) costs. Cost cutting should not be accomplished at the expense of later operation costs.

3.2 Alternative Process Technologies

The scope of this paper is limited to the prevailing needs of the CEE countries, i.e., to provide basic, mostly secondary treatment to most communities and industries. The design has to meet moderate effluent standards at moderate costs. This includes new as well as rehabilitated plants. Some examples of innovative thinking rather than a complex essay of the subject are presented.

Simple plant retrofitting

a) *Emscher tanks*

In the size category from 500 to 5 000 PE a number of older plants consists of simple pre-treatment (screens, grit chambers), Emscher (Imhoff) tanks, trickling filters and final clarifiers. Combined, i.e., primary and secondary, sludge is digested and stored (for 90 to 120 days) in the sludge compartment of the Emscher tank. The technological advantage of Emscher tanks is a relatively constant temperature in the sludge compartment throughout the year because they are generally built deep below the ground surface. Sludge is well digested and can be used in agriculture when desired by farmers. The Emscher tank technology was recently upgraded by using enamelled circular steel tanks surrounded by a ring-shaped aeration volume.

Two examples of Emscher tank retrofitting are shown in Figure 3. Part 1 is the conventional Emscher tank, Part 2 is the retrofitting proposed by Manczak in Poland years ago. Part 3 is retrofitting to an UASBR (Upflow Anaerobic Sludge Bed Reactor) by Batek (1993) in the Czech Republic.

Retrofitting for extended aeration requires a new sludge storage tank (bin, silo). The retrofitting to UASBR is used for industrial wastewater pre-treatment. There are several options for final treatment, either separate or in combination with communal sewage.

b) *Trickling filters*

Numerous trickling filters (with solid media) have been upgraded by adding a top layer of plastic media or replacing the solid media by the plastic one. In the USA, technologies called TF/AS (Trickling Filter/Activated Sludge) or similar RF/AS, BF/AS, TF/SC, ABF are widely used (Daigger and Buttz, 1992).

c) *Activated sludge systems*

Small and medium size extended aeration activated sludge systems without primary settling are extremely popular. As shown in Tab. 2, energy consumption can be high but until recently, energy use was not of much concern. Applications include oxidation ditches, other systems with mechanical aerators (brush type and vertical aerators), continuous flow and sequencing batch reactors. Several companies have been introducing control of aeration by a timer or by a microchip with remarkable savings in energy, amounting to 30 to 40% savings of the original energy consumption.

The systems without primary settling rarely suffer from sludge bulking. Also, they exhibit significant simultaneous denitrification. With the above mentioned aeration control and submersible low-speed mixers, denitrification is well controlled.

Several, rather unsuccessful, attempts have been made to extend this technology to larger plants. Two factors seem to be limiting:

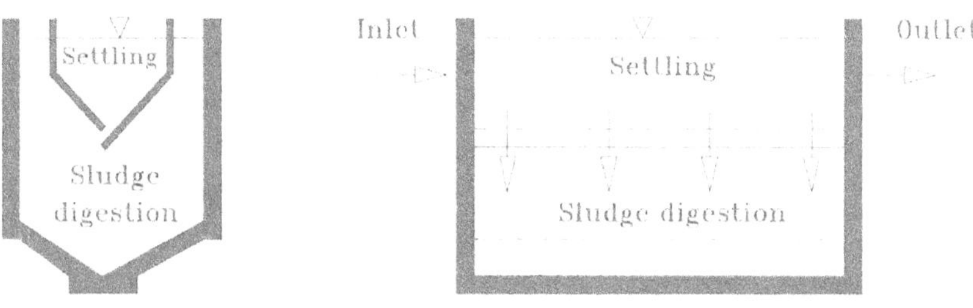

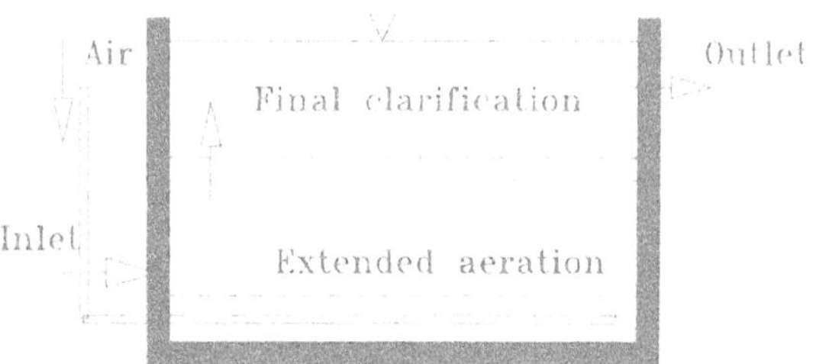

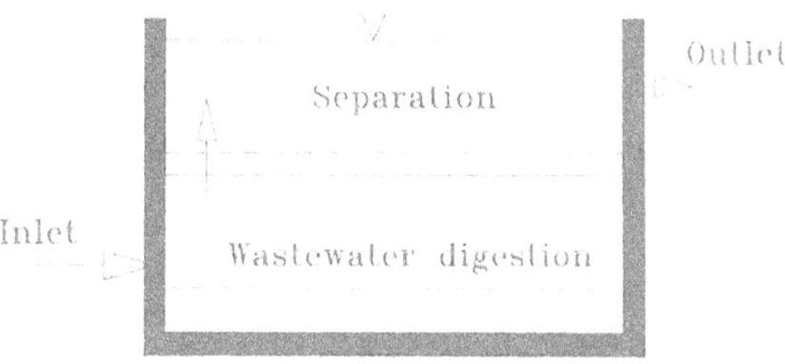

Figure 3. Two types of Emscher tank retrofitting

- Without primary settling, large quantities of grit may deposit in aeration tanks.
- If anaerobic sludge digestion is considered, another drawback appears. The combined sludge does not produce enough biogas to maintain the digestion system self-supporting in energy. Energy balance depends on several factors: start-up, aeration tank temperature, the degree of sludge thickening prior to digestion and on the digester parameters.

More information on similar systems is necessary.

Several new less expensive technologies have been submitted in new projects recently. Figure 4 shows a simplified sketch of the solution which from time to time has appeared in various countries.

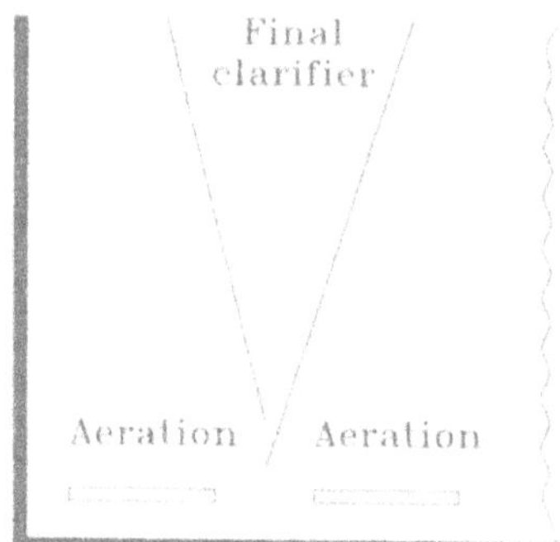

Figure 4: Inserted final clarifier

The idea depicted on Figure 4 is sound. Conventional clarifiers are designed for thickening of both the effluent and the recycled flow, Q_o+Q_r. The smaller is the recycled flow, the higher has to be the underflow (recycle) concentration and the larger clarifier area is required. Theoretically, inserted clarifiers can operate with less thickening and thus require a smaller area which is based on clarification and not on the thickening criteria. Practical implementation is not easy, mainly because of hydraulic problems. Performance of plants of this type will have to be carefully evaluated.

The RDN technology is popular at medium size and large plants in the Czech Republic. Several modifications of the basic configuration, shown in Figure 5, have been developed. The main advantages are: good sludge bulking control, increased aerobic sludge age in small tank volummes, smaller internal recirculation, good overall economy for moderate ammonia and nitrate standards (ammonia nitrogen 3 -5 mg/l, nitrate nitrogen depending on the BOD/N ratio in the wastewater).

The basic RDN configuration can be modified by using more than three compartments as shown in Table 3 wherein shaded areas indicate anoxic/anaerobic compartments. In all modifications derived from the RDN process, sections DS1 and DS2 are anoxic/anaerobic selectors. The internal cycle is always introduced into section DE2. The DE3 section can be operated flexibly, either oxic or anoxic.

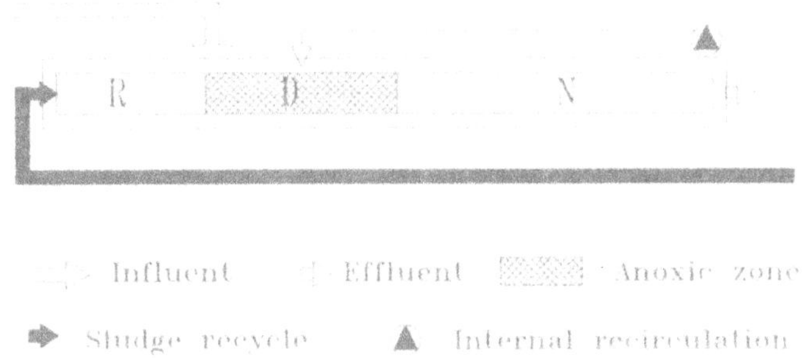

Figure 5: The RDN process

Table 3 Process alternatives operational in the same RDN tank configuration.

Volummes	m^3	Influent/Recirculation fractions				
Process		DRDN	RDN	AnRDN	RAnDN	MLE
DE1	7 500	0,2/0	0/0			1/1
R	15 000					
DS1	4 000	0,8/0	1/0	1/0	1/0	
DS2	3 500					
DE2	40 000	0/1	0/1	0/1	0/1	
DE3	20 000					flexible
Nitrification	120 000					
Total	210 000					

Table 4 shows the results of mathematical modeling (IAWQ Activated Sludge Model No.1, calibrated by long-term pilot plant operation) of five competitive processes for a plant with a maximum dry weather flow 4 m^3/s. Four variations of the RDN process are compared to the conventional pre-denitrification, MLE system. Though the target ammonia concentration (2 mg/l) is below the optimum of the standard RDN process, several modifications are superior to the conventional MLE process. Another significant feature is suppression of bulking which cannot be yet satisfactorily predicted by models. It is known from experiments that RDN and AnRDN are superior even in that respect. This example also indicates the trend to designing plants which can be operated flexibly, in order to cope with changing conditions, mainly temperature, flows and composition of wastewater.

Table 3 Results of mathematical modeling of several competitive processes

Process	Effluent NH_4-N	Effluent NO_3-N	Minimum HCO_3'	Internal recirculation	Nitrification volume
	mg/l	mg/l	mmol/l	%	m^3
DRDN	1,0	10,1	1,6	100	120 000
RDN	2,4	11,3	1,1	100	120 000
AnRDN	1,4	11,9	1,6	100	140 000
RAnDN	1,6	10,5	1,4	130	140 000
RAnDN	3,1	10,1	1,4	100	120 000
MLE	2,5	10,7	1,7	100	120 000
MLE	1,3	9,4	1,7	200	140 000

4. CONCLUSIONS

An appropriate regulative system for wastewater discharges forms the backbone of the water pollution abatement policy. The Czech system, adopted in 1991, introduced emission (effluent) standards supported by ambient (water quality) standards. Emission standards are a tool oriented towards basic clean-up of the country's rivers while ambient standards are meant to be the tool for final tuning of the needs above the basic clean- up. The system seems to be functioning well. Some observers indicate their fear of radicalization which could lead to modifications of the standards and adopting too stringent additional regulations, such as increased charges, etc.

Low cost technology is undoubtedly at the periphery of interests of the world scientific community. There seems to be a general feeling that low cost technology is something technically backward, not worth spending much energy. This is, of course, totally wrong. Not much technical skill is needed if money is superfluous. That has been the case for several countries where state and other subsidies of municipal treatment plants were abundantly available.

Urgent needs to improve water quality of the receiving water bodies and, at the same time, lack of finances in many countries will hopefully add some momentum to research and development on low cost technologies. Several examples of low cost wastewater treatment processes have been introduced. As long as funding in low cost technologies continues to be negligible and interest of the scientific community lukewarm, not much progress can be expected.

REFERENCES

Batek, J. (1993) Personal communication

BCEOM (French Engineering Consultants) (1992) Prefeasibility Study of the Oder/Odra River Basin - Component of the Baltic Sea Environment Programme. For the Commission of the European Communities and the European Investment Bank.

Bhamidimarri, R. (editor) (1988) Alternative Waste Treatment Systems. Elsevier Applied Science, London and New York.

Czech Statistical Office (1992) Results of the general census, Prague

Daigger, G.T. and Buttz, J.A. (1992) Upgrading wastewater treatment plants. Technomic Pub. Co., Lancaster, USA.

Fleckseder, H. (1993) Treatment of Urban Wastewaters in Central and Eastern Europe - What Criteria for Evidence to Select the Route to Go. Manuscript (private communication).

Helmer, R. (1987) " Socio-economic development levels and adequate regulatory policy for water quality management," Wat. Sci. Tech. 9:257-272.

HYDROPROJEKT (Czech Engineering Consultants) (1991) Assessment of the Financial and Capacities and Requirements for Construction of Municipal and Industrial Wastewaters Treatment Plants (in Czech). For the International Elbe Project.

Institute of Hydrology and Meteorology (1991). Water quality in streams, Prague 1991.

Kos, M. (1993) Personal communication.

Ministry of Forestry and Water (1986) Data on operation of municipal wastewater treatment plants. Prague

APPENDIX 1

Czech effluent standards

Indicators I

1. Sewage and municipal wastewaters

Population	BOD mg/l		COD mg/l		TSS mg/l		NH4-N mg/l		Ptot mg/l	
equivalents	2005	2005	2005	2005	2005	2005	200			
<50	80	40	-*	-	65	50	-	-	-	-
<500	60	50	-	-	55	40	-	-	-	-
<5 000	50	40	170	135	45	35	-	20	-	-
<25 000	45	35	150	120	35	30	25	15	-	5
<100 000	35	30	125	105	30	25	15	10	5	3
>100 000	30	25	110	90	25	20	10	5	3	1,5

Pollution load PE = 60g BOD/inh.d

All values are maximum values in an 8 h flow non proportional (equal volume of each sample) composite sample

* values not yet published
BOD analyses with nitrification inhibition

2. Industrial and special wastewaters

Contains a large set of emission standards for a number of industries.

Indicators II

Contains mainly verbal characterization of the desired stream water quality

Indicators III

Ambient - Water Quality Standards - a list of 55 priority pollutants.

Example of the first 15 indicators:

No.	Indicator	Units	Values for rivers with major drinking water uptakes	Values for other rivers
1.	Dissolved oxygen	mg/l	min. 6	min. 4
2.	BOD	mg/l	4	8
3.	COD permanganate	mg/l	8	20
4.	COD dichromate	mg/l	20	50
5.	Hydrogen sulphide and sulphides	mg/l	ND*	0.02
6.	Acidobasic reaction	pH	6.0 -8.5	6.0-9.0
7.	Total dissolved solids	mg/l	500	1000
8.	Total iron	mg/l	0.5	2.0
9.	Total manganese	mg/l	0.2	0.5
10.	Ammonia nitrogen	mg/l	0.5	2.5
11.	Free ammonia	mg/l	ND	0.5
12.	Nitrite nitrogen	mg/l	0.02	0.05
13.	Nitrate nitrogen	mg/l	3.4	11
14.	Organic nitrogen	mg/l	1.5	3.0
15.	Total phosphorous	mg/l	0.15	0.4

*ND - not detectable

CHAPTER 10

WASTEWATER TREATMENT PROCESS DEVELOPMENT IN CENTRAL AND EASTERN EUROPE - STRATEGIES FOR A STEPWISE DEVELOPMENT INVOLVING CHEMICAL AND BIOLOGICAL TREATMENT

Mogens Henze[1] and Hallvard Ødegaard[2]

1. INTRODUCTION

Wastewater management strategies are needed for Central and Eastern Europe. Economic limitations prohibit an immediate upgrading of all wastewater treatment to the standards of Western Europe and North America. Consequently, the choice of optimal wastewater management strategy then becomes very important. One can ask a question whether the historical development of wastewater treatment in Western Europe during the last 60 years is optimal for Eastern and Central Europe or whether other alternatives are feasible. Today a broad spectrum of wastewater treatment methods is available, which allows to remove pollutants to varying efficiencies. Treatment plants now being built in Western Europe are based on high removal efficiencies and have very high marginal costs per unit pollutant removed, as well as high overall costs per unit pollutant removed.

This paper will discuss the optimum use of the limited financial resources in Central and Eastern Europe. The plants to be built must be designed in a way that allows for cost-effective upgrading to higher removal efficiencies at a latter stage. In the distant future they will probably need to be upgraded to EEC effluent criteria standards. However, it has to be pointed out that the step-wise development of wastewater treatment has not been the optimal route in Western Europe, therefore, this chapter approaches the problem without assuming this *a priori*.

---

[1] Institute of Environmental Science and Engineering, Technical University of Denmark, Lyngby, Denmark

[2] Division of Hydraulic and Environmental Engineering, Norwegian Institute of Technology, Trondheim, Norway

NATO ASI Series, Partnership Sub-Series, 2. Environment – Vol. 3
Remediation and Management of Degraded River Basins
Edited by V. Novotny and L. Somlyódy

2. WASTEWATER MANAGEMENT

The choice of strategy for wastewater management will have to be influenced by the type of settlement that the wastewater handling system is to serve. Obviously the strategy for serving a small group of houses will have to be different from that of serving a metropolitan area. In order to focus the discussion, only strategies for cities with wastewater load greater or equal to 10,000 population equivalent (PE) will be dealt with in this paper. Cities in the population range of 10,000 -100,000 cover probably the major part of the Eastern and Central European population. Additionally, there are some metropolitan areas with treatment plants of more than 100,000 PE. In the subsequent sections we will discuss the strategic management of wastewater from cities (10,000-100,000 PE) and metropolitan areas (> 100,000 PE).

2.1 Basic Considerations

Wastewater collection system. The wastewater collection system in Eastern and Central European cities is normally the combined sewer system.

Treatment requirement. The tendency in Eastern and Central Europe is to use the EC effluent standards. When considering the different levels of technological development, this might not be optimal, neither from an economical nor from an ecological point of view.

Treatment technology. The treatment technology to be chosen will to a greater extent be governed by local conditions and costs, than by effluent standards since these will be more or less the same everywhere.

Industrial discharges. The influence of industrial discharges is substantial in city wastewater. For bigger cities, however, the industrial wastewater from one factory is mixed with the wastewater from many other factories and with domestic wastewater. The result is, that municipal wastewater has a reasonably stable composition from day to day.

Sludge handling. For many cities, sludge handling represents the greatest wastewater management problem. In some countries the possibility to use sludge on farmland is limited because of the abundance of cattle manure. Sludge may also have too high a content of heavy metals to be allowed on farmland. There is a trend towards more comprehensive sludge treatment, for instance drying or incineration of sludge, to a cost that may be higher than that of purifying the wastewater. In these situations the magnitude of the sludge production is a key factor and technologies that can reduce this should be asked for.

Treatment plant operation. Operation of city wastewater treatment plants is like operating process industries. Many cities have shown that there is a big potential for bringing the cost per m^3 of treated wastewater down by using optimal management strategies.

Environmental nuisances. In many cases, treatment plants for cities are now situated within living areas as a result of ever increasing expansion of the cities. The environmental nuisances (smell, noise, traffic, psychology, etc.) has to be considered as an important factor.

Energy consumption. The use of energy must be considered as a priority since energy is a key cost factor. The production of biogas from anaerobic stabilization and the use of heat pumps for energy recovery should be considered.

In wastewater management from very large metropolitan areas, the question is often how several treatment plants should be operated in relation to each other and the metropolitan wastewater collection system. It appears that treatment plants larger than 1 mill. PE are difficult to control. Large metropolitan areas ought to be divided into several smaller wastewater management areas - also in order to spread the risks of failure.

2.2 Wastewater Characteristics

Wastewater characteristics vary considerably from place to place. In order to establish a basis for discussions in the following parts of this paper, the composition shown in Table 1 is considered typical for the cases dealt with. The lower values in Table 1 represent a poor sewer system (combined system) or a high per capita water consumption, while the upper level represents a good separate sewer network.

During heavy rains, storm water dilutes the original sewage water and considerable amounts of this mixed wastewater reaches the receiving waters through combined sewer overflows. In cities with efficient wastewater treatment, combined sewer overflows represent the major pollutional load to the receiving waters. Planning for the reduction of the total load by analyzing storm flow patterns and by using real time control systems is one of the big challenges in wastewater management in Western Europe.

Table 1. Typical Raw Wastewater Characteristics for Eastern and Central European Cities

Component	Concentration g/m^3 (mg/l)	Component	Concentration g/m^3
SS	150-300	Total N	30-60
SS settleable	100-200	NH_4-N	20-40
COD total	350-700	Total P	5-12
COD soluble	150-300	PO_4-P	4-8
BOD total	150-300	Fecal coliforms	10^{11} (#/m^3)
BOD soluble	60-120		

Control systems for integrated operation of sewer systems and wastewater treatment plants are needed (Harremoës et al., 1993). Because of the situation described, the characteristics of the wastewater may change considerably from one city to another. The treatment strategy has to be based on these local characteristics.

3. WASTEWATER TREATMENT METHODS FOR EASTERN AND CENTRAL EUROPE

There are numerous wastewater treatment methods in use today. Some have a long history, others recent. The experience shows that many new methods will not stand up against time and after ten years or so they might be forgotten. This makes it a difficult task to come up with a set of relevant treatment methods for long term strategy as the one approached in this chapter. Some of the methods included in the subsequent analyses have limited practical use at present, but they have been included to give perspective, and because they are believed to have a future.

We shall shortly describe the wastewater treatment methods considered most relevant for the Eastern and Central European cities. Table 2 shows the expected effluent quality for the various treatment methods mentioned.

Table 2. Expected Effluent Quality for Wastewater Treatment Plants Larger than 10,000 PE in Eastern and Central Europe for the Various Treatment Methods Discussed.
(The effluent quality is based on a raw wastewater composition as shown in the Table 1)

Process	**BOD** g/m³	**Total N** g/m³	**Total P** g/m³	**Fecal Colif.** #/m³	**DO** g/m³
Raw wastewater	250	48	12	10^{11}	0
Mechanical	175	45	10	$3 * 10^{10}$	0
Chem high load	150	45	4	10^{8}	1
Chem low load	100	40	1.5	$2 * 10^{7}$	1
Biol high load	100	40	10	$4 * 10^{7}$	2
Biol low load	30	40	10	$2 * 10^{7}$	3
Biol/chem simultaneous precipitation	30	35	1.5	10^{7}	2
Biol/chem pre-prec.	15	35	1.5	10^{7}	2
N-removal activ. sludge	15	15	1.5	10^{7}	3
N-removal bio-film	10	12	1	10^{7}	5

3.1 Mechanical Treatment

Mechanical treatment (see Figure 1) involves normally settling of the coarser particles in the wastewater. In view of what is achieved by mechanical treatment (see Table 2), the cost/benefit ratio for such plants is high. It is therefore recommended, that mechanical treatment should only be erected in Eastern and Central Europe as the first step in development towards more advanced treatment plants.

3.2 Chemical Treatment

In this chapter we have chosen to differentiate between two types of chemical treatment of wastewater:

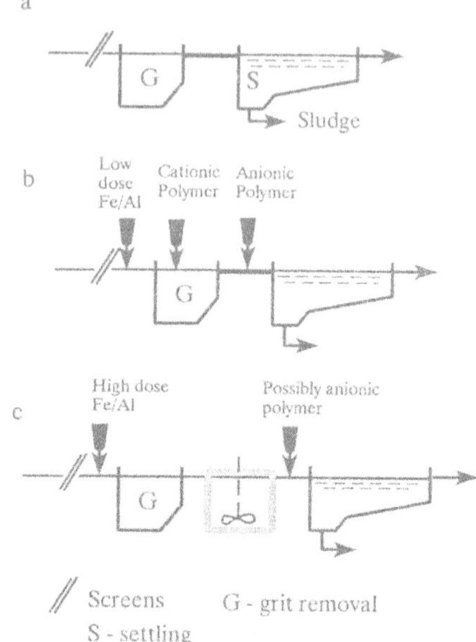

Figure 1.
Traditional mechanical and chemical treatment process units

a. Chemically enhanced mechanical treatment (called CEMT in the USA)

b. Primary precipitation (often called direct precipitation)

Chemically enhanced mechanical treatment plants are to a limited extent used in the USA and primarily in order to improve the treatment efficiency (for BOD and SS) of a mechanical plant.

Primary precipitation is used extensively in Scandinavia in order to remove phosphates in addition to organic matter. It is used either as the only treatment step, which is very common in Norway, or as a pre-precipitation step in a chemical/biological treatment plant. Far more data and experience are available with respect to primary precipitation than to chemically enhanced mechanical treatment. The latter is still under development, and its influence on the following treatment steps has not been investigated in detail. In Table 2, chemically enhanced mechanical treatment is referred to as high load chemical treatment while primary precipitation is referred to as low load chemical treatment.

3.2.1 Chemically enhanced mechanical treatment plants

In CEMT-plants (see Figure 1b), chemicals are primarily added in order to improve separation of suspended solids and enhance removal of organic matter. A metal coagulant is dosed prior to the existing mechanical settling tank (for instance in the grit chamber). Since the intention of the process is primarily to coagulate suspended matter, the dosage of metal coagulant is low (for instance < 50 g $FeCl_3/m^3$).

Flocculation is obtained by addition of an organic polymer after the metal coagulant. If insufficient coagulation is obtained with the low metal coagulant dosage, a cationic polymer has to be added as additional coagulant. As flocculant an anionic polymer is mostly used.

3.2.2 Primary precipitation plants

In primary precipitation plants (see Fig.1c) a metal coagulant is added prior to the flocculation tanks at a relatively high dosage (for instance, 150-250 g $FeCl_3/m^3$), since the main goal in this case is to remove phosphorous in addition to suspended organic matter. Flocculation is normally carried out by the use of mechanical stirring, but may be improved by the use of an anionic polymer as well. Floc separation is normally carried in settling tanks, but flotation is sometimes used.

A three-valence salt of iron or aluminum is normally used as a coagulant, in fewer cases lime can also be used. The commercially available coagulants are often mixtures of metal-chlorides and sulphates. Recently, pre-polymerized coagulants have been used to an increasing extent. The dosage of coagulant is normally controlled by the wastewater flow. In cases of low alkalinity wastewater dosage control based on wastewater flow with an overriding pH control, is used. Typical dosages are in the range of 100-200 g $Al_2(SO_4)_3/m^3$ or 150-250 g $FeCl_3/m^3$. Normally 2-4 flocculation tanks in series are used. An anionic polymer may be used as flocculant aid in cases where mechanical flocculation is unsatisfactory or an improved floc settling velocity is needed.

Sludge production in primary precipitation plants is almost a linear function of the coagulant dosage. When using aluminum sulphate, the average sludge production is about 0.25 kg SS/m^3 of water treated or about 30 kg SS/(PE*year) (Ødegaard, 1992). Energy consumption is very low in chemical plants, but the cost for chemicals may be considerable. Sludge from chemical treatment plants has a high organic matter content and, therefore, has a high potential for gas production. There is no indication that the aluminum or iron contents have any adverse effect on biogas production.

3.3 Biological Treatment

Biological treatment can be accomplished using many alternative unit processes, the two main ones being activated sludge (dispersed growth) and biofilm (attached growth) systems.

3.3.1 Activated sludge plants

These plants can be designed either for high or low loading (low and high sludge age - ranging from 1 to 30 days). The low loaded activated sludge plants are designed for nitrification (aerobic sludge age 15-25 days). Under very cold winter conditions, the nitrification efficiency may be partly reduced. The low loaded plants are often combined with chemical treatment through simultaneous precipitation (addition of a coagulant to the activated sludge tank). As a chemical treatment alternative, pre-precipitation can be used ahead of the activated sludge plant. The surplus sludge from these plants is stabilized in the activated sludge tanks, and digestion is not needed. High load activated sludge plants do not nitrify and do not stabilize the sludge, therefore, the surplus sludge must be stabilized separately. The high load processes need a closer process control than the low load processes.

3.3.2 Biofilm treatment plants

The biofilm plants can be designed as trickling filters, submerged aerated filter and rotating biological contactors. Most of the biofilm reactors require pretreatment in the form of primary settling or coagulation/precipitation. Biofilm processes will normally be designed for a medium to high organic load. A low load must be applied, however, if nitrification is needed. Biofilm reactors require less residence time than activated sludge reactors. This makes them more compact, but also more sensitive to peak pollutional loads in terms of effluent quality. Biofilm-based systems are, however, less sensitive than activated sludge systems to hydraulic peak loads with respect to sludge separation. Surplus sludge from biofilm processes is normally collected in a settling tank following the biofilm tank. Collected sludge is important with respect to the effluent quality, but not for the continued function of the process.

3.3.3 Biological/chemical treatment

Biological and chemical treatment processes may be combined using three basic alternatives: Pre-precipitation, simultaneous precipitation and post-precipitation (chemical treatment before, simultaneous to or after biological treatment).

3.4 Nitrogen Removal

Nitrogen removal can be performed both in activated sludge and biofilm processes. A prerequisite of biological denitrification is an organic carbon source, which can be applied either from the wastewater itself, or from added industrial chemicals like methanol or acetate. In some cases industrial wastes can be used for this purpose. There are many process alternatives for biological denitrification. Pre-denitrification is recommended when raw water contains a sufficient amount of easily biodegradable organic matter for the degree of denitrification needed. In this process, denitrification is placed ahead of nitrification, and a recycle- or an alternating process mode brings the nitrate from the nitrification step to the up-front denitrification step (recycle process or alternating process). When an external carbon source has to be used because of too little easily biodegradable organic matter in the raw wastewater, post-denitrification is the typical solution. In this process, nitrification in the first step is followed in series by denitrification as the second step. The two alternatives mentioned herein may be designed both for biofilm processes and for activated sludge processes. The choice between the two types will depend on the overall economy, and in some cases land space available, because the biofilm process has the smallest land space requirement. The surplus sludge has properties similar to traditional biological treatment with or without simultaneous precipitation.

Biological denitrification may also be used in order to improve the performance of (or prevent) problems in nitrifying treatment plants. Nitrification removes a considerable amount of alkalinity, but a denitrification process will theoretically restore half of it. The high energy (air) consumption coupled to nitrification, is also partly regained by denitrification and finally, rising sludge in activated sludge secondary settlers due to denitrification in the bottom of the settler can be avoided (Henze et al, 1993).

3.5 Sludge Handling and Disposal

Sludge handling at a treatment plant is strongly influenced by its size. Medium range treatment plants, 10.000 - 100.000 PE, will offer an increasing range of cost effective sludge treatment methods with increasing population load. A method which is too expensive for a small treatment plant, for example, sludge digestion, may become for bigger plants an economically attractive solution. For plants larger than 100.000 PE, all sludge treatment methods, inclusive heat treatment and incineration at the treatment plant, will be alternatives worth taking into account in the design of the plant.

4. SELECTION OF LOCAL WASTEWATER TREATMENT STRATEGY

The choice of wastewater treatment technology will have to depend on the type of settlement and on a series of local factors.

4.1 Regionally Based Effluent Criteria

To an increasing extent effluent criteria have been unified all over Europe. The effluent criteria given in the EC Directive on Municipal Wastewater Treatment (CEC, 1991) have been especially leading in this direction. Earlier, the basis for the criteria was the pollution situation of each individual receiving water as specified by the (ambient) water quality criteria. Many countries, even within the EC, still use this principle, but now in addition to the EC Directive (effluent) criteria. The most sound approach to the setting of effluent standards, is the one based on local receiving water quality. In practice, however, this approach is extremely time consuming and unreliable (see Chapter 1). Due to endless discussions, improvement of receiving water quality has been postponed for years or decades. The use of minimum standards, in areas needing immediate wastewater treatment action, is therefore recommended. Even if the EC Directive of Municipal Wastewater Treatment exists, we find important that for Central and Eastern Europe the effluent standards are regionally based. This is necessary if one is seeking the optimum treatment strategy from an economic and environmental point of view and is especially important in the transition phase, between the present situation and a future situation where EC standards may be adopted. There are also reasons to make the standards dependent upon the size of the plant. Experience shows that the smallest plants (on-site or village plants) do not reach the same effluent quality as the larger ones. The expected levels of effluent quality for the various treatment methods discussed later were given earlier in Table 2.

4.2 Sludge Disposal

Many of the pollutants that are removed during wastewater treatment, can be found in sludge. Depending upon the amount of industrial wastewaters that are mixed into the municipal sewage, sludge may contain undesirable high amounts of metals and organic micropollutants. In Table 3, typical ranges for metals in European wastewater sludges without heavy industrial load are given.

The possibilities for final disposal of sludge are of utmost importance for the choice of wastewater treatment strategy. There are only three final disposal possibilities:

a. Land-fill (disposal on land)
b. Use in agriculture or other use (use on land)
c. Incineration and disposal/reuse of ash (disposal in the air)

Table 3. Metal content (mg/kg DS) in a typical municipal sludge without heavy industrial load.

Component	Mechanical	Chemical	Biol/Chem	Biol/Chem
Cadmium	5-10	4-8	5-10	6-12
Copper	100-200	80-160	150-300	200-400
Chromium	50-100	30-60	75-150	100-200
Lead	200-300	150-300	175-350	200-400
Mercury	4-10	2-6	3-8	4-10
Nickel	30-80	20-60	20-60	30-100
Zinc	1500-3000	1000-2000	1500-3000	2000-4000
Nitrogen (g N/kg DS)	20-40	12-25	20-50	15-40
Phosphorous (g P/kg DS)	5-10	20-50	15-40	20-50
Potassium (g K/kg DS)	2-4	1.5-3	2-4	1.5-3
VDS/DS (%)	60-80	40-60	50-60	40-60

Different disposal possibilities require different sludge treatment before disposal. Land-fill dumping creates a considerable nuisance (odor, smell, secondary water pollution, etc) in many places, and the land available for such dumps is becoming increasingly more limited. If, however, such wasteland exists (closed mining areas for instance), land-fill dumping may be the optimal sludge disposal solution from an economical point of view, since pretreatment may be limited to sludge dewatering.

Use of sludge in agriculture or other uses as for instance top soil for green lawns (golf courses, finalization of solid waste dumps, high-way embankments, etc), is a much more sensible disposal method. Municipal wastewater sludge is a good soil conditioner in most cases, because of its humus content and its content of valuable nutrients (P and N). Limitations for the beneficial use of sludge are, however the content of undesirable components (i.e., metals, organic micropollutants and bacteria/viruses) and the availability of land on which the sludge may be disposed and/or used.

In many countries, the content of undesirable components has lead to general elimination of sludge use in agriculture. This is unfortunate from an ecological point. Sludge should be used on farmland, provided that the contaminant content of the sludge is well monitored and controlled and is within acceptable limits. It is possible to define these limits, and they should be set low in order to be sure that the use of sludge does not have any adverse effects on crops.

The knowledge about the content and consequence of organic micropollutants (CHC, PCB, PAH) is much less comprehensive than the knowledge about heavy metals. At this stage, however, there does not seem to be any severe threat from such substances as long as the discharge of industrial effluents to the municipal sewer that contain such components is controlled. Very seldom, raw effluents from chemical industries are co-treated with municipal sewage and this should be avoided if the sludge is to be used on farmland. Depending on which kind of agricultural use is considered, sludge should be disinfected in order to prevent microbial pollution by pathogenic microorganisms. In regions where there is an excess of organic waste, especially in the form of cattle manure, sludge use on farmland is difficult. There may be lack of land in the region, requiring soil improvement that the sludge may represent. This is especially so in metropolitan areas where the amount of sludge produced is far higher than the need for sludge on farmland within reasonable distances.

The third sludge disposal option, disposal in the air by sludge incineration, is therefore becoming more and more used in spite of the large cost associated with this disposal method. As demonstrated in Table 4, the final disposal method may dictate the sludge treatment method and hence also, to a certain extent, the wastewater treatment strategy.

4.3 Co-treatment with Industrial Wastewater

In most towns, local industries discharge their wastewater into municipal sewers and municipal wastewater treatment plants co-treat both municipal and industrial wastewater. This may be advantageous or disadvantageous, depending on the local circumstances. Industrial wastewater should be pretreated in such a way that the industrial wastewater does not harm the municipal wastewater treatment process. The discharge from food processing industries (dairies, breweries, slaughter-houses, etc) increases organic loading and leads to a requirement of larger tanks in a biological treatment plant for organic matter removal. In a treatment plant based on pre-denitrification, however, this extra organic matter may ease the nitrogen removal process. In a chemical treatment plant, the extra load of soluble organic matter will deteriorate the treatment efficiency, as well as requiring a higher chemical dosage.

Table 4. Relationship between final sludge disposal and sludge treatment method

Final disposal method	Sludge treatment methods
Landfill disposal	Thickening and dewatering (> 20 % DS)
Farm-land use	Thickening, stabilization, disinfection, dewatering (> 20 % DS)
Incineration	Thickening, dewatering, drying, incineration (> 30% DS)

4.4 Upgrading Existing Plants.

The lifetime of a wastewater treatment plant is seldom more than 20 years. The tanks, normally made of concrete, have, however, at least the double life time. Because of both ageing and use of new standards as well as because of increasing load, treatment plants have to be upgraded.

4.4.1 Upgrading mechanical plants

The easiest way of upgrading a mechanical plant is to introduce addition of coagulants, transferring the plant into a chemically enhanced or a primary precipitation plant. When planning for a future upgrading, one has to be aware of the need for dosing equipment, chemical mixing possibilities, introduction of flocculation tanks and sufficient area of the setting tanks. When using a metal coagulant alone, without a polymer as flocculant, the area of the settling tank in a chemical plant has to be bigger than in a mechanical plant. By proper use of polymer, however, the overflow rate in the upgraded chemical plant may be as high as in the original mechanical plant.

Primary precipitation plants will produce more sludge than the chemically enhanced mechanical plants, but the treatment results will be better and this type of plant may be expected to be easier to operate. When upgrading a mechanical plant in order to reach a BOD-removal higher than about 70%, a biological step has to be introduced. In most cases, this will imply extending the plant with a biological step.

4.4.2 Upgrading chemical plants

When a chemical plant is to be upgraded, it is either because the hydraulic load is too high or because the removal efficiency with respect to organic matter or nitrogen is too low. In the first case, many primary precipitation plants have successfully been upgraded by the introduction of a proper organic polymer as a flocculant, thereby increasing the acceptable hydraulic surface load of the setting tank.

When improved organic matter or nitrogen removal is the goal of the upgrading, an additional biological step has to be introduced. In this case, the existing chemical plant is used either as a pre-precipitation step or as a post-precipitation step. When using pre-precipitation, the size of the biological step will be considerably reduced as compared to a situation where post-precipitation is used.

This solution is recommended, therefore, when only improved removal of oxygen consuming matter is to be obtained. However, if N-removal is to be achieved, pre-precipitation changes the C/N-ratio in the raw water unfavorably. This means that an external carbon source has to be used in the pre-precipitation, post-denitrification alternative. This is economically favorable, when the wastewater is relatively dilute in the first place, but may be unfavorable in the case of a wastewater with a high carbon/nitrogen ratio.

When upgrading a chemical plant to a biological/chemical plant based on post-precipitation, biofilm processes may have an advantage, since there will be no need for an extra separation step between the biofilm reactor and the post-precipitation step.

4.4.3 Upgrading biological plants

When a biological treatment plant is to be upgraded, it is normally to include nutrient removal (P and/or N). The simplest way of including P-removal in a traditional mechanical/biological treatment plant, is to introduce pre-precipitation or simultaneous precipitation. Pre-precipitation reduces the organic load on the biological step considerably, normally sufficiently to achieve nitrification. Depending upon the existing volume of the biological tanks, even full N-removal might be achieved by the introduction of a single sludge post-denitrification step. When simultaneous precipitation is chosen, the sludge age will be reduced and process conditions for nitrification precipitation may be used also when the existing biological plant is based on a biofilm process.

4.5 Operation and Sensitivity to Failures in External Supplies (Energy, Chemicals, etc)

Chemical treatment plants require considerably more attention than mechanical ones. The most important operational issue, is control of the coagulant dosage. If the dosage of coagulant fails, the chemical plant is reduced to a mechanical one. It is therefore of utmost importance that the supply of coagulants and flocculants is reliable and that the chemicals are correctly dosed. Chemical treatment plants recover very quickly (within 2 hours) from operational problems. Biological plants require normally less control equipment than chemical plants. On the other hand, the biological process itself may be more difficult to operate. This is especially true for activated sludge plants, where separation of the activated sludge is the critical part. Both chemical and biological plants are vulnerable to energy supply failures, chemical plants because of the dosing pumps, and activated sludge plants because of air supply as well as pumps. The consequence of a major energy supply failure (say a day or two) is much more severe in biological plants, because the biomass will deteriorate. It may take weeks to restore the operation back to what it was before the failure. In a chemical plant such a failure will only represent a problem as long as the failure continues.

4.6 Area Requirements

Chemical plants are compact and require only about 5-10 % more space than a mechanical plant. Biological plants based on activated sludge require 2-3 times the area required by a mechanical plant. This makes it cheaper to cover chemical plants than biological ones. Over the last years, very compact biological processes based on biofilm reactors have been developed, and the combination of chemical precipitation and biofilm processes may have area requirements for nutrient removal (N and P) which for the whole plant is less than 40 percent of the area needed for a process based on activated sludge.

5. ECONOMY

In order to select the optimal wastewater treatment strategy, the overall economy for the various processes has to be considered. Some cost calculations are presented below. They are estimates which, however, allow for comparison between the treatment methods. It is emphasized that the actual costs may and will vary from one location to another location. The data presented here are based on cost analyses carried out in Scandinavia and Central Europe. Typical process combinations of particular interest to the development of wastewater treatment in Eastern and Central Europe, have been evaluated.

All the cost data presented here have been treated in a similar way. It is emphasized that one should not take out single cost elements and discuss their magnitude, without discussing all the other elements in the analysis as a whole. Cost data may be misinterpreted if the basis for the cost evaluation is not well understood, and therefore a brief explanation of this basis shall be given.

5.1 Basis for Evaluation

Table 5 gives the specific load and wastewater composition assumed in the following considerations:

The cost data given, are related to design flows that are equal to: Qdesign = k* Quality average, where the k-factor varies with the size of the plant.

In the cost analyses data for a treatment plant of 100.000 PE, k=1.2 has been used. In the cost data tables (Tables 7 and 8), the raw wastewater composition is assumed to be related to an average specific hydraulic load of about 400 *l*/PE*d (last column in Table 5). This is a typical average total daily specific load in Central and Eastern Europe used for design purposes which also includes infiltration water (for instance as the sum of 250 *l*/PE*d wastewater and 150 *l*/PE*d total infiltration).

Table 5. Wastewater specific load and composition used for cost calculations.

	Specific load g/ (PE*d)	Wastewater composition, g/m^3 at 250 *l*/PE*d	at 400 *l*/PE*d
BOD_5	62.5	250	150
SS	62.5	250	150
Tot P	3.0	12	7.5
Tot N	12.0	48	30

Five different groups of process combinations have been considered:

1. Mechanical treatment
2. Chemical treatment

3. Biological treatment
4. Biological/Chemical treatment without N-removal
5. Biological/Chemical treatment with N-removal

For each of the groups (except for that of mechanical treatment), two process situations have been considered (for example high and low treatment efficiency). The different process solutions have been chosen because they are considered to be interesting for Eastern and Central Europe. Table 6 shows the treatment efficiencies that may be obtained with the different process alternatives.

Table 6. Effluent concentrations for various treatment alternatives (Wastewater: 400 l/(PE*d))

Treatment process	BOD_5 g/m³	SS g/m³	Tot P g/m³	Tot N g/m³	Sludge produced g/m³
Raw wastewater	150	150	7.5	30	-
Mechanical	105	60	6.5	25	75
Chemical					
a. CEMT[1]	75	30	2.5	23	200
b. Prim prec[2]	45	15	0.8	21	300
Biological					
a. High load[3]	45	30	5.3	23	125
b. Low load[4]	20	15	5.3	21	125
Biol/chemical					
a. Sim prec[5]	20	20	1.0	20	160
b. Pre-prec[6]	10	15	0.5	20	330
Biol/chem/N-rem					
a. Predenitri/ sim prec[7]	10	10	1.0	10	180
b. Postdenit/ pre-prec[8]	5	10	0.5	5	330

[1] Chemically enhanced mechanical treatment (Chemical high load)
[2] Traditional primary precipitation (Chemical low load)
[3] Biological high load - Activated sludge with sludge load = 0.5 kg BOD_5/kg SS*d
[4] Biological normal load - Activated sludge with sludge load = 0.2 kg BOD_5/kg SS*d
[5] Biological/chemical - Simultaneous precipitation in normally loaded activated sludge plant
[6] Biological/chemical - Pre-precipitation followed by normally loaded activated sludge plant
[7] Biol./chem. incl. N-removal - Predenitr./simult.precip. in activated sludge plant (sludge age of 13 days)
[8] Biol./chem. incl. N-removal - Pre-precip. followed by biofilm process with post-denitr. (external C- source)

Design criteria used for wastewater treatment processes are based on common European criteria, and can be found elsewhere (Ødegaard and Henze, 1992). Three alternatives for sludge handling have been economically evaluated:

a. Dewatering only

b. Anaerobic stabilization and dewatering

c. Dewatering and incineration

5.2 Cost Calculations for Wastewater

In Table 7 the unit cost (US $/m^3) is given as the sum of capitalized unit cost (12% interest, 20 year life time of equipment) and operation and maintenance cost, for treatment plants of 100,000 PE. The costs are given for a specific hydraulic load of 400 *l*/PE*d, similar to the wastewater composition assumed in

Table 7. Costs of wastewater treatment plants (100,000 PE)
(Capitalized, US $/m^3, Q = 400 *l*/PE*d)

Treatment process	Capital cost	O & M cost	Total
Mechanical	0.110	0.050	0.160
Chemical			
a. CEMT[1)]	0.117	0.075	0.192
b. Primary precip[2)]	0.133	0.083	0.217
Biological			
a. High load[3)]	0.150	0.083	0.233
b. Low load[4)]	0.180	0.083	0.263
Biological/chemical			
a. Sim. precip[5)]	0.150	0.108	0.258
b. Pre-precip[6)]	0.167	0.125	0.292
Biological/chemical, N-removal			
a. Pre-denitrification /sim.precip[7)]	0.283	0.158	0.441
b. Post-denitrification /pre-precip[8)]	0.225	0.175	0.400

1) Chemically enhance mechanical treatment (Chemical high load)
2) Traditional primary precipitation (Chemical low load)
3) Biological high load - Activated sludge with sludge load = 0.5 kg BOD_5/kg SS*d
4) Biological normal load - Activated sludge with sludge load = 0.2 kg BOD_5/kg SS*d
5) Biological/chemical - Simultaneous precipitation in normally loaded activated sludge plant
6) Biological/chemical - Pre-precipitation followed by normally loaded activated sludge plant
7) Biol./chem incl. N-removal - Pre-denitr./simult. precip. in activated sludge plant (sludge age of 13 days)
8) Biol./chem. incl. N-removal - Pre-precip. followed by biofilm process with post-denitr. (external C- source)

Table 6. The costs given in Table 7, do not include sludge handling, only sludge storage (thickeners or aerated storage depending upon the size of the plant).

6. STEP-WISE DEVELOPMENT OF WASTEWATER TREATMENT PLANTS

Treatment plants may be developed in a step-wise manner. This may be particularly important for Eastern and Central Europe, since in many cases it may be impossible from an economical point of view to construct and operate treatment plants according to EC standards within a foreseeable future. Typically, three situations dominate in Eastern and Central Europe today:

a. Mechanical treatment is in operation
b. Biological treatment (typically based on activated sludge process) is in operation.
c. No treatment facility is in operation.

In the following section, scenarios based on these three initial situations will be briefly discussed.

6.1 Unit Cost for Removal of Pollutants by Various Processes

When establishing a strategy for step-wise development of wastewater treatment in a region, the unit costs for removal of the pollutants are an important factor. In Table 8 unit costs have been calculated to illustrate how they change with the process and the degree of removal. It can be seen from this table that the marginal costs are high for removal of phosphorus and nitrogen. When building a new treatment plant with no existing one in place, several possibilities exist. When upgrading mechanical treatment plants, primary precipitation offers rather cheap removal of organic matter. The figures from the table will be used in the discussion below of the various possibilities for step-wise development of wastewater treatment, illustrated in Figure 2 a-k.

6.2 Step-wise Development of an Existing Mechanical Treatment Plant for 100,000 PE.

In Figure 2a the existing plant is shown. The first step will be to upgrade the plant to chemical treatment, either as chemically enhanced mechanical treatment (see Figure 2b) or as primary precipitation (see Figure 2c). By the use of organic flocculants, the sludge settling rate will be high enough to avoid expansion of the primary tank surface area. Both plant types will have low additional investment, as seen in Table 7, but the operational cost will increase due to the chemicals added. If phosphorous removal is a high priority, a chemical primary precipitation plant should be chosen.

Table 8. Unit costs in US $/Kg of pollutant for step-wise development of wastewater treatment plants. (Numbers in parentheses are the percentage of the total cost attributed to the removal of the given pollutant).

Step:	BOD	Phosphorus	Nitrogen
No → Mechanical	4 (100)	-	-
No → Biological, high load	2 (100)	-	-
No → Chemically enhanced	1.3 (50)	19 (50)	-
No → Primary precipitation	1.0 (50)	15 (50)	-
Mechanical → Primary precipitation	0.4 (50)	4 (50)	-
Mechanical → Biological, high load	1 (100)	-	-
Biological → Biological/chemical	0.3 (5)	10 (95)	-
Biological/chemical → Nitrogen removal	-	-	10 (100)

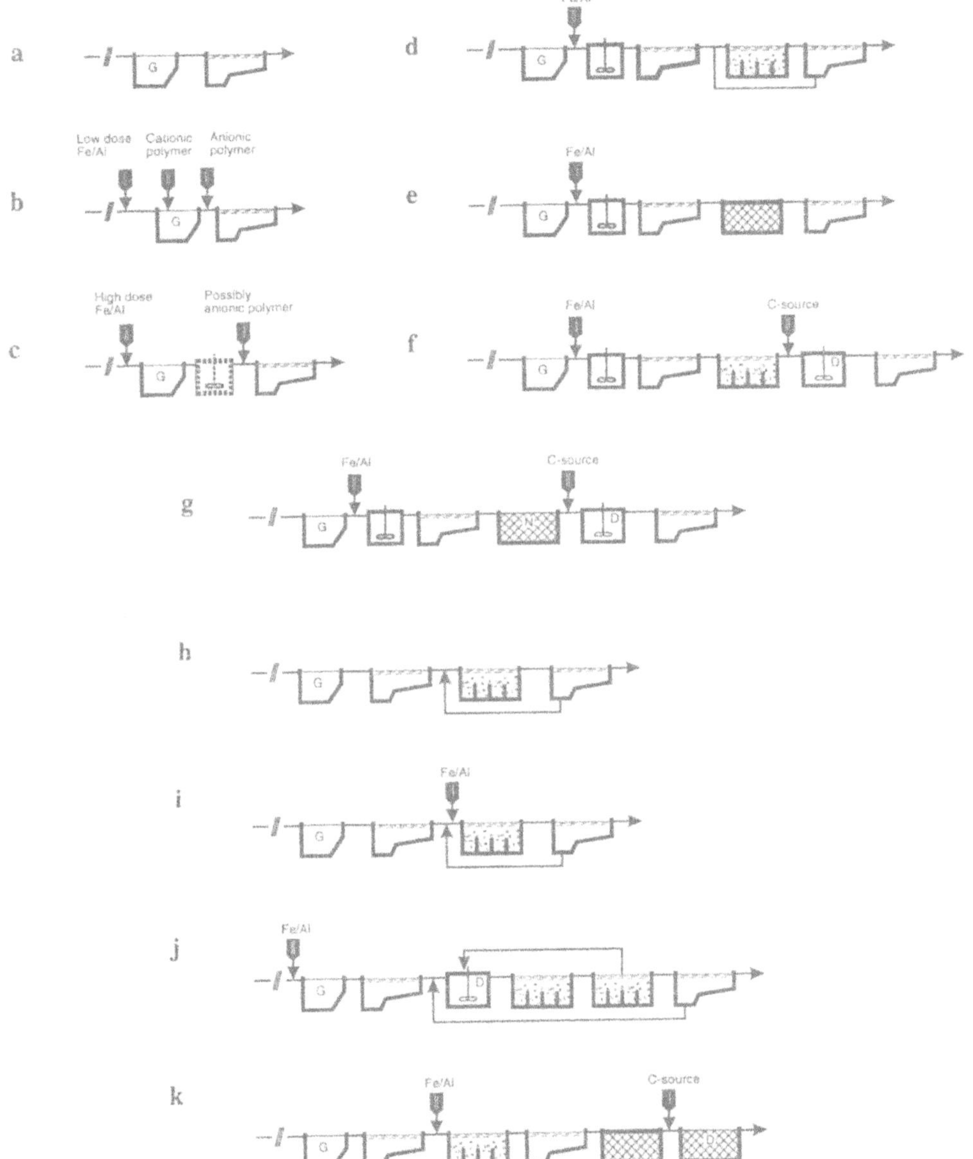

Figure 2 Upgrading of wastewater treatment plants

It is important that only coagulants with a low heavy metal content are used in order to prevent excessive heavy metal concentrations in the sludge.

If better removal of oxygen-consuming matter (BOD and NH_4^+) is required due to local receiving water conditions, a biological treatment step has to be introduced in addition to the chemical one. This may most conveniently be done by the introduction of a biological step after the chemical one (pre-precipitation). Both activated sludge (Figure 2d) and biofilm processes (Figure 2e) may be used for the biological step.

Upgrading from chemical/biological to chemical/biological including N-removal is the final step in order to reach advanced standards, as those in Table 2 or the present EC standard. If an activated sludge process exists, a one-sludge pre- or post-denitrification process would be recommended (Figure 2f). A carbon source would have to be added, either as an external carbon source (methanol, ethanol, organic industry waste, etc.) or as hydrolysate made from processing the sludge from the pre-precipitation step. If a biofilm process exists, a two-sludge post-denitrification system based on separate biofilm reactors for nitrification and denitrification would have to be recommended (Figure 2g).

6.3 Step-wise Development from an Existing Biological Treatment Pant for 100,00 PE

An existing biological plant can be with or without primary settling. In the case of Eastern and Central Europe most plants will have primary settling, as shown in Fig. 2h. Phosphorus removal can be included by either pre-precipitation (Figure 2i). Pre-precipitation will, in addition to removal of phosphorus, also reduce the organic load to the activated sludge step. In cases where overload of the activated sludge process is a problem, pre-precipitation is an obvious solution. Sludge production will increase significantly by this alternative.

Simultaneous precipitation, Fig. 2i, will be an alternative, either when the activated sludge treatment plant can accommodate the extra sludge production, or when the activated sludge tanks are expanded.

Upgrading from biological/chemical to nitrogen removal can be accomplished with a biofilm process or a process based on activated sludge. An activated sludge process will mean that a pre-denitrification tank will have to be added in front of the existing activated sludge tank, see Fig. 2j. If nitrification was not included in the existing plant, then the activated sludge tank has to be increased in volume by a factor of 2-3. Depending on the composition of the wastewater, there might be a need for adding an extra carbon source to the denitrification process.

The biofilm alternative is shown in Fig. 2k for a case where simultaneous precipitation exists. If the existing plant does not include nitrification, then a nitrifying and a denitrifying filter can be added. In this case a carbon source is needed for the denitrification process. The nitrogen removal part of the process will be unchanged if the existing plant has pre-precipitation.

6.4 Step-wise Development from no Treatment to Efficient Removal of BOD, Phosphorus and Nitrogen

The situation of no existing treatment allows for an optimal step-wise development of the treatment plant. There stages for the development would be as follows.

Stage 1: No treatment to chemical treatment.

From Table 6 it is seen that primary precipitation (chemical, low load) and high load biological treatment yield the same effluent quality with respect to BOD. In addition, primary precipitation removes phosphorus efficiently. The total cost effectiveness is higher for precipitation than for high load biological treatment as seen in Table 9. Thus primary precipitation (Figure 2c) will be the first stage in a step-wise development. Alternatively, chemically enhanced mechanical treatment could be used but this does not seem to be the optimal first step in a step-wise treatment plant scenario. The reason is that this process is as complicated as primary precipitation and it is more expensive per kg of BOD and phosphorus removed.

Stage 2: Chemical to chemical/biological treatment.

The primary precipitation plant can be upgraded with biological treatment, either high or low loaded, depending on whether nitrification is needed. This plant lay-out is shown in Figure 2d and 2e for activated sludge and biofilm processes respectively.

Stage 3: Chemical/biological treatment to chemical/biological with nitrogen removal.

The upgrading of this process follows the pattern discussed earlier and shown in Fig. 2f and 2g.

7. DISCUSSION

Optimal wastewater development must be based on local factors. This means that general recommendations given here might be changed due to local conditions. The approach using unit costs per kg of pollutant removed as a guide for the development of a strategy is healthy on a regional basis.

The fact that identical amounts of organic matter removed in chemical and biological treatment processes are of a different nature, should neither be forgotten nor misused in selecting the best development strategy. The organic matter in the effluent from a chemical treatment process degrades at a faster rate than that from a conventional biological treatment process, resulting in faster oxygen depletion in the receiving water bodies receiving chemically treated effluents. On an overall basis both effluents will give identical total oxygen losses in the receiving waters. For many large rivers in Central and Eastern Europe, the difference in oxygen depletion rate has no impact on water quality, but for smaller rivers it might influence the quality and thus the selection of optimal treatment alternative. The analysis made here illustrates, that the historical development of wastewater treatment in the 20th century is not the natural optimal step-wise development today. Existence of many alternative treatment methods allows for more cost-effective management in the future development of wastewater treatment in Eastern and Central Europe.

8. CONCLUSIONS

Based on the results presented in this chapter the following general conclusions can be drawn with respect to the strategy of wastewater treatment plant development in Central and Eastern Europe:

- Strategy for development should not just mirror the one used during the last century in Western Europe and North America.

- It is recommended to build many wastewater treatment plants with medium removal efficiency rather than a few plants with high removal efficiency.

- The new generation of wastewater treatment plants in Eastern and Central Europe should at first focus on removal of organic matter and phosphorus.

- Effluent criteria should be based on local or regional conditions and, for the new generation of Central and Eastern Europe treatment plants, should not mimic those presently applied in the EC. Less stringent criteria should be established for a specific transitional period.

- Removal of nitrogen from wastewater is very expensive and has in most cases only minor influence on the quality of the receiving waters as long as there is no overall efficient removal of organic matter and phosphorus.

- Treatment plants should be planned for the possibility of convenient upgrading for nitrogen removal at a later stage.

- The upgrading strategy should be as follows:

  No treatment exists. Primary precipitation is the most cost effective first treatment step in cases where no treatment exists or in cases where an unfinished mechanical or mechanical/biological treatment plant exists. A chemical treatment plant can, without loss of investment, be upgraded to a chemical/biological treatment plant with or without nitrogen removal.

  Mechanical treatment exists. Existing primary treatment plants can be upgraded to chemical treatment plants and then later be further upgraded.

  Biological treatment exists. Existing biological treatment plants can be step-wise upgraded with pre-precipitation or simultaneous precipitation and finally with nitrogen removal.

- By upgrading and building of new treatment plants, the amounts of sludge will increase considerably. The sludge should preferably be used as a soil conditioner and fertilizer in agriculture. However, this requires good control with the metal content. Some of the sludge will contain excessive metal concentrations to allow for agricultural use. This sludge must be disposed at landfills or incinerated.

- In order to reduce the metal content in wastewater sludge, source control must be established for industrial effluents. The experience from Western Europe is that it may take many years before the metal content is reduced to a level that will allow for agricultural use of the sludge.

Nevertheless, this strategy should be followed in order to improve sludge quality in the long run.

- o The wastewater treatment processes recommended herein are not high-tech processes when compared to even small industrial enterprises. In spite of this, all efforts should be made to select uncomplicated process alternatives for wastewater as well as for sludge treatment.

- o It is recommended to develop wastewater treatment strategies on a regional basis. This will be cost effective and will make it possible to build up regional expertise and production of hardware for the treatment processes. Singular treatment plants of star quality will be a waste of money and will not develop a local expertise and a local wastewater industry.

9. SUMMARY

Wastewater management in Central and Eastern Europe is a choice between two basic alternatives. One is to build a limited number of high efficiency treatment plants and the other to build a larger number of plants with less efficiency. This paper discusses these options based on technical and economical considerations. The conclusion is that a step-wise development with chemical precipitation as the first step and biological treatment as the second step, in general is the recommendable solution.

REFERENCES

CEC (1991) Council of European Communities. Directive concerning urban wastewater treatment (91/271/EEC). Official Journal L135/40

Harremoës, P., Capodaglio, A. G., Hellström, B. G., Henze, M., Jensen, K. N., Lynggaard-Jensen, A., Otterpohl, R. and Søeberg, H. (1993) "Wastewater treatment plants under transient loading performance, modelling and control," Wat.Sci.Tech. 27(12):71-115.

Henze, M., Dupont, R., Grau, P. and de la Sota, A. (1993) "Rising sludge in secondary settlers due to denitrification," Wat.Res. 27:231-236.

Henze, M. and Ødegaard, H. (1994) "An analysis of wastewater treatment strategies for Eastern and Central Europe," Pap. presented at the Biennial Conference of IAWQ, Budapest, Hungary also publ. in Wat.Sci.Tech.

Ødegaard, H. (1992) "Norwegian experiences with chemical treatment of raw wastewater," Wat.Sci.Tech. **25**(12):255-264.

Ødegaard, H. and Henze, M. (1992) "Evaluation of alternative municipal wastewater management strategies," World Bank Report, Copenhagen, December 1992.1

CHAPTER 11

FACTORS AFFECTING WATER QUALITY OF (LARGE) RIVERS -PAST EXPERIENCES AND FUTURE OUTLOOK

Part I: Present Views and State-of-the-Art

Hermann H. Hahn[1] and Neithard Müller[2]

1. INTRODUCTION

1.1 Water Quality Changes of Surface Water Systems as Described in Water Quality Modeling

Man tends to generalize specific and real time experience into abstract reports and formulae. One such instrument for generalization is the formulation of mathematical models to reproduce observations and to extrapolate such observations into other frames of time and location. Effects of pollution and other factors upon water quality of (large) rivers are such examples, i.e., observations are so numerous and, if not reported in a more generalized form, the information becomes so heterogeneous if not contradictory, that it might be difficult to formulate a consistent concept of quality changing factors and of the processes which they initiate. Therefore, even very early quantitative attempts of water quality assessment and control have used the instrument of mathematical modeling to describe dominant factors that control or change river water quality and to quantify their effects.

It seems useful, therefore, to discuss the changes in water quality of larger river system and their potential controls, in those terms that are used in water quality modeling for such systems. This appears more appropriate since other contributions in this book also address water quality modeling.

---

[1] Institut für Siedlungswasserwirtschaft, Universität Karlsruhe, Germany

[2] Weber Engineering, Pforzheim, Germany

NATO ASI Series, Partnership Sub-Series, 2. Environment – Vol. 3
Remediation and Management of Degraded River Basins
Edited by V. Novotny and L. Somlyódy

2. PRESENT VIEW - THE SITUATION OF THE RIVER NECKAR AND ITS RELEVANCE FOR OTHER RIVER SYSTEMS

2.1 The Neckar System

Neckar hydrology and hydrography: The river Neckar is located in the center of the State of Baden-Württemberg. Its headwaters are in the South-East slopes of the Black Forest, making this water body in the upper third of its course a more rapidly moving strea and, therefore, well aerated hill-region type of river. Then it enters the heavily settled and industrialized area of Central Württemberg where it was channelized and converted into a Federal waterway. It continues as a slow moving stream to its confluence with the Rhine River at Mannheim. Half of the Baden-Württemberg population (about 3 million inhabitants) resides in the drainage area of the river. Most of its drainage area is in the leeward direction of the Black Forest, which explains the low average precipitation of this region. The total length of the river is about 300 km, which also includes a larger stretch for the old river bed and a slightly shorter one for the parallel Federal waterway. Its average discharge at the last free river gauging station, before the begining of the system of locks is roughly 40 m^3/s. The ratio of a defined floodwater (Q_{95}) to a defined low flow (Q_5) is significantly larger than that of other German rivers.

Predominant uses: The previously mentioned high density of population and industrialization of the Neckar's drainage area makes it one of the most intensively and heavily used rivers. The uses of the river include:

- Navigation as the Federal waterway with large annual tonnage,
- Cooling water extractions for numerous conventional and some nuclear power plants,
- Water supply for industry and stand-by water supply for municipalities,
- Recipient for wastewater from all inhabitants and industries in the area[3],
- Irrigation water for agriculture and
- Recreational uses such as fishing and boating.

Due to early discernible users' conflicts, resulting in a degradation of water quality, the potable water is now supplied to the largest degree possible from elsewhere: the most significant amount is

[3] This includes numerous combined sewer overflows from municipal sewer systems into the Neckar.

delivered from Lake Constance and to a lesser degree from the Danube and the Upper Rhine rivers. Decreasing intensities of use can be observed in the area of water supply while there is an increase in the "non-consumptive" recreational uses (compare also next paragraph on changes in water quality).

2.2 Past and Present Water Quality

It is not surprising that the intensive and multidimensional use of this river system has brought significant stresses for all ecological functions. Very early, readily noticeable and visible pollution could be registered. Then, roughly coinciding with the industrial recovery after World War II, alarming deficits in the oxygen balance could be seen - initially indirectly through repeated severe events of fish-kill. (Critical oxygen deficits as reason for these fish - kills were later confirmed by analyzing river oxygen concentrations both by sampling and laboratory measurements and by on-line electrochemical instruments - Figure 1). This means that Dissolved Organic Carbon (DOC) as well as reduced nitrogen compounds were identified as critical loading parameters. Sources responsible for such discharges inlcude municipal sewer systems, municipal treatment plants and combined sewer overflows. Relatively early the water quality of the river system was also evaluated in terms of the quality of river sediments. Sediments analyses showed two major problem areas:

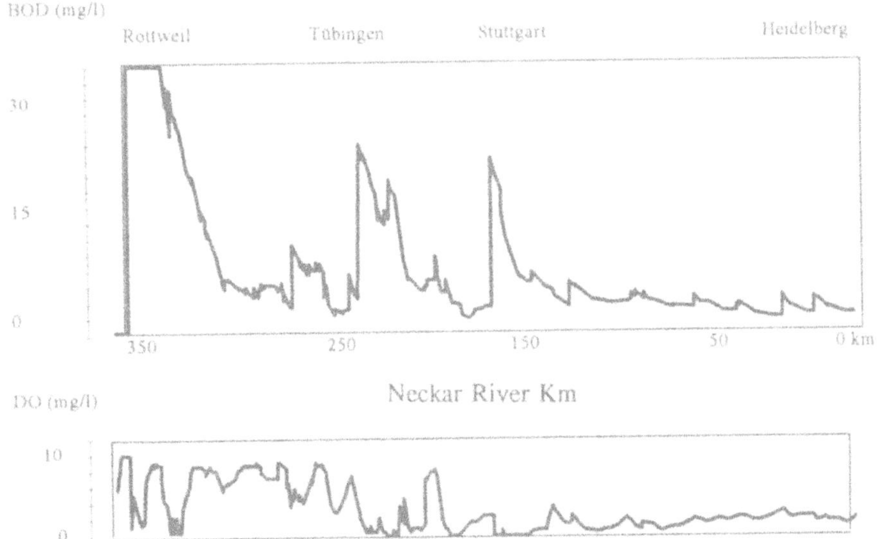

Fugure 1. BOD and DO profilrs of the Neckar River for least favorable boundary conditions - 1975 loads

(1) elevated sediment accumulation (from surface erosion and combined sewer overflows), and

(2) sediment pollution predominantly with reduced nitrogen compounds (less permanent) and heavy metals (permanent and severe - Figure 2). The latter pollution phenomenon was attributed foremost to industrial discharges.

Today the directly visible and indirectly registered pollution, i.e., pollution that causes stress to the oxygen balance, have been reduced to a rather reassuring level (Figure 3 for the oxygen deficit and Figure 4 for the heavy metals loading of the sediments). Analyses of bi-monthly collected samples and results from strategically placed monitors for pH, pO_2, T, conductivity and turbidity show that all acute pollutants are reduced below the levels that are perceived to be critical. Accumulation, long-term and long-range pollutants are not yet assessed regarding their ultimate effects.

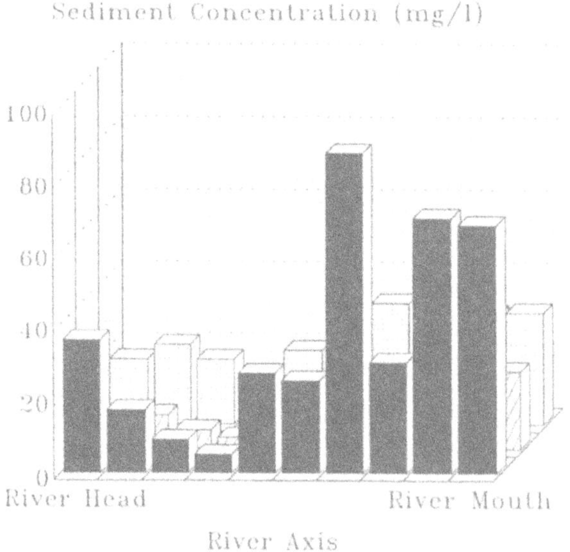

Figure 2. Heavy metals in Neckar sediments in 1970-1973 (after Förstner and Müller, 1979)

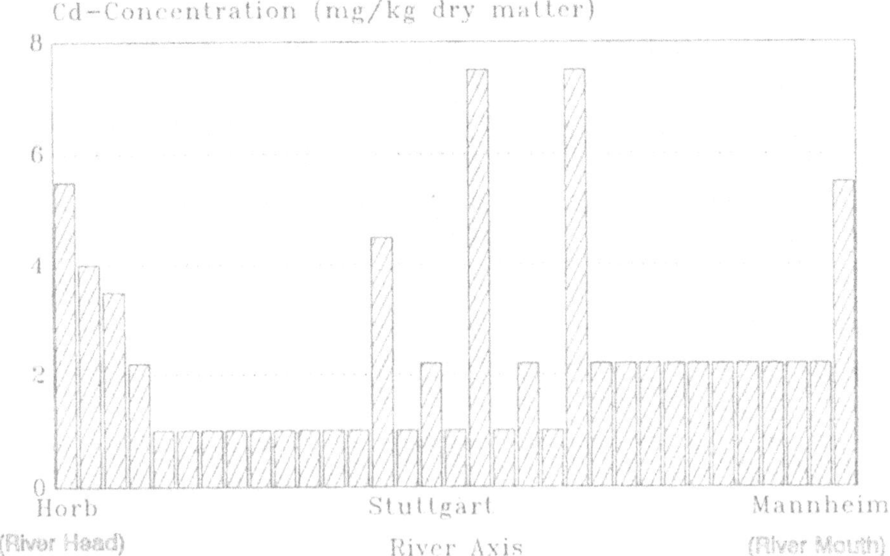

Figure 3. Average oxygen deficit at the mouth of the Neckar River (Anonymous, 1992)

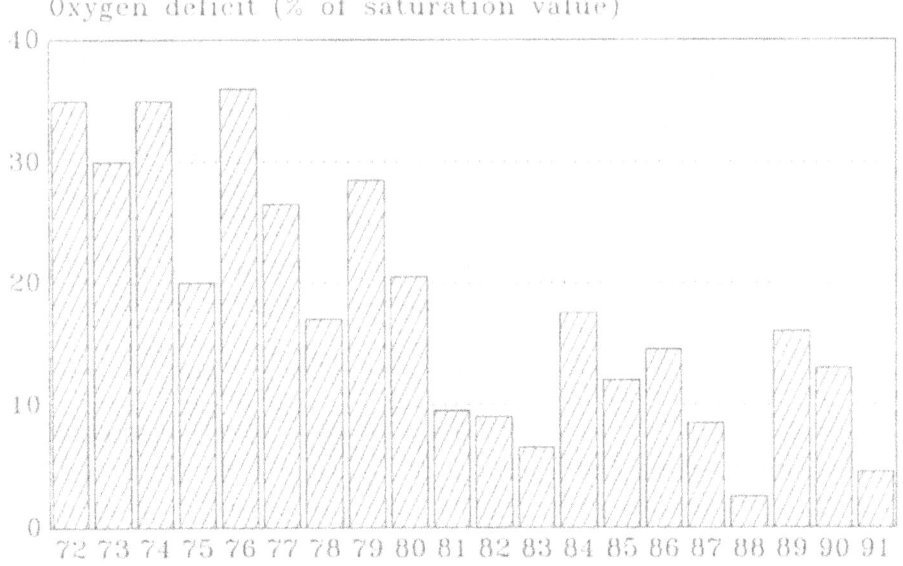

Figure 4. Cadmium content of the Neckar sediments in 1991/1992 (Anonymous, 1992).

2.3 Results of Remediation Efforts

Extensive and costly efforts of planning, building, and operating municipal wastewater treatment plants - and subsequent efforts of planning and building stormwater retention tanks (with subsequent routing of stored water masses to the central treatment plant) to alleviate pollution resulting from combined sewer overflow, were rewarding. Observations of dissolved oxygen concentrations significantly improved and fish that had been absent before were breeding again and could be caught. In efforts parallel to this development, raw industrial pollution was curbed either by improvements of central wastewater treatment or through tightened requirements for indirect discharge into the sewer system. This was reflected in a slow but confirmed "recovery" of the river sediments: erosion, sediment input and reduced pollutant input have quantitatively lowered average pollutant concentration over time (Figure 3). Long-range effects, i.e., effects upon downstream river systems and long-term reactions, for instance, accumulation of pollutants in food-chain systems, were not routinely monitored. In all, it could be stated that by the mid to late 80's the Neckar had recovered from an oxygen-depleted, sediment-polluted river to a habitat of medium to high quality for fish and fowl. This was confirmed by water authorities, fishery associations and recreational users. In an unplanned/un-intended way, dredging the Federal waterway also helped the recovery. Introducing dilution water was contemplated and then rightly rejected.

Future problems of the Neckar Basin: It is not exaggerated to say that most point-sources discharging into the Neckar (or its tributaries) have been identified, registered, controlled and subsequently monitored. Thus, the less visible diffuse (nonpoint) sources are the focal point of the next steps in curbing Neckar pollution. There are three major areas of concern:

(a) Non-controllable diffuse use of polluting substances and subsequent discharges into the Neckar, such as household chemicals - this problem can only be solved through changing use patterns, i.e., by substitution with non-phosphate detergnets.

(b) Dangerous input from industrial processes, such as heavy metal compounds or anthropogenic organic compounds (frequently categorized as "priority pollutants" or anthropogenic pollution) - this type of pollution is addressed by rewriting the (Federal German) water laws, demanding that wastewater containing dangerous substances (accumulating, non-degradable, cancerogenic, teratogenic, mutagenic) is treated with the best available technology.

(c) Non-point input from agriculture and settled areas such as nitrogen fertilizers from arable fields or lead from traffic areas; for this type of pollution there are not yet any proven controls. It is thought that curbing this input by offering financial incentives to the farmer for reducing N-fertilization and, for instance, further intensified storage of stormwater runoff with subsequent treatment at decentralized stations or at the central treatment plant may provide a solution.

2.4 Specificity of the Neckar Case Study

In terms of hydrography the Neckar is in comparison with other German (or European) rivers, a smaller to medium size river, draining a smaller area. However, population density and industrialization is high, making this river potentially a highly loaded one. Besides this, other German rivers differ in:

(a) the higher average precipitation within the drainage area;

(b) the less extreme low flow to high flow ratio, i.e., showing on an average a more uniform discharge pattern (and thus having less frequently a "sediment-moving" or sediment-cleaning flood event);

(c) mostly negligible input of water from other drainage areas; and

(d) in the fact that they have longer free flowing river stretches and not, as the Neckar, a sequence of locks and dams with stretches of low flow in between.

Finally it should be pointed out, that nearly over the whole river stretch where barging is possible there are two Neckar channels, the old and natural channel and the new Federal waterway. The latter one carries the main discharge of the whole system.

In terms of use: Indirectly the specificity of the Neckar users' pattern has already been described in the previous section. Two types of users should be specifically mentioned since they affect water quality to a large extent and therefore control other potential uses: (1) discharge of (treated) domestic and industrial wastewater from 3 million inhabitants or 6 million population equivalents (including the effects of combined-sewer-overflow from the large settled area) and, (2) the higher number of large-scale power plants requiring large amounts of cooling water. This increases noticeably the Neckar river water temperature.

Such diverse uses have caused re-orientation of satisfying the demand for potable water from other sources (as said above): i.e., the Neckar is not used as much as other German rivers for water supply purposes. A very significant user of the Neckar is commercial navigation which has little direct effect upon the water quality but - through the concurrent construction/operation of 26 dams and locks has altered the main characteristics of this formerly free-flowing river into one of a chain of 26 smaller impoundments. This causes more sedimentation, in particular of fine materials, changed boundary conditions for the oxygen balance and in this way affects other users. It seems to further define recreational use of this river which increases the demand for and public willingness to improve water quality. Use pattern of other (German) rivers might include larger municipal and industrial water supply demand, larger irrigation demand, possibly still commercial fishing and flood protection structures.

In terms of resulting water quality (past and present): The very high level of oxygen depletion during low flow periods of the Neckar has at this time not been observed in other German rivers. As compared to the river Neckar they were - with few exceptions - less burdened with (degradable) municipal wastewater if one calculates this on a loading per unit discharge basis. Nevertheless, intensification of conventional biological wastewater treatment was also planned and put into effect in other river basins. This has lead to a significant improvement with time in the oxygen balance and related water quality parameters. Comparing the Federal River Water Quality Maps (Gewässergütekarte) over the years one can see that classes[4] IV and III disappear and class II (on a saprobic index scale: alpha-meta-saprobic) emerges in nearly all river basins. Other water quality problems, such as high nitrogen and phosphorous loading (N from agriculture and/or industry, P from domestic sources, agriculture and possibly industry) have existed in nearly all basins and have received increased attention because of their long-range effects manifested in the major drainage system such as the North Sea. There are differences between different rivers, depending on the their use pattern, but nitrogen and phosphorous loading has to be considered a problem in nearly all German river basin systems. Other problem areas, such as high salt concentration or heavy metals pollution or organic priority pollutant input are not nearly as uniform for all river systems. The relative degree of pollution of river sediments, i.e., the more lasting type of pollution, depends on the prevalence of less or non-degradable pollutants such as heavy metals or complex organic material and the silt content and siltation tendency of the river. The sediment contamination today

---

[4] Based on the modified Kolkwitz - Marson's saprobic system classification.

exhibits the largest difference in water quality between German river basins.

2.4.1 Neckar remediation measures and other concepts for remediation of (German) rivers

Of the conceivable remediation measures (Figure 5) only four were applied extensively in the Neckar river system.

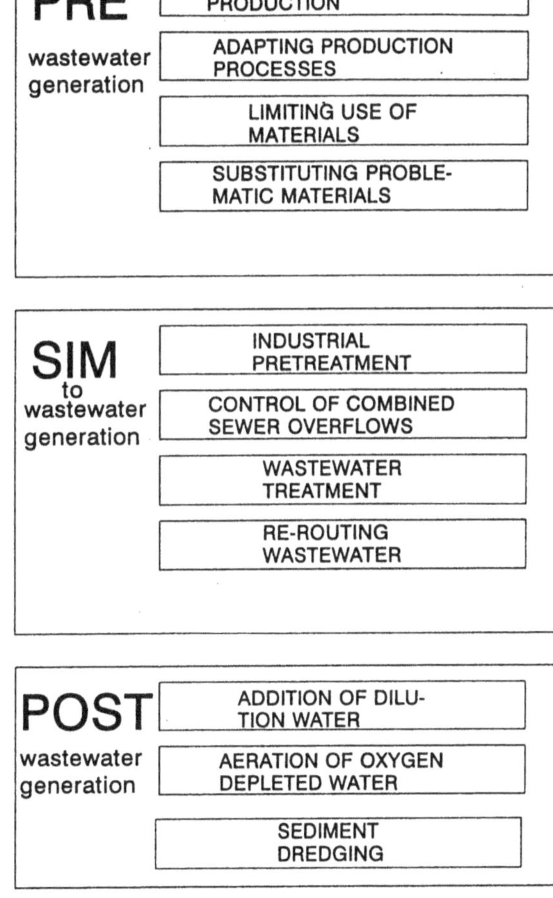

Figure 5. Summary of feasible remediation measures for river basins.

(a) In terms of *PRE-wastewater-generation measures* the following options were taken:

* the administrative regulation of the substitution of phosphorous in detergents

(b) Using the instruments of *control SIMULTANEOUS to the wastewater collection treatment and discharge* (today frequently labeled as "end of the pipe" control measures), Neckar authorities focused on:

* reducing point-sources' load through intensified wastewater treatment,
* curbing storm water related pollution by constructing runoff storage and additional treatment capacity.

(c) Resorting to *POST-wastewater-discharge measures:*

* river sediment dredging has been performed on an annual basis (predominantly for the purpose of keeping the Federal waterway operable).

The same remediation measures have been applied in other German river basins. Sediment dredging is done only if shipping is endangered. (Exception is the Ruhr valley where river impoundments had been conceived and constructed to enhance sedimentation and therefore require regular dredging). Post-discharge measures, such as dilution or river aeration have not been applied to any (long-term) extent. Pre-discharge measures might even be more frequent in the near future, concentrating on the removal of specific wastewater constituents. These control and remediation measures appear to be the only area where further development is possible and needed.

2.5 Future Developments of River Basin Planning and Management in Germany

River basin management had a long tradition in Germany. The founding of the Ruhr Valley Authority in 1922 as an administrative board of great independence and competence proved to be a very effective instrument for the integral planning and operation of installations for water quality protection in the Ruhr area. The great success of this type of solution of river basin planning and management questions has led to the subsequent establishment of some other 40 river basin associations. In recent years, however, no river basin associations have been founded; it appeared to be of less interest due to the fact that researchers, practitioners and administrators focused predominantly on the development of solutions for specific wastewater sources, i.e., treatment plants. This might have implied the forming of regional management agencies if it was proven necessary in terms of technical or economic boundary conditions. There were even motions against the forming of a Neckar River Authority using the arguments that this might delay the development of wastewater treatment facilities. Further more voluntary cooperation of communities, supported by financial incentives from the State government might have been held up for reasons or redistribution of central funds.

Today a high level of wastewater treatment has been accomplished, consequently, it is necessary to ask whether more environmental control in this area is necessary and if so, where and for which constituents. These questions can only be addressed rationally in overall planning. River basin management will experience a revival, along with the renewed use of (new and differently conceived) river water quality modeling. The main questions must still be formulated in a more exact form, but they will have to address the following issues:

(1) What pollutants have a noticeable environmental effect even if only over a longer period of input, considering even lower doses of polluting material?

(2) Where are the most important sources of such pollution?

(3) What is the most expedient way, i.e., expedient in administrative, technical and economical terms, of controlling this pollution?

(4) To what extent should the pollutants be controlled (i.e., returning to the long-term and global analyses of the type indicated in issue (1))?

3. FUTURE OUTLOOK - WHICH FACTORS ARE STILL RELEVANT AND WHICH HAVE EMERGED AS NEW ?

Science usually advances most successfully by learning from weaknesses and faults of previous analyses. Thus, the preceding discussion of specific questions in the field of water quality description and prediction serves as the background for the following suggestions on necessary new developments and extensions of existing work.

3.1 Enlarging Engineering Know-how Contained in River Water Quality Models

Having described the Neckar River Basin analysis, as well as discussing its relevance for other river basins we already indicated that there are significant phenomena affecting water quality which have not yet been considered in available analyses. Reasons for this could be that they were not included in the formulation of existing river models - or that they cannot yet be accommodated in (existing) water quality models.

However, such remarks should not lead to the impression that water quality modeling had not been used so far in the development of strategies for river basin sanitation. Since the early seventies of the twentieth century each river basin authority or each regional water authority which had to cope with potential changes in river water quality developed water quality modeling capacities. Yet in Germany there was no centralized or synchronized development with comprehensible documentation. This means, that each model has been mostly used only by its developer. And if other agencies needed modeling facilities they rather developed their own instrument than transcribing an existing one. It appeared as if specific water quality models were only useful tools to those who had developed them. And this did not necessarily enhance the trust in this instrument nor further its broad use by all water resource planners.

Short comings of water quality modeling in terms of technical and scientific content of the model as well as documentation must be corrected by improving the engineering knowledge base of these instruments in order to facilitate pollution control planning and quality remediation possibly through the application of mathematical modeling. For illustration the following (non-complete) list of short-comings shall be briefly described:

- Enlarging the list of input parameters that are relevant for river water quality such as to accommodate non-point sources or remote loading.
- Up-dating current less satisfactorily formulated building blocks for specific physical, chemical and biological processes (the phenomena of sedimentation and erosion might serve to illustrate this fact: the present form of these simulation blocks does not allow a very realistic representation of water quality changes).
- Including effects of more recently developed engineering controls such as inclusion of riparian flood plain management, ground-water recharge, etc.

For most (German) river basins the following can be shown: With the noticeable reduction in the loading from non- or partially treated domestic and industrial sewage streams the effects from non-point sources, such as combined sewer overflows or area erosion from agriculture, become more significant. First, these input sources - possibly also more removed from the actual river bed and characterized in socio-graphic terms - have to be evaluated in quality parameters that differ from traditional wastewater analyses. Therefore such sources are not yet well documented. Second, existing methods for the testing of the efficiency of remediation or control measures will have to be adapted, including above all water quality modeling, which does not yet include these decision options. For instance time variable loading of a system with highly adsorptive lead compounds interacting with varying loads of particulate inorganic and organic solids has not been considered. These compounds might react with traditional wastewater constituents, e.g., in the process of river self-purification. Such phenomena must be quantified and included into computational routines.

Sedimentation and erosion have been previously described as the most frequently occurring processes in rivers. Their effects are directly registered in the accumulation of river sediments, or in their disappearance. This has been "controlled" in a very pragmatic way by either attempting to withhold particulate matter to a larger degree at point-sources (the first step in conventional wastewater treatment was and still is a sedimentation step) or by regularly dredging the river bed. Yet, there are indirect effects upon water quality from such particulate solids transport or deposition at unpredictable locations. It is known how to formulate sedimentation as a physical phenomena; the same holds (to a less perfect degree) for the process of erosion. Yet, in all the available physical formulations there are process-characteristic parameters, such as for instance the instantaneous river bottom shear velocity, which is not routinely calculated in available water quality models. Thus, the processes of sedimentation and erosion, even though they are included into various simulation routines, are not yet described with necessary precision as function of time in order to describe or predict transport and deposition of such dangerous pollutants.

Developing remedial measures should also include an understanding that we can no longer expect to live in a completely natural environment; we cannot escape living in an environment influenced by man. There is no way of going back in development. We can begin to understand processes and reactions characteristic

- for regions of alluvial forests
- for inundation areas,
- for groundwater recharging areas,
- for more complex river side arm systems,
- etc.

and we can begin to develop an understanding for control and management, in particular also in terms of the necessary dynamics of necessary control measures. It is better to be in a position to "act" and to have an idea of the consequences of the action, than to "leave it to chance". It is necessary, after accepting the existence of structural changes in river systems to define purposes of such a multi-objective system anew. New concepts will have to address, amongst others, the following issues (listed below without the claim of completeness nor necessarily in an hierarchal order):

- Flood Protection
- Environmental Conservation
- Groundwater Resources
- Energy Production
- Urban and Industrial Sewerage Discharges
- Navigation
- Others

and new design and operation rules for each of these sometimes conflicting uses of the water body will have to be developed such that a compromise can be attained in satisfying these properly weighted goals. It has to be borne in mind that boundary conditions will change in terms of hydrological constraints, socio-economic developments and even social preferences that will make it necessary to react flexibly, with adjustable operational rules.

3.2 Including Additional Non-Chemical Water Quality Affecting Aspects

In the past, water quality changes have been looked at as isolated or "controlled" events. Thus the more realistic scenario of superposing multiple and not adequately identified processes was not considered. More realistic scenarios would consist of:

- Combining in a probabilistic manner water quality affecting processes such as the least or less favorable conditions for water quality can be described.

- Analyzing long-term effects of pollution of river systems and the corresponding reaction of the system, commonly referred to as "long-term simulation".

- Coupling one river system with another more remote one that could exert a significant effect upon the basin under consideration.

Water quality planning decisions for river basins are usually made for boundary conditions that describe in a very definite and unique way the situation that is thought to be realistic for the question at hand. Furthermore, these boundary conditions are selected such that a "worst" case is assumed. However such scenarios are generally confined to one level where the specific questions is thought to be relevant. Whether other boundary conditions might affect or might be affected, i.e., whether less likely scenarios might lead to an even "worse" situation, is not analyzed or investigated. There is consequently a need to combine least-favorable boundary conditions in river basin decision making or in water quality modeling that supports this decision making. To answer this demand the analysis of historical and potential boundary conditions must be extended. On the basis of such information one will have to select a probabilistic set of boundary conditions, possibly for repeated draws on the combination of data. It is also possible to answer this demand by generating a continuum of computed water quality records through long-term simulation (see next paragraph).

It is known that the water quality status of time 'n' will influence the status of time period 'n+1' as an initial condition. It will affect water quality possibly as much as all those physical, chemical and biological phenomena, that constitute the mathematical model. In the past, such boundary conditions have been defined in a deterministic way, frequently on the basis of historical events. Now, such boundary conditions are no longer (pre-)set in a rather arbitrary way but result from a logic coupling of consecutive analyses: The result will

(a) become a more realistic picture on water quality changes and

(b) furnish a data-series that can be used in a probabilistic manner.

Such conclusions expressed and discussed in probabilistic terms are essential for today's planning and decision making. Yet, existing water quality models are for the most part not capable of 'long-term simulation'. The numerical and computation effort would presently be too large, which means that such models would have to be rewritten, possibly with less emphasis on describing precisely every conceivable reaction step, but rather allowing longer simulation periods.

Iin view of necessary decisions, today's analyses of river basins are made exclusively for a particular basin or even a fraction thereof. However, in most instances it is appropriate to widen the horizon of analysis, i.e., to increase the regional scope of the analysis. Most recently, with questions of long-range and long-term effects, this widening of the basis of analysis becomes even mandatory. Such increase of the regional scope could be the inclusion of upstream or downstream river systems; this might initially only enlarge the number of computational steps, without increasing the complexity of the problem. More recently it has been recognized that areas geographically remote from the river under investigation might also affect the water quality in a way yet to be described in exact physical terms. Here, other methods of water quality assessment and prediction might be called for (see Part II of this chapter). There might be additional possibilities (and needs) for widening the data base in analyzing the consequences of certain administrative and/or engineering measures envisioned for a river system: frequently political goals or boundary conditions cannot be formulated in those dimensions that are familiar to the water quality modeling expert or that are even accessible to such computational routines. Then variables that are by that definition "incommensurable" must be addressed and evaluated in a synoptic way.

3.3 Increasing the Public and Public Officials Awareness of Water Quality Processes by Improving Comprehension and Presentation of Data

In the past, possible changes in water quality, resulting from anthropogenic actions have been described predominantly in a qualitative or verbal manner. Quantitative assessment appeared to be too complex. Therefore, such effects were not given the same weight or consideration as other quantifiable phenomena. Even today, aspects of changing water quality are neglected in decision making with the same arguments. This may result from the fact that known (and quantified) water quality effects are summarized in complex mathematical water quality models. The computed results do not appear accessible to everybody or are not comprehensible to the decision makers.

If in the future the above described processes which affect water quality are to be included into river basin planning and management then steps have to be taken, to make existing knowledge bases more readily available to all interested. Such steps could entail:

- More clear display of cause and effect of water quality changes in connection with definite engineering (or administrative) measures, such as impoundment or water extraction from river systems.

- Direct comparison of calculated/predicted/anticipated water quality data with observed data by having a direct feed-in of observations.

- The translation of frequently complex scientific representations of water quality changing processes or water quality changes into readily comprehensible graphical or tabular results, enabling even less familiar decision makers to understand cause and effect of decision options.

One possibility of increasing the attractiveness and comprehension of water quality modeling is to develop interactive modeling. The decision maker who requests (computational) data should be included into the process of obtaining the data. This could be accomplished by repeated simulation runs. These would be calculated immediately upon the setting or correcting of a boundary value or a system's parameter by a specialist. Such procedures would emphasize the "experimental" nature of many of these studies. It would de-emphasize the blind belief in computational results that are furnished with an apparent degree of precision that is not supported by assumptions underlying the model structure and model use.

The coupling of real time measurements and computations to water quality changes could be an additional instrument for the improvement of comprehension and acceptability of water quality modeling. An ever growing list of water quality variables can be analyzed in real-time on a continuous basis (for instance through electrochemical methods), or at least close to real time and on a quasi-continuous basis (for instance through auto-analyzers). These data are frequently relayed by telegraphic or wire-less means to central stations where they can be displayed. However, it is not always known how to interpret observed changes or what the cause of such variations in water quality might be. In these instances water quality models with measurement and data evaluation routines could lead to an updating or self-correcting of water quality models.

Finally, the problem of general accessibility of water quality models must be addressed. It seems that in Germany developed and proven water quality models or their updates are only marginally documented. There is seemingly no effort made to help in the transfer of knowledge to potential users: mathematical water quality models are most frequently handled like "proprietory property" of the agency or of the author who developed them. In order to have more professionals contributing to the knowledge of these data bases it appears desirable to have water quality models

(1) developed in a fashion that maximizes comprehensibility and clarity,

(2) documented in a precise and standard form that is accessible to the scientific community, and

(3) included into programs of technology transfer, i.e., making coordinated efforts to spread the understanding and application of these instruments.

4. CONCLUSIONS - WHAT DEVELOPMENTS HAVE ALREADY BEGUN TODAY AND WHAT IS TO BE DONE TOMORROW?

River basin management using synoptic planning and decision making instruments has received new interest due to a re-orientation in pollution control. This leads to renewed interest in and development of water quality models as a powerful instrument for such large-scale and long-term analyses.

The specialists, using or developing (mathematical) water quality models, have responded to this challenge. They have developed new generations of models with the intent of avoiding faults and weaknesses of previous model generations. The familiar BOD model has disappeared and mathematical descriptions and predictions of nutrient concentration scenarios are dealt with today. Non-stationary situations and complex (hydrodynamic) phenomena influencing for instance sedimentation and erosion processes, can be described in first approximation. In many instances these descriptions are still rather crude. This might be illustrated by the fact that in most water quality models that claim to simulate the "behavior" of heavy metals in a river system, these heavy metals are defined as conservative water constituents.

There is still much to be done in terms of developing more comprehensive instruments of water quality assessment without increasing the degree of complexity of such methods. The request for interactive modeling, showing results in less abstracted and more graphically oriented forms has only been answered in part. The notion that traditional water quality assessment could and should be combined with "far-away" socio-economic data has not yet been translated into a readily available tool for pollution control officials. The general accessibility or general distribution of water quality analyzing routines (models) is still far from reality. On one hand this results from the lack of instrumentation that would allow the use and display of non-numerical information. On the other hand this is explained by the still narrow knowledge base of today which does not correspond to the one necessary for such tools.

REFERENCES

Anonymous (1977) "Sanierungsprogramm Neckar", Ministerium für Ernährung, Landwirtschaft und Umwelt, Baden-Württemberg, Heft 5

Förstner, U. and G. Müller G., (1979) Schwermetalle in Flüssen und Seen, Springer Verlag Berlin, Heidelberg, New York

Anonymous (1992) "Umweltdaten 91/92 Baden-Württemberg," Landesanstalt für Umweltschutz, Umweltministerium, Stuttgart

Part II: Future Outlook

Neithard Müller[1] and Herman H. Hahn[2]

1. INTRODUCTION

As was shown by Hahn and Müller in Part I of this chapter, river basin systems, especially for larger rivers are very complex. Almost all who deal with water quality of flowing waters simulate certain properties of the real world system under consideration. This is true today and will certainly continue in the future, keeping the amazing progress of computer hardware in mind. The same is true for water quality analysis of reservoirs. Therefore, this paper focuses on water quality models, since it is the favorite tool of those dealing with water quality in the field of water resources management.

According to the environmental, political, and economical boundary conditions, different priorities, different quality aspects and different working strategies certainly have to be chosen. This contribution will focus on water quality aspects that are urgent in Germany or countries with similar environmental conditions. These have been described to some extent in the preceding paper.

As time goes by, population grows, industrialization proceeds and one can notice changes in the field of water quality investigation. Although the overall amount of pollutants may grow, the effective discharges into the rivers from point inputs are diminishing due to rigorous wastewater treatment legislation and a well functioning governmental system to control compliance with regulations and limits. The diffuse sources are therefore becoming more important and can no longer be neglected. This is true for the classical oxygen demanding pollutants and nutrients, such as phosphorous and nitrogen. But on the other hand there are increasing amounts of dangerous substances that are contributing to the burden of the rivers. Even if wastewater containing these substances were treated with the best technology available, an overall increase would still be noticed in some cases, because these substances have to be considered as a diffuse load that is released all over the catchment area in (very) small quantities, due to

---

[1] Weber Engineering, Pforzheim, Germany

[2] Institut für Siedlungswasserwirtschaft, Universität Karlsruhe, Germany

the activities of many individuals. On the other hand, the river basins and the rivers themselves are being used to a much higher degree than before, because the number of uses have increased and each single use has intensified. Last, but not least, a new attitude towards the environment in general is about to develop, which explains the readiness for a certain amount of renunciation and a self-sacrificing devotion.

There are two ways to improve water quality of a river or the overall situation in a river basin; (1) reduction of pollutant input (direct or diffuse) into the river, or, (2) improvement of the rivers' condition. To achieve a reduction of the input there are basically two methods: 1) avoid the release of a certain pollutant or 2) treatment of the effluent. To change the river itself is somewhat difficult especially since technical restoration measures are becoming less popular today. Furthermore possible actions are limited and often very expensive.

2. INCREASE OF FUTURE KNOWLEDGE BY MODEL APPLICATION OR SCIENTIFIC RESEARCH

If we aim for a position where we can "act" that is to positively influence the situation in a river basin, rather than "leave it to chance" we need to be as efficient as possible. In order to be able to decide which measure might be positive or to determine ranking of several feasible possibilities, one needs to evaluate the effects of these measures, taking the costs into account. The questions that have to be answered have been formulated in detail in Part I (Hahn and Müller). There are different approaches that can be used in the future to find answers to these questions, or at least to get closer to the goal of a better understanding as to what affects water quality. As it was stated this is usually done using mathematical models to simulate certain water properties which are judged for systematically changed boundary or tributary conditions.

2.1 Application of Classical Water Quality Models (Empirical/deterministic Approach)

The most obvious approach is certainly to continue the way water quality modelling has been developing in the past, that is applying deterministic models simulating water quality parameters of a river as a function of river characteristics and input quantities that are fixed usually as a point, sometimes as line sources. Assuming that when new scientific knowledge becomes available models will be modified to include this knowledge. Thus a-priori knowledge is used to improve the models in use today. There are several possible ways to alter existing models or to create new ones from scratch, starting from the latest engineering technology. These possibilities are discussed in the following section. Some of these shall be considered for a new water quality model which is just being developed in Germany (ATV 1994).

2.1.1 Consideration of a greater number of processes

The first method is to add more modules to a model, reflecting additional (and/or new) transformations of water quality parameters. These modules might represent chemical, biochemical, biological, physical or mixed empirically determined processes. The closer one looks at the water quality parameters and the lower the concentrations of pollutants become, the greater the variety of the biocoenosis involved. If in the past one single biomass variable was often sufficient to describe the mass transfer of nutrients and the oxygen balance, different algae and other micro-organisms with their appropriate metabolism should be considered, as well as benthic macroinvertebrates and water plants, to get a closer understanding of the changes in water quality parameters.

2.1.2 Consideration of additional water quality parameters

Another method allowing a closer understanding of the river situation requires simulation of additional water quality parameters, corresponding to the development of science in the field of environmental research. Today water quality models focus on oxygen, algae, biomass, one or several fractions of phosphorous and nitrogen and some conservative constituents. Additional parameters being of interest for further investigations might be organic and inorganic substances or groups of such substances, including for example fungicides, pesticides, chlorinated hydrocarbons, complex forming reagents (NTA, EDTA) and heavy metals. These substances might be of future interest or are already the focus of researchers today due to the following reasons:

- Better analytical methods allowing the tracing of new substances and a lowering of the detection threshold;

- A large number of new substances have been synthesized; many of these are influencing water quality in one way or another;

- Synergistic effects cause the need to consider parameters which on their own were without significant influence;

- Negative influences of classical parameters are becoming less important and thus issues that have been covered before come to light; and

- Some parameters are necessary in connection with additional biological species and their metabolism.

Water quality models that are under construction will certainly be characterized by the potential to simulate several or all of the above parameters. In comparing model results against real-time measurements of water constituents, techniques are developed to detect accidents or to give warnings.

2.1.3 Water use accounting

Furthermore, we will have to discuss the expression "water quality" and how it will be used in the future. When we usually talk of water quality, every individual is thinking of something different. To illustrate this we define the scale of water quality as having three possible states: {good, medium, bad}. This might be a rather rough classification but it is sufficient for this illustration. Imagine now a given water with certain characteristics. Depending of the use each individual has in mind, the assessment will differ although the water is still the same. A water supplier for example has different expectation what should be water quality that is good for water purification as opposed to a fisherman. This is due to the different water quality requirements necessary for certain uses. As the number and intensity of different users rises, this issue is getting more and more important. Therefore, it might not be sufficient in the future to simulate certain water quality parameters, but to consider the constituents concentrations in association with the demands of the people in the river basin and possibly those downstream.

2.1.4 Necessity of transforming environmental information in the river basin into point-source or line-source concentrations of water quality

When we want to apply water quality models for the impact determination of the activities or the general situation in the river basin a problem arises, because the river quality models do not usually allow for spatial data input but are normally designed for point or line input. Therefore it is necessary to transform the situation in the river basin into a set of sources, distributed along the river axis. For each lateral source, a quantity of water is usually required along with a set of values, one for each constituent the model will simulate. If we are using a dynamic and non-stationary model, all these values will be required for the time of the simulation, i.e., a time series of water quantities and concentrations. This seems feasible for actual point sources such as wastewater treatment plant outlets. But for diffuse inputs from the river basin that are not close to the river under consideration, this might be very difficult. The user of a model must determine which information describing the situation in the catchment area should be considered. Following this, a chain of casual connections must then be set up that links a certain environmental characteristic in the basin with information of one or more point sources. Such a chain is very unlikely to be found. Even if there is enough *a-priori* knowledge that such a chain of deterministic sub-models can be defined, the volume of data necessary is too great to be handled or not available at all. Another difficulty arises from the large number of inaccuracies occurring in such a chain. This leads to uncertain results of the water quality model.

2.1.5 Implementation of classical water quality models' further development

Classical water quality models might therefore be further developed by adding new modules, corresponding to the additional processes depicted above. This addition is the easier if it has been foreseen and allowed for when the model was created through the use of modular techniques. This is more likely if a model is newer, as this strategy was followed only recently. A modular structure of a model helps greatly because the addition of a new process may affect some other modules, for example, via variables describing constituent concentrations or physical, chemical or biological state attributes. Such a network of interaction is easier to handle if a well-defined interface between two modules exists. The modules can also include elements that previously have not been included and whose influences had been considered through the input data only. Examples for such modules might be qualitative or quantitative groundwater or surface runoff models. Another possibility is the improvement of the algorithm solving the governing equations. This can concern numerical errors, e.g., numerical diffusion, or the transition to non-stationary and/or dynamic algorithms.

Of course a further possibility consists in extensions of the hydraulic submodel. Usually water quality models use relatively limited routines to depict the hydraulic situation of a river. When we want to take meshed systems into account including for example lateral channels, improved algorithms have to be applied. This aspect might be very important for rivers that are left undisturbed and hence are very likely characterized by many secondary river beds, possibly connected with the main stream on one or both sides, following the actual discharge and thus water level. Such hydraulic model improvements are necessary especially when dynamic aspects are important.

The easier a model is to handle, the more it will be used. In comparison to other models that are similar in performance. Besides this evident effect there is another issue to be recognized. A model that is easy to use will certainly be used more often for the determination of effects that will result due to a variation of the input data. Today, an easy-to-use model will most likely have a user interface enabling interactive execution. Consequently it has to include tools or routines for graphical result visualization since it is proven that this is the most powerful way, in terms of time to interpret the results, in terms of reliability (because it is easy and fast to check plausibility) and because people are more motivated. A very helpful tool, favorable to be used for assessing different scenarios, is a means for automatic or at least user-supported parameter estimation. Water quality models are often very parameter intensive and even for models with very few parameters, difficulties in the determination have been reported by Masliev and Somlyódy (1994).

2.2 Application of Innovative Model Concepts

When we are developing models for the system under consideration, using methods as illustrated in the paragraphs above, there arises a problem with increasing complexity. This problem was described by Zadeh (1973) "In general, complexity and precision bear an inverse relation to one another in the sense that, as the complexity of a system increases, our ability to make precise and yet significant statements about its behavior diminishes until a threshold is reached beyond which precision and significance (or relevance) become almost mutually exclusive characteristics." Thus, there seems to be a certain limit where we have to look for other means of investigation, other than the deterministic-style model we have used so far.

No additional processes

We do not know all the processes involved in the sphere of influence of a river and many of those we are aware of are not fully understood. Therefore, it is in principle not possible to create a model that covers every possible effect and due to an enormous amount of data necessary to operate such a model would be impractical anyway. One strategy in the future might consequently consist of disregarding a model made of stand-alone modules, each representing a certain process and coupled together to form the model. Instead we may apply methods representing the system as a whole with implicit internal connections.

No consideration of isolated water quality parameters

As it was shown above, the number of parameters to be considered is constantly increasing. This trend will certainly continue for a while. The question that comes up in this context is: if we want to simulate all these parameters and do not really know which ones are the most important, how can we determine when we should stop taking new parameters into account? Each new generation of parameters is getting more numerous. The first water quality models took basically two parameters into account: Dissolved Oxygen and Biochemical Demand (BOD) concentrations. The next generation included nutrients among others, phosphorous and nitrogen. It was necessary though to use several forms of these substances (for example two P- and four N-fractions) and thus clearly raising the number of parameters. When we now talk about toxic substances or heavy metals or any other group of today's water quality parameters, the number of single constituents is even greater. The disadvantage is that for each parameter we have to modify the old model, influencing other parts of the model. We can develop models of the classical type to whatever extent we wish but the day will come when a new substance must be considered and we will again be without a model to use.

For a number of parameters it is still controversial whether they should be investigated using deterministic methods. Chemical Oxygen Demand (COD), for example, is used in some water quality models. Such a model has equations that describe the change of COD concentration although there are great differences in what actually causes the oxygen demand. This might be one reason why it is sometimes very hard to calibrate a given model for a certain river system.

As point-source influence is diminishing diffuse sources are getting more important

If we try to look into the future, we will certainly have to deal more and more with diffuse rather than point sources. Tributaries represent less input to the overall load of a river due to the continuous increase in treatment. The extent, to which this is the case, varies from substance to substance. Figure 1 and 2 illustrate this fact for the load of phosphorous and nitrogen of the river Main (Bayerisches Staatsministerium für Landesentwicklung und Umweltfragen, 1994). The P-load from point sources has been diminishing for 20 years due to the substitution of phosphorous in detergents and increasing wastewater treatment. Nitrogen has only begun diminishing recently as a result of very strict nitrogen standards for wastewater treatment.

On the other hand, the emissions of small quantities of substances that negatively influence water quality by every individual are increasing. We know little or nothing about the spatial and temporal

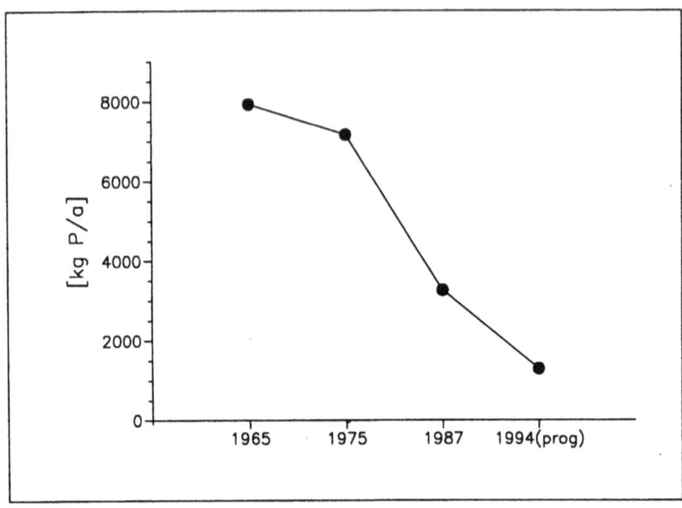

Figure 1: Phosphorus load of the river Main from wastewater treatment plants and rainwater overflow.

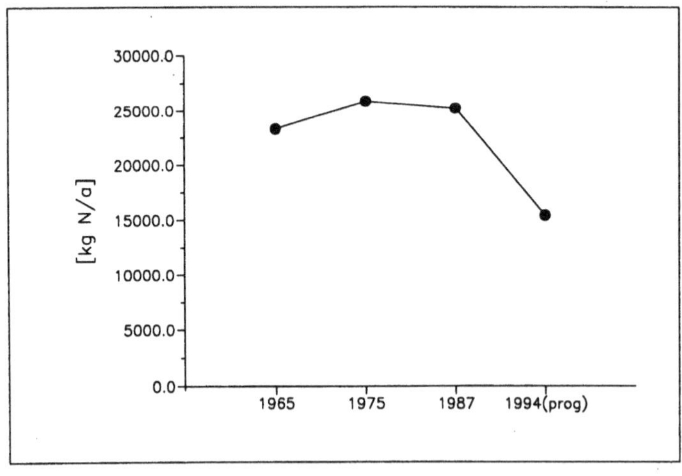

Figure 2: Nitrogen load of the river Main from wastewater treatment plants and rainwater overflow.

distribution of such emissions. It would, therefore, be necessary to use other, more exact deterministic methods to take such aspects into account.

Creation of river basin models though extension of classical water quality models

The first idea that comes to mind when thinking of models representing complete river basins, is the extension of existing models. It would be good if it were possible to use the well-known water quality model by adding some pre-processors, representing the sequence of effects that lead to an influence on the water quality. Such pre-processors would translate the activities in the catchment area into a set of input data for the existing model. To obtain enough scientific knowledge to enable us to create such pre-processors, a great deal of research is required, probably leading to models with an enormous data demand.

How to find out which information is important

All the arguments stated so far lead to the recognition, that we will have to use other methods in the long run to find out what influences are the most important irrespective of the future comprehension of the term "water quality". A method allowing for statements about this topic whatever the boundary conditions might be, and whatever water quality aspect will be focused on. The authors believe that successful developments will apply innovative methods that will be discussed and illustrated in this chapter.

2.3 Methods and Methodologies

The methods to be discussed will be divided into two groups. The first group covers all the aspects in connection with new computer technologies and the second group contains statistical approaches.

2.3.1 Use of new computer technologies

Interactivity: Interactivity has been mentioned already in connection with the improvement of classical water quality models. When we are talking about new concepts we have more in mind than a simple, nice and easy to understand visualization of results. We could extend the interactivity to a degree where the user of a model can get information of every relevant variable in the model at any time and the possibility to intervene at any point he wants. Thus he can investigate what will be the results of certain changes. This can be done using object oriented programming techniques. The user could select a set of items he might be interested in and the state of these items might be displayed in different windows or frames in some way or another.

GIS: Another tool indispensable for future models is some kind of Geographic Information System (GIS). The moment we wish to deal with spatial information, we have to use such a system. In this way we can handle the huge amount of data following from transition to a two-dimensional approach. The application of a GIS allows for the necessary activities to handle spatial information, e.g., overlaying of different characteristics or cookie-cutter-like operations. A database for the storage and administration of spatial data is also included as well as some means for two- or even three-dimensional visualization. The apparent contradiction to the claim for models with less data demand can be resolved easily. The absolute amount of data is certainly greater, but the time required for obtaining such data is about the same, regardless of the actual number of values (within certain limits) of each data type. In classical water quality models we need one value of many different data types; to obtain this information requires more work because each value has to be taken care of and there is no volume discount.

A GIS might represent a stand alone water quality model for river basins that can be used to judge the influence of spatial information in the catchment area. Using numerical capabilities and visualization might lead to some conclusions on connections between water quality aspects and certain characteristics of the catchment area. This method requires an investigator to analyze the processed and visualized data. The investigator should be an expert and using the tremendous capabilities of the human eye can often determine connections by looking at maps that show the spatial data of interest and the river (system). Such maps can show single attributes or aggregated information (Fedra, 1991).

It has often been reported that the human visual faculty could be used with success for similar tasks, in which the interpretation of colored areas was necessary. In connection with sound technical knowledge this is a possibility that leads to useful results. The quality aspect studied could be displayed color coded along with the spatial data. This procedure is illustrated in Figure 3. The area shown represents the catchment area of the river Neckar, illustrated in the preceding Part I. The average nitrate-concentration is represented by colors red to blue. High concentrations are represented by red colors, low concentrations by blue colors. The color of the catchment area represents the relative amount of the area used for agriculture. The lighter the brown shade the less intense is the use. A trend can be seen that the nitrate concentration in the Neckar River is higher where agricultural use is high and vice versa.

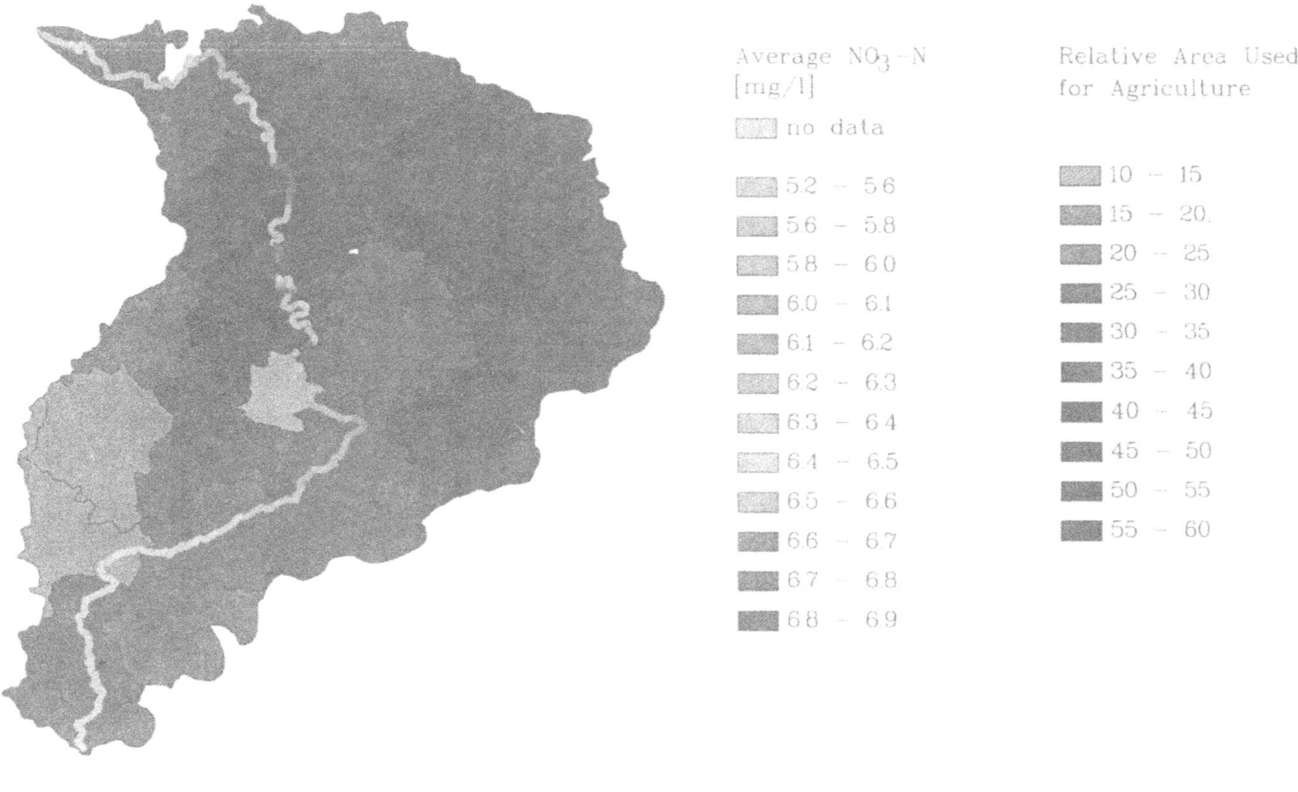

Figure 3. Intensity of agricultural land use in the catchment of the River Neckar and nitrate concentrations of the river.

Spatial-characteristic data in the river basin can be obtained from several sources. Unlike the data representing the boundary conditions for classical water quality models, this information is to some extend stored in more or less central, usually governmental and sometimes commercial, databases. Such databases will be accessible using networks (such as INTERNET) and hence the process of data collection will be shortened and less difficult, even though the absolute amount of data may increase. The advantage, compared to the data collection process so far, is the fact that using central databases and networks it is no longer necessary to gather all the different information needed from many different organizations and locations, with different logical and physical formats. Of course access to and handling of such databases must be improved, along with the acceptance and expanding of appropriate networks.[3]

Artificial Intelligence: Another emerging technique used for the purpose of improving water resources planning is Artificial Intelligence. Even though the image of Artificial Intelligence, usually used in the form of Expert Systems or Knowledge-Based Systems, decreased in the late eighties, it is slowly increasing again and is reaching a more appropriate level without the exaggerated expectations that existed a decade ago, when such techniques became available.

When we are talking about artificial intelligence in the field of water quality management, we will mostly deal with Rule-Based systems. Such systems have already been used for the support of planners and it is certainly a technique that will become more popular in the future. Fields of application are widespread, but the success will be most likely for problems that can be characterized as follows: basic scientific knowledge is available (experts are available and willing to publish their knowledge); connections between elements of a system are numerous but each single link is rather simple; the network of these basic knowledge bits is complex. In other words, although everything necessary to solve the problem is available, the solution or the way to achieve it is not obvious.

Presently the creation of systems using artificial intelligence is to some extent limited by the so called bottleneck of Expert Systems, which is the knowledge connection and the insufficient power of computers. The second argument might not seem to be too important, as computer technology is advancing rapidly, however, it has to be kept in mind that the size of rule-bases is rising too. To reduce the hampering effect of the first argument, tools have been developed for assistance, but they have only a limited capability so far.

[3] For example, the U.S. Environmental Protection Agency has developed a water quality data base STORET that can be remotely accessed and data can be downloaded by a computer with a modem.

Fuzzy Logic: The next method we want to discuss is using fuzzy approaches. Fuzzy logic uses a mathematical concept allowing for fuzzy judgments instead of a strict assignment to one set. Thus statements such as "water quality is strongly affected by the wastewater treatment plant runoff" can be easily modelled. This concept allows for modelling of knowledge that is not exact. Therefore, it is suitable for water quality management where we often only have rough ideas and measurements that are often characterized by large errors. This is frequently the case in water quality management because in most cases parameters or measuring methods are used that have been developed for another purpose and do not suit exactly the needs for the water quality simulation.

Fuzzy logic approaches have been introduced in the field of environmental systems and in water quality modelling recently (Bardossy, 1993). First applications have proven the applicability of this method. Jensen (1992) and Müller (1994) experimented with such methods to simulate the water quality due to information concerning the surrounding area like population density, municipal area, agricultural area, area of forests, number of companies, number of wastewater treatment plants, SO_2-emission and amount of yearly rainfall.

A relatively simple method of fuzzy modelling uses rules as well. Herein rules are composed of fuzzy elements, like fuzzy numbers, and the operators linking these elements also have been extended to the fuzzy concept. A method exists in establishing rules that connect different items of input information, described for example using triangular fuzzy numbers, and map them onto another fuzzy number representing some amount of change of water quality (Figure 4). The fuzzy "and" in the rules can be implemented in several different ways, for example using operators like "Einstein-Sum" or "Algebraic-Sum" (Kaufmann and Gupta, 1985). Several rules together, each representing an unknown process of transformation, are aggregated to describe overall change of water quality along a certain river segment. This aggregation can be done using an averaging approach, transforming the fuzzy number to a real number describing the change in water quality at the same time (Jensen, 1992; Bardossy and Disse, 1993).

To set up the necessary rule base, a relative weight has to be defined for all triangular fuzzy numbers used to assign the measured value of information. This can be done with setting up of rules containing parameters that are determined in a process that keeps continuous track of them, using sets

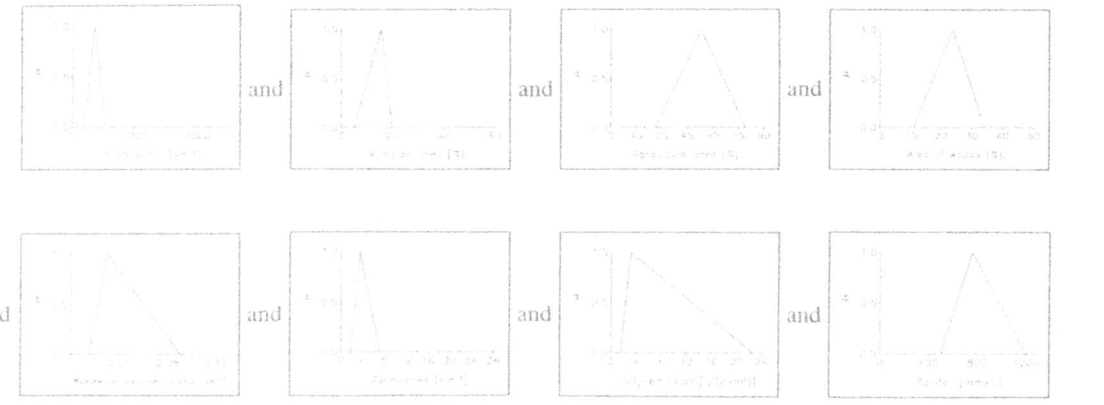

Figure 4 **Fuzzy rule.**

of data containing water quality and corresponding load information. Such a gauging process will be easier if more data sets are available. The computer power requirement necessary to determine the parameters of a relatively small problem is fairly big, therefore a Monte-Carlo simulation can be used and, thus, the time required could be reduced significantly. Once the parameters of the rules are fixed, the model represents the real system. By interpreting the rules it seems possible to derive knowledge of the connections of the system.

The first tests only used data of the relative close surroundings of a river. This led to results that are not sufficiently representative of the real world. Two aspects should be considered in the future for similar investigations: when using area information we must not completely disregard point information like main tributaries, and the size of the catchment area taken into consideration must not be too small.

Fuzzy models have so far been more successful when fewer independent variables were considered and more sets of data had been available for gauging the rules. Thus future investigations should start with relatively well-known systems, that can be considered as one system.

Learning Algorithms: Learning algorithms, like neural networks are propagated, whenever systems have to be modelled for which not much is known about the internal circumstances. These algorithms, due to their name, create the impression that the user no longer needs to know anything about the problem because the algorithms will learn on their own how the problem has to be solved. However, this is not the case. It is always required that the user supervises the algorithm and makes certain decisions that are necessary to control the learning capabilities of these algorithms. With the ever growing speed of computers, the potential for these methods is rising and in the future, this technology will be used more often.

2.3.2 Statistical Models

Classical statistic: For future investigations there will be an increasing number of applications using statistical or stochastic methods. This is due to the fact that we often do not understand functional deterministic description of the processes and there are also errors involved. If there is a large amount of data available, statistical methods are certainly appropriate. As a first step to an integral river basin model there are hydrological methods available now which describe the temporal spatial distribution of rainfall in a catchment area (Bardossy, 1994). The next step for a river basin model necessarily requiring statistical modelling will be the groundwater flow. There are certainly good deterministic models available for specific cases but for integral groundwater investigations of larger river basins, including transport in the unsaturated zone, too much data is required to allow strictly deterministic modelling. For the water quality aspects in the narrow sense of the river itself, the necessary amount of data is usually not available. This is the reason why statistical methods have been used only in special cases.

Self-organizing algorithms: If classical statistical methods fail or are not applicable because there is not enough data available, we can use self-organizing algorithms to create a model. Such algorithms are getting more popular and this trend will certainly continue in the future. One representative of this class of methods which has already been applied for different water quality aspects is the Group Method of Data Handling (GMDH) (Ivakhnenko, 1968; Ivakhnenko and Müller, 1984; Farlow, 1984). This method determines a model of optimal complexity, describing changing values of a quantity due to variations of a number of variables. There is no *a-priori* information needed other than to determine the relevant input data. It was used for investigations on nitrogen transformations (Duffy, 1974; Rohde, 1992) and also for river water quality analysis taking spatial data into account, including socio-economic information available from an official statistical, meteorological, geological and hydrological information in the catchment area (Müller, 1994). It was applied for the river Neckar which was described earlier.

The data of the Neckar river basin was transposed onto the river, using a subdivision of the catchment area published by German water authorities. Each catchment area of main tributaries is repeatedly subdivided, as shown in Figure 5. This subdivision represents the topology of the catchment area. The use of this system of subdivisions can replace the application of a Digital Terrain Model. The resulting brook catchment areas ($6x10^{-2}$ to 120 km^2) have been considered to be sufficiently small to be used as a representation of the topologic structure of the river basin. Every area is assigned a number. These numbers can be used to determine which is the next area upstream or downstream of a certain area considered. The section of the river, where a certain part of the catchment area contributes in one way or another, was determined using these numbers, moving downstream until the river Neckar was reached.

The areas to which the socio-economic information are related are rural districts. Information of geology etc. obviously is also not necessarily based on areas with administrative borders. The boundaries of the data-relevant division of the catchment area differ from those of the subdivided catchment area. It was therefore necessary to proportion the information of the corresponding part of the data-related region onto the respective small catchment area.

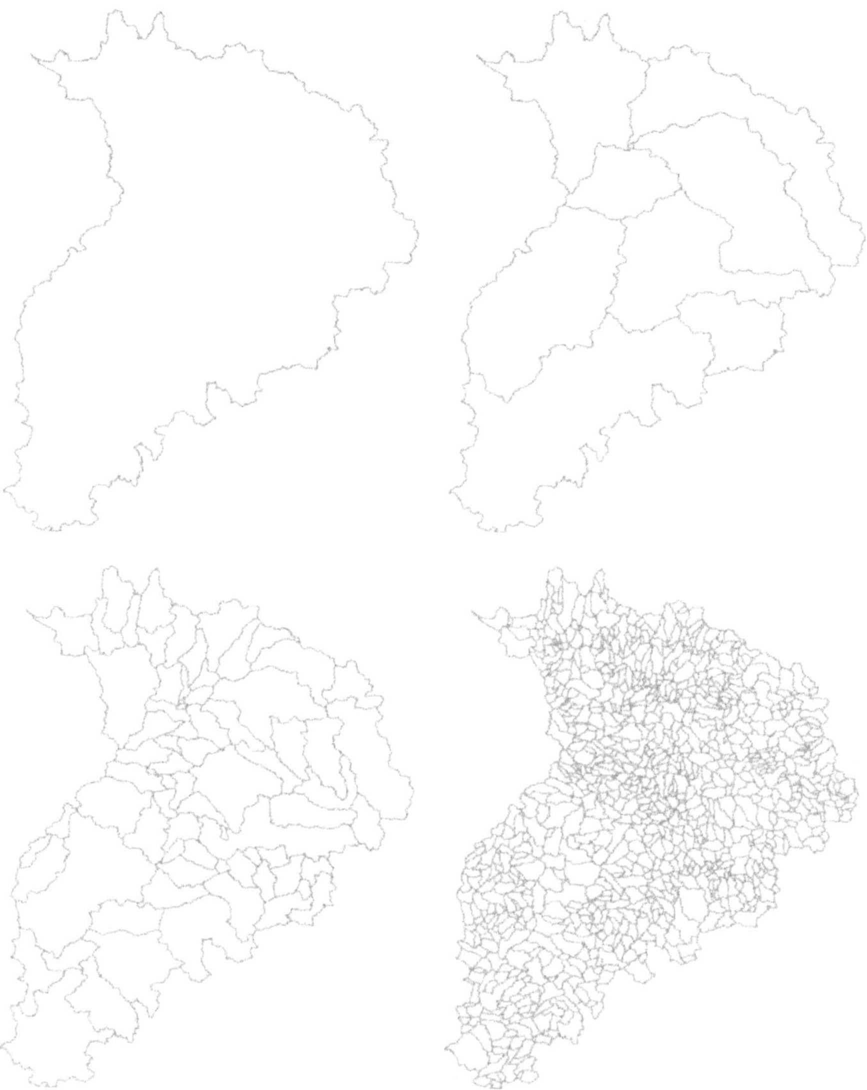

Figure 5. Progressive subdivision of the basin are on the River Neckar

In order to reduce the amount of information, parts of the river basin have been accounted for by introducing the corresponding tributary instead of its catchment area (Figure 6). Tributaries have been considered as point sources if the average dry water discharge was more than one tenth of the River Neckar's discharge at the mouth of the tributary. The area of investigation is illustrated in Figure 7.

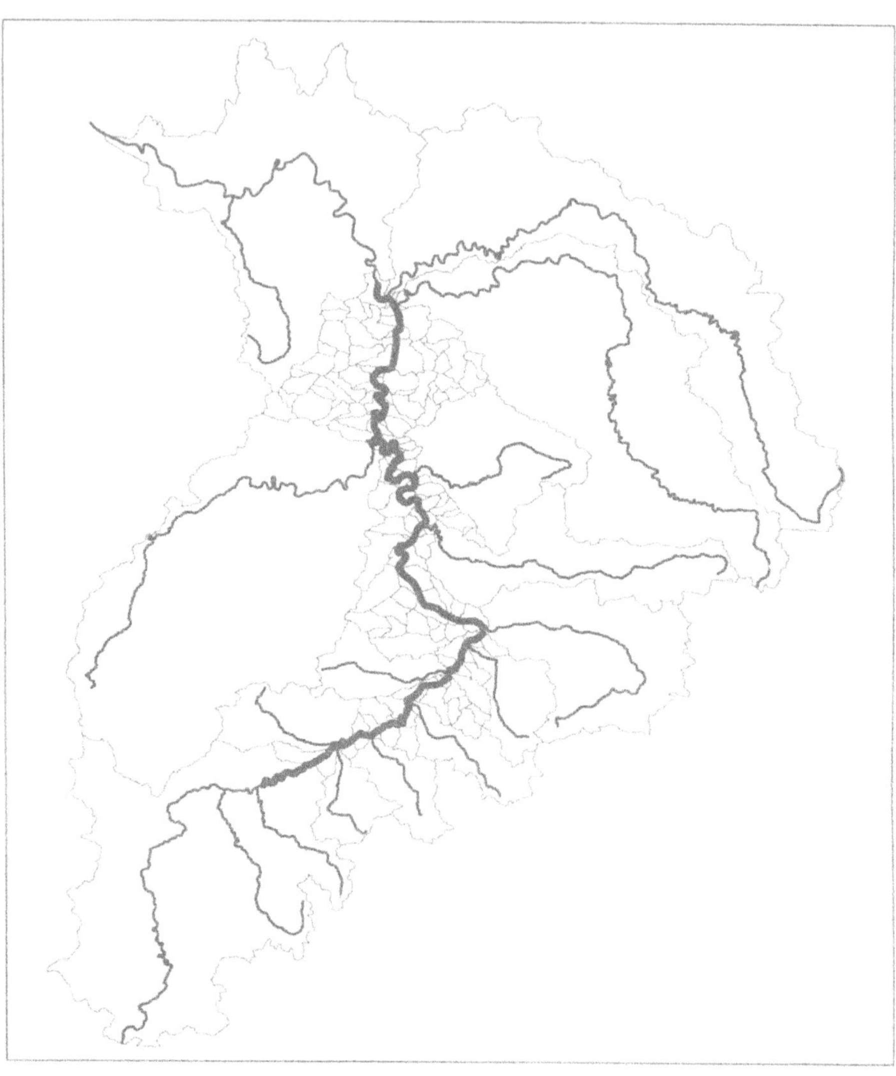

Figure 6. Basin area of the River Neckar with main tributaries

Figure 7. Resulting area of investigation

This method could be used to determine the intensity of influence of certain information on water quality, without using any specific knowledge, except for the selection of relevant information. The ultimate goal, which would be a method determining a transfer function between any number of data representing the model input, and the water quality, representing the model output, could not be achieved. But the method proved applicable to determine an order of preference of the input information items with a very small original data base. It could be shown for example, that the BOD-load from wastewater treatment plants is no longer a significantly influencing quantity for the water quality of the river Neckar. A far greater influence can be attributed to agricultural use, e.g. cattle feedlot or pasture intensity or amount of surface used for farming. These results agree with the expert knowledge, but have been established exclusively with the use of raw data describing the situation in the catchment area and the water quality resulting from it.

The algorithm can be applied to situations where only a small amount of data is available, where other statistical methods cannot be applied. With some further development of the approach a wider field of application can be assured.

REFERENCES

Abwassertechnische Vereinigung (ATV) (1994): Jahresbericht 1993, Kölnische Verlagsdruckerei GmbH, Köln.

Bardossy, A. (1994) "Stochkastische Modelle zur Beschreibung der raum-zeitlichen Variabilität des Niederschlages.," Mitteilungen des Institues für Hydrologie und Wasserwirtschaft der Universität Karlsruhe (TH).

Bardossy, A. and Bogardi (1993) The use of Fuzzy rule-based models for the description of environmental systems. In: Informatik für den Umweltschutz. (Jaeschke, A. et al.)

Barsossy, A. and M. Disse (1993) "A fuzzy-rule based model for infiltration," Water Resources Research **29**(2):373-382

Bayerisches Staatsministerium für Landesentwicklung und Umweltfragen (Hrsg.) (1994): Wasserwirtschaftlicher Rahmenplan Main, München.

Duffy, J.J. (1974): Identification of Soil Nitrogen Transformations In: An Agricultural Ecosystem. Dissertation, Washington University, Saint Louis.

Farlow, St. J. (ed.) (1984) Self-organizing Methods in Modelling, Marcel Dekker, New Yoork

Fedra, K. (1991) "Smart Software For Water Resources Planning And Management." In: Loucks, D.P.; da Costa, J. R. (eds.): Decision Support Systems. NATO ASI Series G: Ecological Sciences, Vol.26, S. 145-172. Springer, Verlag, Berlin.

Ivakhnenko, A.G. and J.A. Müller (1984) Selbstorganisation von Vorkersagemodellen. VEB, Berlin

Ivakhnenko, A.G. (1968) " Group Method of Data Handling - A Rival of the Method of Stochastic Approximation," Sov. Autom. Control, Vol. B, No 3, pp. 43-45

Jen sen, R. (1992) Verwendung von soziologischen Daten zur Beschreibung der Gewässergüte mit Hilfe von unscharen (fuzzy) Ansätzen. Diplomarbeit am Institut für Siedlungswasserwirtschaft der Universität Karlsruhe (TH).

Kaufmann, A., and M.M. Gupta (1985) Introduction to Fuzzy Arithmetic. Theory and Applications. New York: Van Nostrand Reinhold, New York, N.Y.

Masliev, I; and L. Somlyódy (1994) Uncertainty Analysis and Parameter Estimation for a Class of River Dissolved Oxygen Models, Working Paper - International Institute for Applied Systems Analysis, WP-94-9, Laxenburg, Austria, IIASA.

Müller, N. (1994) "Gewässergütemodellierung von Flieβgewassern unter Berücksichtigung qualitativer, quantitativer, flächenhafter und sozioöknonomischer Informationen," Schriftenreihe des Instituts für Siedlungswasserwirtschaft der Universität Karlsruhe (TH), Band 70.

Rohde, A. (1992) Nährstoffhaushalt und Stickstoffdynamik in der ungesättigten Bodenzone. In: Plate, E. (Hrsg.) (1992): Weiherbach-Projekt "Prognosemodell für die Gewässerbelastung durch Stofftransport aus einem kleinen ländlichen Einzugsgebiet". Mitteilungen des Institutes für Hydrologie und Wasserwirtschaft der Universität Karlsruhe (TH).

Zadeh, L.A. (1973) Outline of a New Approach to the Analysis of Complex Systems and Decision Processes. IEEE Transactions on Systems, Man & Cybernetics SMC-3, p. 28-44.

CHAPTER 12

RIVER BASIN WATER QUALITY MANAGEMENT STRATEGIES IN THE CENTRAL EUROPEAN REGION: AN EXAMPLE OF THE NITRA RIVER (SLOVAKIA)

L. Somlyódy[1,2], I. Masliev[1], and M. Kularathna[1]

1. INTRODUCTION

Water quality management in most Western countries is based on the effluent quality standards, leading to uniform emission reductions at all sites. The development of such a policy is a simple task, and both the ambient water quality and the costs directly follow from these standards. Enforcement is also straightforward. This system is widely known as "end-of-pipe control". In fact, the actual impacts and costs are often of little interest or unknown in advance. It is generally assumed that receiving water quality will be "good" if stringent effluent criteria were selected and money was available to realize the strategy (i.e., society is willing to pay for a safe environment). The choice of technology is also a side-effect of this system since standard values are most frequently set on the basis of a few (or one) well proven technologies (e.g., secondary activated sludge biological treatment in the U.S.).

Following this policy, Western countries achieved remarkable improvements in the state of the water environment during the past two decades or so. However, it was not a cheap process. For instance, control and management of municipal emissions alone necessitated expenditures amounting to 1% or more of the GDP in several countries for the above period.

The situation is rather different in many Central and East European (CEE) countries, including

[1] International Institute for Applied System Analysis, Laxenburg, Austria

[2] Department of Water and Wastewater Engineering, Technical University of Budapest, Hungary

NATO ASI Series, Partnership Sub-Series, 2. Environment - Vol. 3
Remediation and Management of Degraded River Basins
Edited by V. Novotny and L. Somlyódy

Slovakia. The per capita costs of addressing rather serious water pollution problems can be estimated to be several thousand US$ which can exceed the annual specific GDP (GDP in these countries is roughly 1/5 to 1/10 of Western European values). The economies of the CEE countries are presently in a rather poor state. National debts are high, production is still declining, inflation rates are significant, and unemployment in some CEE countries reaches 15%. Increasing prices and decreasing salaries, bread-and-butter worries, and the need to restructure the economy as fast as possible puts environment at a low priority for both the public and various governments. One or two percent of the GDP are simply not available for municipal emission control (and due to the difference in per capita GDPs between the West and East, a much larger investment with respect to income, would be needed to have an equally rapid improvement as in Western countries). Economic recovery will unfortunately be a slow process; it suffices to refer to the slow development of Portugal or Spain - which were in a comparable situation to some of the CEE countries 10-15 years ago - or to the fact that doubling the GDP assuming a 5% growth rate would require about fifteen years.

Environment in many countries of the region is in poor shape, since for several decades policy makers virtually neglected environmental issues. Money is not available to correct the situation immediately. Therefore, a question can be raised of investigating various emission control strategies, having in mind the present scarce financial resources. One of the options is to adopt the multi-stage control strategy, answering the most pressing needs immediately and ensuring transition to sustainable development practice when money becomes available (Somlyódy, 1993). The potential of saving exists in two directions mostly: implement a non-uniform emission control strategy where the main objective would be to achieve acceptable ambient quality of the receiving river, and broaden the range of alternative technical solutions to include non-traditional cost-effective waste removal and upgrading methods such as chemical enhanced primary and secondary treatment. Both of these approaches were utilized simultaneously in the Nitra River basin study (Somlyódy et al, 1994).

In a broader sense of legislation, the issue is to add efficiency as a critical element under the present conditions of CEE countries, to the underlying principles of Western legislation such as equity, uniformity and easy-to-enforce statues. The implementation of an "efficient" legislation for the short run (say the next 10-20 years) is certainly not an easy issue and requires a proper institutional setting. This is beyond the scope of this chapter. Here we solely wish to demonstrate some of the technical tools as a pre-requisite which can be applied to develop cost-effective strategies which can be further extended later on.

The development and comparison of these strategies (based on effluent standards alone or mixed standards) is the major objective here, with reference to the Nitra watershed in Slovakia. The work to be outlined is a central element of a comprehensive policy oriented river basin research program with the involvement of the Water Research Center (Bratislava) and the Váh River Basin Authority (for details the reader is referred to Somlyódy et al. (1994). Other components of the study include a survey of various emissions, the development of wastewater treatment alternatives and upgrading strategies for most of the municipalities (Somlyódy et al., 1994), longitudinal water quality profile measurements and their evaluation (see Masliev and Somlyódy, 1993; Masliev et al., 1994), and the application of water quality simulation models from simple dissolved oxygen models to a more complete nutrient cycling model QUAL2E (see Breithaupt and Somlyódy (1994) for the latter) which are used to assess and compare various control policies.

The policy analysis is based on the dynamic programming approach (DP), which is a versatile and powerful method providing a least-cost policy solution subject to a set of ambient constraints. Numerous uncertainty elements are inherent in the process of decision making on a river basin scale. Therefore, uncertainty and risk analysis were incorporated into the study. Two methods were used: regret analysis based on a set of probable scenarios and direct Monte-Carlo simulation. Both of them used probability distributions of model parameters. The latter were obtained in a model calibration study (Masliev and Somlyódy, 1993).

The chapter is organized as follows: First, the situation in the basin is outlined with respect to the water quality management problem. Second, water quality simulation models used are outlined. Water quality management models are discussed in the third section of the chapter. The fourth section is devoted to various treatment strategies considered as viable control solutions for individual dischargers. Different control strategies developed with a focus on municipal emissions and oxygen balance are discussed in the Section 5, while the role of industrial emissions is demonstrated by sensitivity analysis. The influence of parameter uncertainty (of the traditional dissolved oxygen model employed at this stage) on policies developed is the objective of the analysis in Section 6 using regret analysis methodology. Another approach to uncertainty analysis, a Monte-Carlo *a posteriori* simulation, is discussed in Section 7. Policy recommendations are summarized in Section 8, and Section 9 contains conclusions of the study.

2. EMISSIONS AND WATER QUALITY SITUATION IN THE NITRA RIVER BASIN

The Nitra River is a tributary of the Váh which enters the Danube in Komarno, downstream from Bratislava where the Danube forms the border between Slovakia and Hungary. The length of the river is about 170 km. The mean stream flow near the mouth is 24 m^3/s, while a typical August low flow condition is characterized as being less than 3 m^3/s. The watershed area is 5100 km^2 and the basin population is 650 000 inhabitants (about 40% of them live in rural areas). The region is highly industrialized. Metallurgical works, meat processing and sugar beat factories, tanneries, chemical plants etc. can be found at various locations. The level of water re-cycling and industrial wastewater treatment is low. Agricultural non-point source pollution is negligible. Eleven larger municipalities are sewered and have treatment plants (with a total population of 350 000). The level of public water supply as well as sewerage in these municipalities is close to or above 90%. In rural settlements, however, only a small portion of the population is supplied by potable water, and wastewater treatment is practically non-existent. The level of wastewater treatment is below 50%. Treatment plants, most of which use the activated sludge process without nitrification, are on average 20 years old. Several plants are overloaded by 100% or more, operating at biochemical oxygen demand (BOD5, which will be denoted hereinafter as BOD) removal rates between 60% and 70%. The excess flow above the design capacity receives generally only primary treatment. A map of the watershed illustrating the levels of municipal wastewater treatment is presented in Figure 1. The level of industrial wastewater treatment is shown in Figure 2.

The water quality of the river is one of the poorest in Slovakia. Parameters of the oxygen balance and other chemical components comply only with quality Class IV-V requirements, according to the national classification system (where V class indicates the worst quality). For instance, BOD values can exceed 30 mg/l, and dissolved oxygen (DO) is sometimes below 1-2 mg/l (Somlyódy et al., 1994). Total phosphorus (TP) can reach 2 mg/l, while ammonia nitrogen (NH4-N) can be around 7 mg/l. High algae biomass levels can be observed at the downstream stretch of the river (chlorophyll-"a" values of 150 mg/m^3 are not rare), where travel time is sufficiently long.

Municipalities contribute about 70% of the total emission of traditional pollutants (organic materials, phosphorus and nitrogen) of which 30-50% is due to industrial discharges (see Figs. 1 and 2). Industrial contamination is characterized, among others, by high arsenic concentrations (the origin of which is yet unknown) and high conductivity.

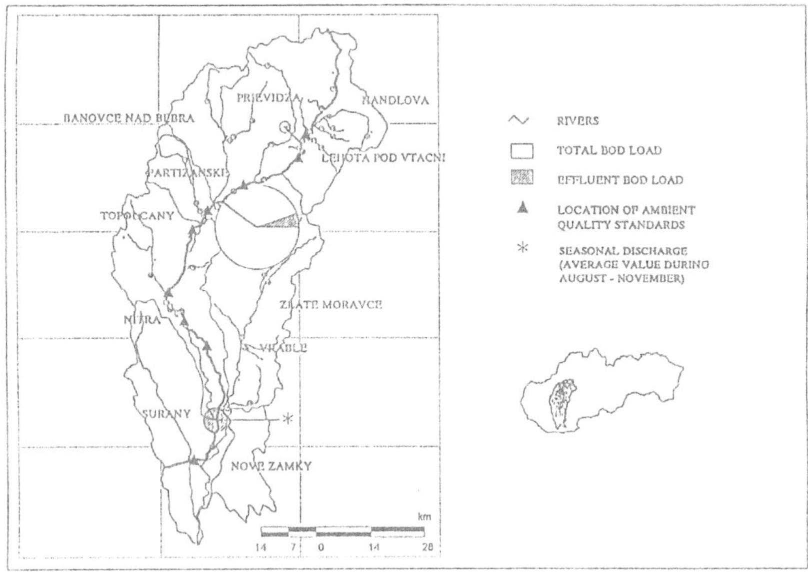

Figure 1. Municipal wastewater treatment in the Nitra basin

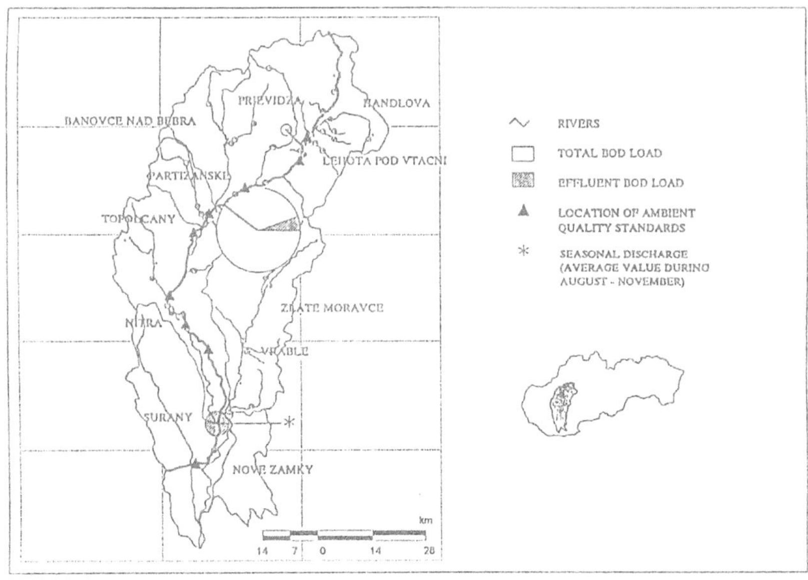

Figure 2. Industrial wastewater treatment in the Nitra basin

At present, the primary use of the Nitra River is for waste disposal (as apparent from what was outlined above) and for water abstraction for industrial and irrigation purposes. For the latter purpose, small dams have been built, mostly at the downstream part of the river.

The economy of the country and the region is in transition. In spite of increasing water prices, domestic water consumption has not yet been reduced significantly. The same statement also applies for industries connected to municipal wastewater treatment plants (WWTPs). Thus, the flow and load of WWTPs have been reduced only slightly during the past three years. In contrast, larger industrial plants outside of urban areas underwent significant changes. A large portion of them were fully shut down. Some others show about 50% discharge reduction and only a few operate in an unchanged fashion. In the short run, this trend is likely to be continued, while longer term alterations are difficult to predict.

A new effluent standard system was enacted in 1992 by the former Czech and Slovak Republics as a basis for future legislation. This system leaves open the possibilities to also incorporate ambient water quality criteria. The procedure realistically distinguished two stages - before and after 2004. For a 100 000 population-equivalents or larger municipality the standards are as follows: BOD=30 mg/l, NH4-N=10 mg/l and TP=3 mg/l. The corresponding limits are slightly more stringent as of 2005.

3. MODELING EFFORT

3.1 Water Quality Simulation Model and Design Conditions for the Control Policy

The most severe water quality problem in the Nitra River catchment is presently organic pollution and subsequent dissolved oxygen depletion during low-flow periods. Therefore, the water quality models used for the study describe a variety of oxygen and organic pollution factors. For policy purposes, the traditional Streeter-Phelps model and its modifications were tested (Masliev and Somlyódy, 1993). Different model versions were calibrated to the longitudinal water quality observation performed in August 1992 in the framework of the policy study. At the end, the usage of the three-state variable, linear version model was decided upon, incorporating carbonaceous and nitrogenous biochemical oxygen demand (CBOD and NBOD), as well as DO. The key parameters of this model are the decay rate for CBOD and NBOD, respectively (assumed to be identical, K_d) and the reaeration coefficient K_r. The estimated mean decay rate (K_d) was around 0.8 l/d, which is higher than the recommendation of the literature. The explanation is twofold. First, the presence of partial biological waste water treatment only and, second, the small water depth. The mean reaeration coefficient (varying longitudinally depending on the hydraulic parameters) was about 2.0 l/d, leading to a realistic K_r/K_d ratio.

A sensitivity analysis showed the dominating role of the reaeration coefficient on the DO profile. The statistical distributions of parameters were available from a Monte Carlo parameter estimation performed by Masliev and Somlyódy (1993). The coefficient of variation was about 15% for K_r, and this value will be employed for the deterministic multiobjective analysis. The ongoing research activities that are not described here include: stochastic extension of the present approach, the incorporation of more complex, non-linear water quality models, scheduling of the implementation of treatment strategies, and the systematic evaluation of various management model formulations in a broad sensitivity study framework.

Results of the August 1992 experiment (Masliev et al., 1994) performed under critically low-flow conditions served as the basis for the (single) design scenario (including the derivation of parameters of the DO model). Background pollution of the tributaries and the temperature profile (showing an increase downstream from 19°C to about 25°C) were obtained from the observations. For the generation of design emissions for municipalities, annual average values available for the period 1990-1992 were used in addition to measured ones. Industrial emissions were kept at their 1990 level, mainly as a worst-case scenario and to demonstrate the role of changing industrial emissions (as observed recently) by a sensitivity study. The optimal policies to be developed will focus on municipal emissions. Industry will be considered without analyzing the costs, due to lack of reliable information on industrial discharge control measures.

3.2 Water Quality Management Model

Simulation models, optimization models, or a combination of these two can be employed to determine the "optimum" or the satisfying set of wastewater treatment alternatives for river basins. These methods assist a decision maker in reaching a solution which is optimal (or near-optimal) with respect to some predefined goals. Often there are economic and water quality goals to be considered. The most obvious economic goals are the cost-minimization and proper distribution of costs (in space and time). Water quality goals are usually expressed by quality standards for the receiving waters and/or for wastewater effluents.

In order to select an "optimal" set of wastewater treatment alternatives by simulation, a river model has to estimate the consequences of a variety of feasible alternatives. The number of feasible combinations of treatment alternatives in a river basin increases rapidly as the number of treatment plants and treatment levels increases. Consequently, a large number of simulations are required to select an "optimal" set of alternatives. In order to incorporate the uncertainty of various inputs and parameters, many more simulations in a Monte-Carlo fashion would be needed. This will increase the required number of simulations to an impractical level. Nevertheless, simulation offers the possibility for a more detailed representation of a river system. Simulation models in water quality management include those presented by Warren and Bewtra (1974), Orlob (1982), and Thomann and Mueller (1987). The role of BOD dischargers in the Nitra river basin was analyzed using a simulation model by Koivusalo et al. (1992). A state-of-the-art review of water quality simulation models, including a discussion on decision support systems, can be found in Somlyódy and Varis (1992).

When considering the optimization techniques, linear programming (LP) has been one of the most widely used techniques in water quality management (Loucks et al. (1967), ReVelle et al. (1968), Bishop et al. (1974), Biswas (1981), Burn and Lence (1992)). LP solves a special type of problem in which all relations among the variables are linear. This requirement should be fulfilled by the objective function as well as by the constraints of the model formulation, which is rarely met in practice. The wastewater treatment cost functions as well as the "transformation functions" that relate waste discharge to the river water quality are generally nonlinear.

Dynamic programming (Bellman, 1957) is an optimization method for a multistage decision problem (in general, not necessarily linear). It decomposes a problem with a sequence of decisions into a sequence of sub-problems each having one or a reduced number of decisions. These subproblems are solved recursively, by considering the sub-optimal solution(s) of one subproblem as input(s) to the subsequent subproblem. The selection of optimal wastewater treatment alternatives in a river basin is such a sequential decision problem, involving sequential decisions in space as well as in time. Spatially, the decisions are to be taken for a series of locations in a river basin. Due to the downstream-only propagation of pollutants in a river system, the water quality at a particular location in a river is fully determined by the water quality at the immediate upstream discharge/control point (or by several discharge/control points in the special case where the considered location is below confluence). Similarly, in planning the investments over the planning horizon, the decisions are to be made at a sequence of points in time. The decisions made at one time point directly affect those to be made at the next time step.

These special sequential attributes of the problem make it suitable to be solved by a dynamic programming approach. Applications of dynamic programming (DP) for water quality management problems have been reported by Newsome (1972), Hahn and Cembrowicz (1981), and Cardwell and Ellis (1993). The approach of dynamic programming was selected for the presented study as well, due to generality and robustness of the methodology.

In current application of dynamic programming to the water quality management decision making, the river system was subdivided into a number of reaches that were further divided into "stages". The river network has been defined by the interconnection of different reaches. A "stage" was considered as a part of the river from a point immediately upstream of a "point of action" to a point immediately upstream of the next "point of action" downstream. A "point of action" can be: a wastewater discharge, an abstraction point, a measurement point, a point with a prespecified water quality standard, a weir or one of the artificial points introduced to maintain a generalized computational procedure.

The policy model has to select the optimal treatment strategy from a set of feasible alternatives for each discharge. Consequently, the task of the policy model is to choose the optimal configuration by selecting one treatment alternative for each discharge. In other words, the policy model has to make a yes/no (1/0) decision for each treatment alternative at each municipal discharge. As only one alternative will be selected for each location, all but one alternative at a discharge point will receive "no" decisions.

The other inputs to the policy model include the water quality goals or standards defined by legislation. With these inputs, the policy model makes use of the water quality simulation model estimations to identify the appropriate least-cost policy. The outcome of the policy model is characterized by a treatment configuration, resulting water quality in the river, and the associated costs.

The computations of DP start from the most upstream point of the river system. Using Bellman (1957)'s principle of optimality, the DP calculations are performed stage by stage, proceeding towards the most downstream point of the system. During computation, the current decision (treatment alternative), the previous state (at the previous stage) and the cumulative cost are recorded for each allowable water quality state at each stage. Having reached the most downstream point to be considered, the optimal solution can therefore be traced-back upstream (Somlyódy et al, 1993).

Due to numerous uncertainties inherent in the process of water quality management decision making, it is important to account for them in the management model. The procedure of Monte-Carlo parameter estimation incorporates all the model uncertainty into the probability distribution in the parameter space (Masliev, Somlyódy, 1993). In order to estimate the robustness of the developed policies with respect to uncertain factors, an approach based on multiple scenarios (reflecting alternative hydrologic, meteorologic and pollutant loading conditions) was presented by Burn and Lence (1992). Similar "regret analysis" was used also in this study to explore how the derived policy is affected by uncertain water quality model parameters, showing the gains and losses in terms of water quality (and risks associated) and costs. Additionally, an a posteriori direct Monte Carlo simulation estimates possible variations of water quality and provides probabilistic conclusions on standard violations (Section 7).

4. MUNICIPAL WASTE WATER TREATMENT ALTERNATIVES

For all the WWTPs in the region, analyses were performed to develop feasible treatment alternatives. Each alternative is characterized by the flow, the effluent quality expressed by DO, BOD, SS, NH4-N, nitrate nitrogen (NO3-N) and TP. From of view point of the expenditures, the important characteristics are investment cost (IC), operation and maintenance cost (OMRC) and total actual cost (TAC) (the latter depends on the former two and interest rate on capital). The alternatives were obtained by considering existing units (primary sedimentation tank, aeration basin, final clarifier and sludge processing), their various upgrading possibilities, and the design of new treatment plants. The average project life time was assumed to be 20 years (longer for new plants and shorter for upgraded ones). In addition, a 10% interest rate was employed to obtain total annual costs.

Feasible treatment alternatives at each WWTP were assigned numbers (treatment levels) starting from 0 (no-treatment was indicated by Level 0, while the most expensive solution by the highest level, see below). The number of levels ranged between four and ten, depending on the actual situation and technological calculations. The basis of deriving alternatives were as follows:

- Levels 0 corresponds to no treatment, while Level 1 assumes the operation of existing facilities as it is the case nowadays. The demolition of existing WWTPs and the construction of new, advanced plants characterized approximately by BOD of 15 mg/l, TN of 10 mg/l and TP of 1 mg/l effluent quality (corresponding to the most stringent recommendations of the European Community) was considered as the highest "level" alternative (called "Best Available Technology", BAT).

- The increase of the number and size of units (without changing the type of processes) to compensate overloads (depending on technological calculations) resulted in other alternatives.

- Upgrading by adding a low dosage of chemicals before the primary clarifier was considered as an attractive, cost-effective alternative for most of the WWTPs (which leads to a combined chemical-biological treatment, see the chapter by Henze and Ødegaard and also Henze and Ødegaard, 1994). As shown in the literature, the surface overflow rate and BOD removal of primary sedimentation basins can be practically doubled (Morrissey and Harlemann, 1990). Thus, the capacity of overloaded biological treatment plants can be significantly extended (depending on the capacity of the final clarifier and the sludge line). Jar tests performed at various plants in the Nitra Basin justified the applicability of this method (Murcott and Harlemann, 1994).

- The combination of the above upgrading options is also possible. Similarly, the re-shaping of existing plants and the construction of future WWTPs (which can be based on different technological principles, see Henze and Ødegaard, 1994) can also combine different options (which lead to additional levels). For instance the flow of an existing, overloaded facility can be reduced (even to a nitrification operation mode resulting in a "downgrading") to improve the performance. This would require the construction of a smaller new plant than under the condition of pulling down the old one.

Additional alternatives can correspond to present or future effluent standards in Slovakia, and thus, a comparison to other strategies is a straightforward task. Examples for this can be given with regard to one of the two largest WWTPs in the basin located in Nove Zamky. The design capacity is slightly above 10 000 m^3/d, while the present flow is close to 30 000 m^3/d. Table 1 displays the feasible treatment alternatives and approximate cost estimates for the Nove Zamky treatment plant. The BAT would cost more than 20 million US$ each. The "cheapest", still effective upgrading which improves the effluent BOD value from about 60-80 mg/l to approximately 20 mg/l would cost about 3-6 million US$, while a combination of a "downgraded" old plant and a new one would amount to 11-14 million US$. Further details on cost-effective treatment strategies and applications to the Nitra River basin are included in Somlyódy (1993) and Somlyódy et al (1994).

The analysis outlined above led to a set of tables for each treatment plant. As displayed in Table 1, the lines of the tables represent alternatives or levels characterized by effluent water quality and costs. To each line a (0,1) decision variable is assigned which expresses the linkage to the policy model.

Table 1. Characteristics of the feasible treatment alternatives for Nove Zamky WWTP.

Alternative	Cost (10^6 US$)		Effluent Concentrations (mg/l)			
	IC	OMRC	BOD	TP	NH4-N	NO3-N
0	0	0	240	11	48	0
1	0	1.0	60	9	40	0
2	3	1.2	15	7	4	30
3	2	1.4	20	1.5	34	0
4	5	1.6	15	1.0	4	30
5	6	1.9	15	1.0	2	17
6	13	1.2	15	7	4	30
7	11	1.6	15	1	4	26
8	14	1.9	15	1	0	19
9	21	2.0	10	0.8	0	10

5. WATER QUALITY CONTROL POLICIES

The policy analyses can be classified into three main groups, depending on the type of water quality standards considered: effluent, ambient, and mixed. From the point of view of costs, two different types of objectives can be considered separately. The first is to formulate a policy with a minimum total annual cost (TAC) of waste treatment. The total annual cost comprises the operation, maintenance and replacement costs (OMRC) and the annual repayment of the investment cost. The second goal was to find a policy which requires the minimal basin-wide investment cost (IC), which stem from the tight budgeting situation in CEE countries including Slovakia.

A summary of the most interesting results obtained with the water quality control policy model using dynamic programming is given in Table 2. It incorporates a number of strategies, listing IC and OMRC costs, water quality extremes for DO, BOD, and NH4-N (outside of locations where ambient criteria were set as constraints), as well as the sum of treatment levels (defined above) for all the sites. The first five strategies are based on effluent standards, and thus, both ambient water quality and costs are direct consequences (no optimization is needed). The second block [(6)-(9)] represents least-cost

Table 2. A summary of different control policies.

No:	Policy*	IC [mil. US$]	OMRC [mil. US$]	DO min [mg/l]	BOD max [mg/l]	NH4-N max [mg/l]	Sum of treatment levels
1	No treatment	0.0	0.0	0.1	34.6	7.9	0
2	Current treatment	0.0	5.7	2.3	30.6	7.7	10
3	Eff: BOD£30, NH4-N£10, TP£3	32.1	8.4	5.4	11.3	2.3	35
4	Eff: BOD£25, NH4-N £5, TP£1.5	35.2	8.7	5.4	11.3	2.3	37
5	Best available technology	95.5	11.1	5.7	11.1	2.1	53
6	DO³3	3.2	3.8	3.0	18.7	4.2	8
7	DO³4	4.0	5.6	4.2	13.8	4.1	11
8	DO³5	15.0	6.2	5.0	13.8	3.2	18
9	DO³6**	33.4	7.7	5.6	11.3	2.2	26
10	DO³5, BOD £10	20.0	7.0	5.0	11.2	3.6	20
11	DO³5, BOD£10, NH4-N £2	29.5	7.1	5.6	11.3	2.2	24
12	DO³5, BOD£10, Eff: NH4-N£10	31.0	7.9	5.6	11.1	2.2	29
13	DO³5, BOD£10, Eff: NH4-N£10, TP£3	33.6	8.5	5.5	11.3	2.3	36
14	DO³5, BOD £ 10, simul: with 50% ind.	20.0	7.0	6.0	8.5	3.5	20
15	DO³5, BOD £ 10, simul: with 25% ind.	29.5	7.1	5.5	8.2	3.4	20
16	DO³5, BOD £ 10, 50% ind.	7.2	6.2	5.1	9.7	3.6	14
17	DO³5, BOD £ 10, 25% ind.	6.2	5.7	5.1	11.4	3.8	12

* Standards are expressed in mg/l

** Constraint violated at one of the standard points

strategies defining ambient criterion alone for DO. The next two lines illustrate the consequences of having additional constraints for BOD and NH4-N, respectively, (corresponding to quality Class III). Policies (12) and (13) are mixed ones, where ambient and effluent criteria are used jointly. Finally, the last block shows the influence of industrial emissions. The first two lines were obtained with the aid of simulation and they show the impact of uniformly reduced industrial emissions on water quality. In contrast, the last two lines illustrate least-cost strategies under the reduction of industrial loads by 50% and 75%, respectively.

From the analysis of the Table 2, the following conclusions can be drawn:

- The range of water quality changes due to control is relatively narrow (e.g., between 11 and 35 mg/l for BOD) since only municipal emissions are incorporated into the optimization model. A portion of emission load comes from small industrial discharges which were considered as background non-controllable pollution. The rest of the industry was treated in a sensitivity analysis framework, setting certain uniform reduction levels. The combination of BAT for municipalities and 75% uniform industrial discharge reduction leads to a maximum BOD concentration (BOD_{max}) of about 8 mg/l and a minimum DO concentration (DO_{min}) close to 7 mg/l (the average saturation value was around 8 mg/l). This result illustrates both the role of industry and that of "background" pollution. The optimal budget allocation would be different from the present one if all the discharges of differing origins were accounted for. In contrast to water quality, the domain of meaningful expenditures is extremely broad, between 3 and 95 million US$. It indicates significant saving possibilities. The longitudinal profiles of ambient DO concentrations corresponding to the strategies (2), (3), (5), (6) and (8) are presented in Fig. 3. It can be observed that the strategies (3), (5) and (8) give rather identical DO profiles, although their investment cost requirements are very different (32, 95 and 15 million US$ respectively).

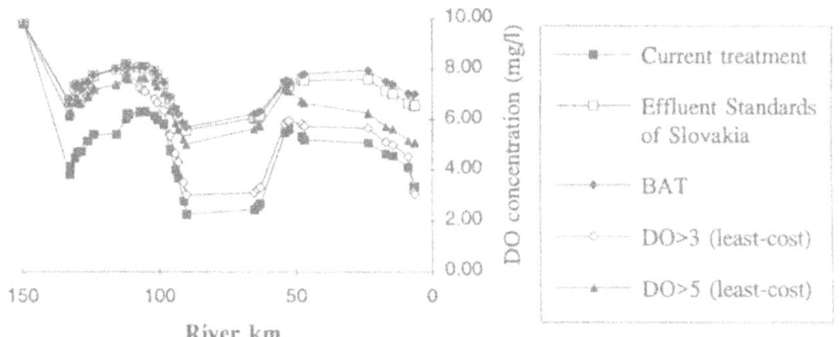

Figure 3. Longitudinal profiles of DO concentration for the policies 2, 3, 5, 6, and 8.

- Four clusters of policies can be identified from the Table. 3-4 million US$ is the smallest investment which leads to measurable improvement in DO (and also in the other two components). The second cluster is formed by policies (8) and (10) in the 15-20 million US$ capital cost domain ($DO \geq 5$ and BOD=11-14 mg/l). The third group comprises strategies (3), (4), (9), and (11)-(13). The investment cost is 30-35 million US$ and the water quality further improves ($DO^{3}5$ mg/l, BOD ~ 11 mg/l, and NH4-N ~ 2 mg/l roughly corresponding to Class III water, while at present Class IV-V is assigned; see Somlyódy et al, 1994). Nevertheless, the marginal cost of this improvement is rather high. It is worthy to note that policies (3) and (4), corresponding to present and future effluent standards in Slovakia are identical from the viewpoint of ambient quality, in spite of the difference in costs. Finally, the last cluster is represented by the construction of new BAT treatment plants. Short term benefits are absent; the capital cost is three times higher than for the previous set, but water quality is practically unchanged (Class III). However, BAT reduces the NH4-N concentrations more pronouncedly than the DO and BOD profiles, due to the associated denitrification process.

- The economic implications of setting standards is strong and non-linear. For ambient water quality standards considering DO alone, the IC shows practically an exponential increase (see Fig. 7) with tightening the minimum DO level. The performance of the strategy corresponding to DO 6 differs only slightly from the BAT strategy. On the contrary, the difference in the investment costs of those two alternatives is 62 million US$. The improvement of DO_{min} from 4 mg/l to 5 mg/l requires 11 million US$ investment. Similarly, the price of specifying BOD as a criterion in addition to DO (not necessarily justified professionally) is a 5 million increase in the cost (see Table 2, lines (8) and (11)). The incorporation of NH4-N has an even stronger influence. The cost implications of setting DO and NH4-N standards are summarized in Fig. 8, considering three different scenarios for the industrial discharges based on their 1990 values (no reduction, 50% reduction, and 75% reduction, of 1990 discharges). It is to be noted that the peculiar low IC associated with the standards $DO^{3}6$, NH4-N£2 in Fig. 5(e) is offset by a higher OMRC cost; indicating a higher amount of total annual cost.

- The reason for the above nonlinear behavior (of the economic implications of setting standards) is the change in the sum of treatment levels and in the spatial configuration. Treatment levels corresponding to each strategy in Table 2, for each location, are displayed in Table 3. As can be seen from Tables 2, although there is hardly any difference between strategies (2) and (7) in the total number of treatment levels, the latter non-uniform policy (see Table 3) is much more efficient. The transition from policy (7) to (8) is achieved primarily by investing in the upgrading of two larger WWTPs, Topolcany and Nitra (Fig. 1), respectively. In contrast to effluent standard based strategies, all the least-cost policies are strongly non-uniform, reflecting longitudinal changes of water quality and the local cost-effectiveness of individual control measures. Mixed policies (e.g., (12) and (13)) show more uniformities

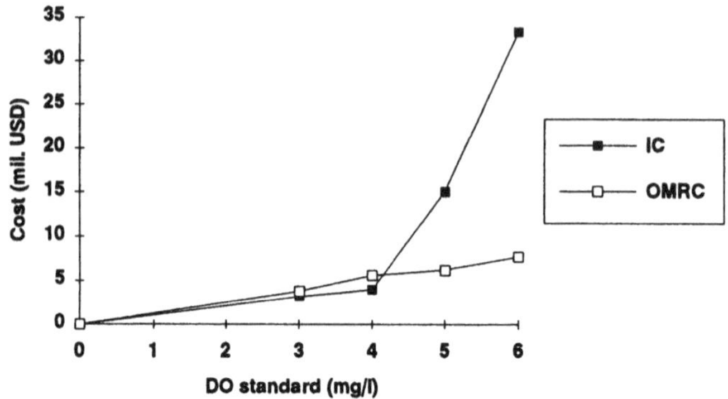

Figure 4. Cost implications of setting different ambient quality standards

Note: BAT will cost 95.5 mill. US$ resulting in DO level close to 6 mg/l

Table 3. Treatment levels corresponding to the strategies in Table 2.

Policy*	Treatment Plant**										
	Ha	Le	Pr	Pa	Ba	To	Ni	Zl	Vr	Su	No
No treatment	0	0	0	0	0	0	0	0	0	0	0
Current treatment	1	1	0	1	1	1	1	1	1	1	1
Eff: BOD£30, NH4-N£10, TP£3	2	2	5	3	3	6	5	2	1	2	4
Eff: BOD£25, NH4-N £5, TP£1.5	2	2	5	3	3	6	6	2	2	2	4
Best available technology	4	3	4	3	5	8	7	4	3	3	9
DO^{3}3	0	0	2	0	1	2	2	0	0	1	0
DO^{3}4	0	0	2	1	3	1	1	0	0	1	2
DO^{3}5	0	0	2	1	3	6	2	0	1	1	2
DO^{3}6	2	1	3	1	3	8	5	0	0	1	2
DO^{3}5, BOD £10	1	1	4	1	3	3	2	1	1	1	2
DO^{3}5, BOD£10, NH4-N £2	2	1	3	1	4	6	5	0	0	0	2
DO^{3}5, BOD£10, Eff: NH4-N£10	2	1	4	1	3	6	5	2	1	2	2
DO^{3}5, BOD£10, Eff: NH4-N£10, TP£3	2	2	5	3	4	6	5	2	1	2	4
DO^{3}5, BOD £ 10, simul: with 50% ind.	1	1	4	1	3	3	2	1	1	1	2
DO^{3}5, BOD £ 10, simul: with 25% ind.	1	1	4	1	3	3	2	1	1	1	2
DO^{3}5, BOD £ 10, 50% ind.	1	0	2	1	3	2	2	0	0	1	2
DO^{3}5, BOD £ 10, 25% ind.	0	1	2	1	1	2	2	0	0	1	2

*Standards are expressed in mg/l **First two letters of the name of the treatment plant (see Fig. 1)

Treatment technologies remove different pollutants simultaneously, but to different extent. The multiple pollutant nature of the problem and the alternative treatment methods is well-reflected by the table. Total phosphorus (TP) plays a less important role in this respect. The influent TP concentration is rather small (due to high water consumption and infiltration to the sewerage), and its reduction to levels set by effluent standards can be realized relatively easily.

- The role of industry is demonstrated by Fig. 5 and the last four lines of Table 2. Assumed emission reductions clearly improve DO and BOD conditions. However, the impact is reflected more pronouncedly by the costs. The "savings" (which in reality depends on the costs of industrial emission control) on the municipal emission control can be 13-14 million US$ (see (10), (16) and (17)), expressing the need to develop an integrated least-cost strategy covering all the discharges of different natures.

- From Table 3, selection of policy (7) or (8) seems to be logical, depending on the budget available (see Somlyódy and Paulsen, 1992 for a similar conclusion from a preliminary analysis of the Nitra problem).

6. THE IMPACT OF THE MODEL UNCERTAINTIES ON THE SELECTION OF CONTROL STRATEGIES: REGRET ANALYSIS APPROACH

Deterministic conditions were assumed for obtaining the least-cost strategies of this study. It means that model parameters and external conditions such as design flow, temperature, background load, etc., were assigned predetermined values for the optimization process. In real life, the situation can differ from the design conditions, so the sensitivity/uncertainty analysis is necessary. In this work, only the model uncertainty (expressed via the parameter uncertainty) was analyzed in more detail in two ways described below. Other external parameters, such as background load, were analyzed by sensitivity analysis (Section 5).

The goal of the first approach was to evaluate the effect of using a parameter which in reality assumes a different value from that taken during design. This implies that a mistake has been done in the design and the magnitude of the mistake, in relation to the actually realized parameter, can be expressed in terms of over-investment or under-investment, together with the associated improvement or worsening

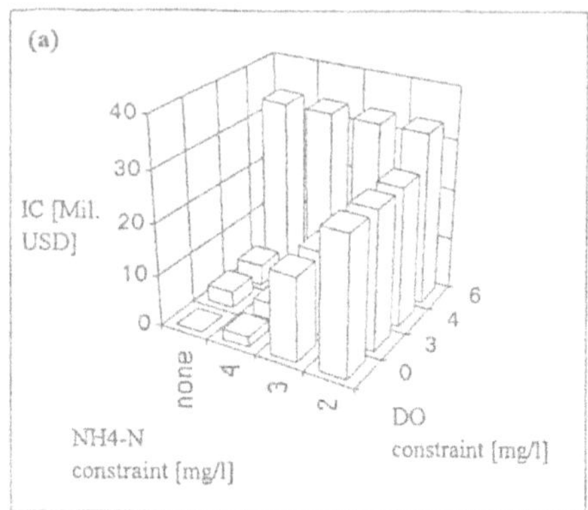

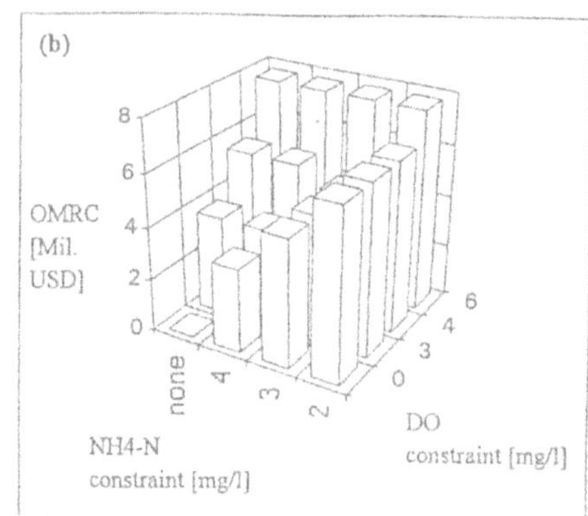

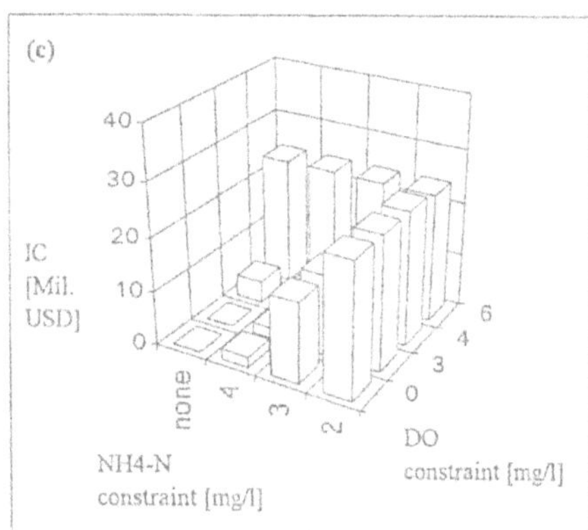

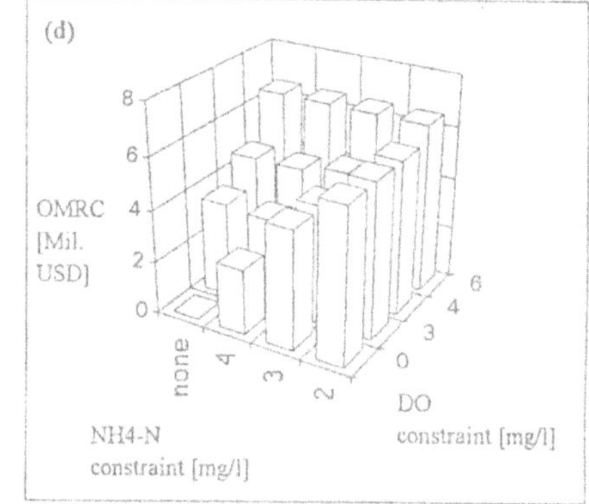

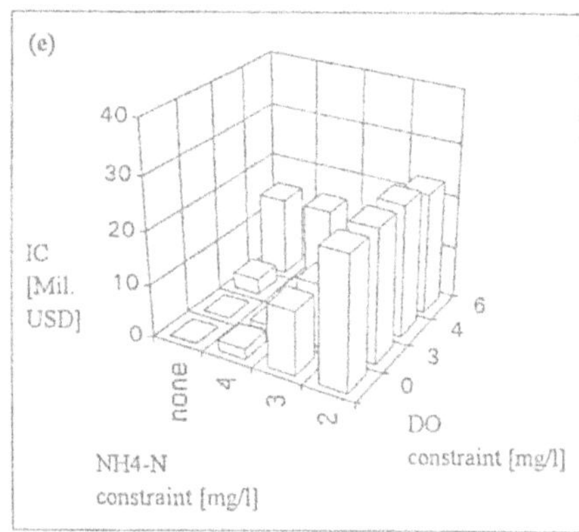

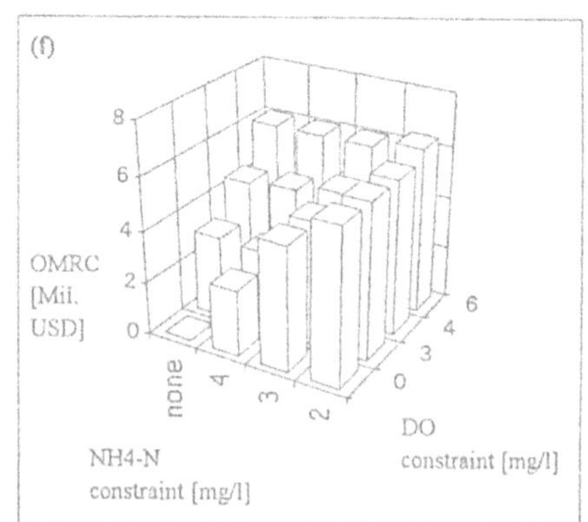

Figure 5. Cost implications of setting DO and NH-4 standards under different industrial discharge scenarios. (a), (b) IC and OMRC corresponding to 1990 discharges; (c), (d) IC and OMRC for a 50% reduction of ind. discharges; (e), (f) for 75% reduction if ind. discharges.

of water quality. This approach to decision making is termed as a "regret-analysis", although the "regret" can be positive or negative (see Burn and Lence, 1992, for a broader analysis). In the original work of Burn and Lence, the stochastic uncertainty of boundary conditions like stream flow and temperature was taken into account, but the approach allows to include any parameters into "design" and "actual" scenarios in a similar way. Here, the uncertainty related to the modelling process was estimated during the Monte-Carlo based parameter estimation procedure (Masliev and Somlyódy, 1993). For policy analysis, the uncertainty in the most sensitive parameter - reaeration rate coefficient - was considered.

The second approach was to estimate the range of water quality indicators under specific treatment strategies. For this purpose, water quality estimation was done by Monte Carlo simulation, and the results lead to probabilistic conclusions on the quality impacts of least-cost strategies. Section 7 describes the Monte Carlo analysis and its results in more detail.

As indicated, the "regret-analysis" estimates the cost and quality implications of not realizing the parameter used for the design. Therefore, this necessitates evaluation of the performance of the least-cost policy under different parameter values. In the policy analysis, the reaeration coefficient was assumed to be 2.0 day^{-1} on the average. The regret-analysis assumes that a deviation of +/- 0.5 day^{-1} from this value is possible, which corresponds to a deviation of +/- 2 times standard deviation. This range would, therefore, cover the most probable range of the parameter.

The difference in the design and realized values will lead to a deviation in ambient quality (the realized quality will be different to the design value), which can be indicated by the minimum DO level. If the deviation in ambient water quality is negative, standards will be "violated", indicating an under-investment, i.e., more investment is needed to compensate for the violation. On the other hand, a positive deviation indicates an over-investment. The over or under investment is the cost difference of least-cost policies corresponding to the two different parameter values (design and realized).

Large negative deviations of ambient DO criteria indicate high vulnerability of the treatment policy decision and, consequently, significant risks. If this risk is to be avoided, one can look for the "safe" solution, taking additional costs. Large investment margins, on the other hand, imply that "safe" policy decisions amounts to high expenses. The regret analysis thus exposes decision risks and helps to determine whether the safe policy is affordable. Figure 6 illustrates the minimum DO concentrations that may occur when the actual reaeration rate is different with respect to design values. Each curve of Figure

6 corresponds to one set of design conditions. The points corresponding to the design conditions are indicated on each of them. The other four points on each curve were obtained by subsequent simulations.

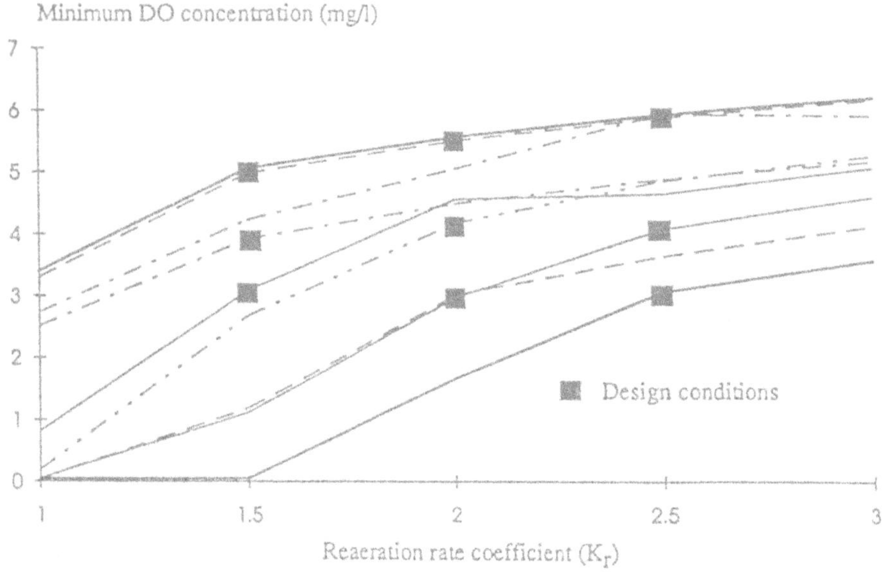

Figure 6. Minimum DO concentrations that may occur when the actual K_r value is different to that considered in the design

For these simulations, the set of "optimal" treatment alternatives found for one particular design condition was assumed to be in operation while in fact design conditions were altered. Thus, the simulations were performed for the four different K_r values, in order to obtain the four other points of the curve. It is observed that the curves of two of the design conditions are coinciding (designs conditions: $K_r=2$, $DO_{min}=5.6$; and $K_r=1.5$, $DO_{min}=5$). That implies the possibility to choose the cheaper one (in this case $K_r=2$, $DO_{min}=5.6$), when considering a selection from those two alternatives.

The summary given in Table 4 outlines the regret analysis for the treatment policies developed for three target DO ambient criteria. The first number in each cell of the table is the difference in IC in million US$. "Missing" investments are indicated by negative signs, and positive figures show "extra" expenses. The second figure is the change of DO criteria in mg/l, negative signs showing the violation of the target ambient standard.

Table 4. Role of uncertainty of the reaeration coefficient on investment cost and the ambient oxygen concentration.

Actual K_r [day^{-1}]	$DO^{3}3$ Design K_r [day^{-1}]			$DO^{3}4$ Design K_r [day^{-1}]			$DO^{3}5$ Design K_r [day^{-1}]		
	1.5	2.0	2.5	1.5	2.0	2.5	1.5	2.0	2.5
1.5	0/0	-1.0/-1.8	-4.2/-3.0	0/0	-17.0/-1.5	-19.5/-3.0	0/0	-16.4/-0.8	-24.2/-2.1
2.0	1.0/1.5	0/0	-3.2/-1.4	17.0/0.6	0/0	-2.5/-1.1	16.4/0.5	0/0	-7.8/-0.6
2.5	4.2/1.6	3.2/0.6	0/0	19.5/1.0	2.5/0.7	0/0	24.2/0.9	7.8/0.5	0/0
IC	4.2	3.2	0.0	21.0	4.0	1.5	31.4	15.0	7.2

The table also illustrates the high sensitivity of investment cost to the value of K_r (the bottom line). A higher K_r leads to increased natural O_2 input to the river water, and thus, a lower level of treatment is necessary.

As can be seen from Table 4, the drop in the DO level caused by the reduction in the reaeration rate exceeds the raise under equal parameter increases. This occurs due to the saturation character of the problem (the upper limit of DO_{min} under municipal emission control is less than 6 mg/l, see Table 2.) Similarly, missing investments exceed overexpenditures at higher DO levels.

For the DO criterion of 3 mg/l, the vulnerability of ambient water quality is high, while possible overexpenditures and additional investment needs are small. This calls for a safe policy decision to invest 4.2 million US$ which guaranties DO above 3 mg/l.

For target DO levels 4 and 5 mg/l, the over and underexpenditures become much higher. The smaller deviations of the DO criteria makes the policy selection even more difficult. Interestingly, the policies aimed at DO level 4 mg/l do not look too attractive and the DO=5 mg/l strategy of 15 million US$ investment is perhaps the best compromise. The additional investment requirement may be rather high if K_r=1.5 day^{-1} is "realized", but the respective improvement in DO level (from 4.2 mg/l to 5 mg/l) is not in proportion with the expenses. Thus, the above policy can be considered as a relatively cheap which (as contrasted to the present DO_{min} level around 2 mg/l) guarantees DO between 4.2 mg/l and 5.5 mg/l (and the "worst" DO levels may be observed under low flow conditions only, occurring for at most one or two weeks in a year).

Finally, DO=6 mg/l strategies (not presented in this paper) do not offer too many interesting features. They are expensive (25-35 million US$, depending on K_{ao}) and safe (DO=5.5-6.4 mg/l, K_{ao}=2 day^{-1}). In addition, over or underexpenditures are smaller (10 million US$) than for the DO=5 mg/l case.

7. THE IMPACT OF THE MODEL UNCERTAINTIES ON THE SELECTION OF CONTROL STRATEGIES: A POSTERIORI MONTE CARLO SIMULATION

The performance of two treatment strategies, under uncertain parameters (coefficients of reaeration, K_r; BOD decay, K_d; and ammonia decay, K_n), were evaluated by *a posteriori* Monte Carlo simulations. The strategies considered were the current treatment and the least-cost (lC) one corresponding to ambient DO 5 mg/l. Although the policy analysis employed a three-component water quality model (DO, BOD and NH4-N), the Monte Carlo analysis was done under three-component as well as two-component (DO and BOD only) model formulations in order to test uncertainty in model structure. The three-component model requires consideration of all three parameters, while ammonia decay rate is excluded in the two-component formulation.

Parameter sets used for these *a posteriori* Monte Carlo simulations were those appropriate for predicting the system behavior under present conditions (Masliev and Somlyódy, 1993). Parameter estimation could be done either jointly, where joint probability distribution of all the relevant parameters is estimated, or marginally, yielding distribution of only one parameter. Identification of joint parameter sets, although more appropriate, require a large number of computations especially if the number of parameters is large. Therefore joint parameters have been identified only with the two-component model formulation, for which a marginal set also has been identified. Parameters for the three-component model were estimated only on a marginal basis.

Accordingly, water quality estimations of the *a posteriori* Monte Carlo analysis also can be categorized into marginal or joint sensitivity analysis, depending on how the parameters were generated.

In the marginal sensitivity analysis, water quality is analyzed with respect to the variations of each individual parameter separately, while joint analysis considers the joint effect. An outline of the analysis is given in Table 5.

Extreme values of DO, BOD and NH4-N in the river (which may occur at different locations), obtained under the least-cost strategy, using three-component marginal sensitivity analysis, are summarized in Fig. 7-9 respectively. The DO plot shown in Fig. 7 is related to the reaeration coefficient. However, estimated DO values are also affected, to a lesser extent, by BOD decay rate and ammonia decay rate coefficients as well; although they are not presented in this paper.

Table 5. Outline of quality estimations of Monte Carlo simulations.

Quality Model	Treatment strategy and the type of analysis	
	Current treatment	Least-cost (IC) strategy (DO5)
Three-component	Marginal	Marginal
Two-component	Joint	Joint
	Marginal	Marginal

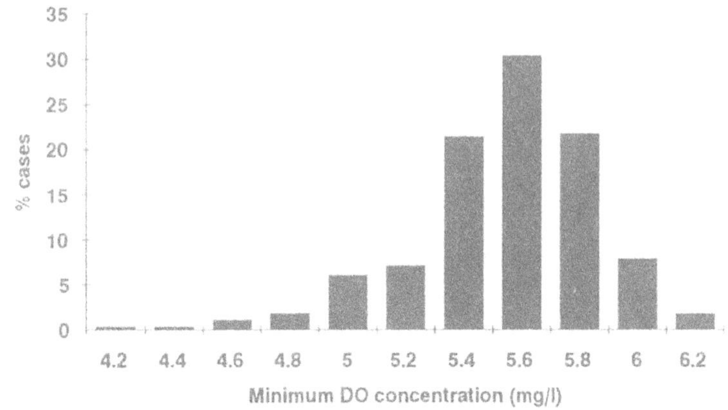

Figure 7. Distribution of the minimum DO concentration in the Nitra River under the least-cost strategy. Deterministic value is 5.0 mg/l (Table 2)

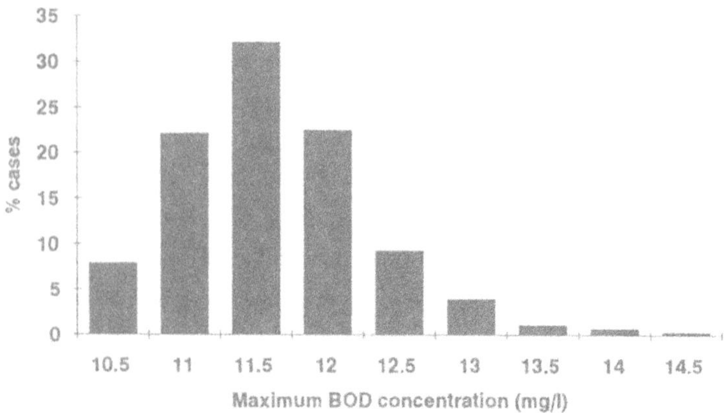

Figure 8. Distribution of the maximum BOD concentration in the Nitra River under the least-cost strategy. Deterministic estimate is 13.8 mg/l (Table 2)

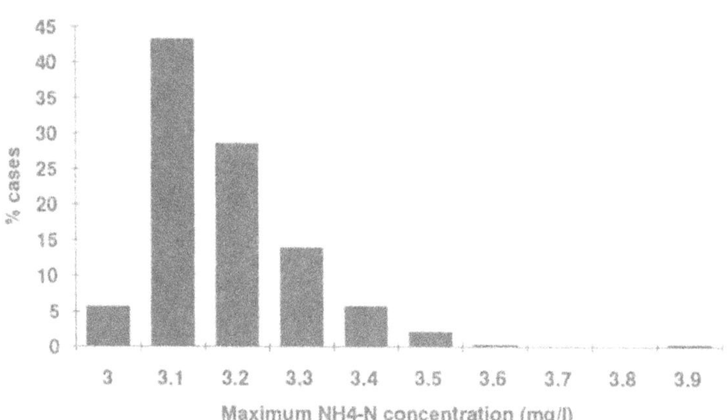

Figure 9. Distribution of the maximum NH4-N concentration in the Nitra River under the least-cost strategy. Deterministic value is 3.2 mg/l (Table 2)

Figures 7 to 9 show that the extreme quality levels are in most cases centered around the deterministic estimates done in the policy analysis (see Table 2). The extreme quality range is relatively small, and even a much larger deviation may not be regarded as undesirable, because the extreme values will occur only at one or two locations within the river basin.

Table 6 presents the mean values and standard deviations of minimum DO levels (for the least-cost strategy), as estimated by the three analyses outlined in Table 5. The results of Table 6, in the case of marginal analyses, refer to the variation of estimated DO with respect to each individual parameter. Minimum DO levels resulting from the current treatment strategy are summarized in Table 7.

Table 6 shows a difference in the mean DO levels estimated by different models. The improvement of water quality resulting from the least-cost strategy can be seen by comparing Tables 6 and 7. It should also be noted that, with the current treatment, the probability of having a minimum DO level which is more than 4 was estimated as zero by each model.

As the relative merits of the methodologies are considered, the regret analysis is more generic in the sense that it can be applied to any scenario-based uncertainty, even to qualitative parameters and subjective estimations. On the other hand, the results of regret analysis are more difficult to interpret and analyze, therefore the number of scenarios should be relatively small. Furthermore, it is based on a notion of "design" scenario or "reference" scenario, which sometimes is not applicable, like in the case where we would like to estimate the effect of the parameter varying according to specified probability distribution. In the latter case, Monte Carlo uncertainty analysis is more appropriate.

Table 6. Statistics of minimum feasible DO levels for the least-cost treatment strategy (estimated by three-component water quality model)

	Parameter of the marginal analysis		
	K_{ao}	K_r	K_n
Mean (mg/l)	5.6	5.2	5.2
Std.deviation	0.3	0.1	0.03
Prob(DO 4)*	100	100	100
Prob(DO 5)	96.4	98.0	100

*probability (%) of having a minimum DO level of more than 4 mg/l

Table 7. Statistics of minimum feasible DO levels for the current strategy (estimated by three-component water quality model).

	Parameter of the marginal analysis		
	K_{ao}	K_r	K_n
Mean (mg/l)	2.9	2.4	2.3
Std.deviation	0.5	0.3	0.3

8. POLICY RECOMMENDATIONS

As mentioned earlier, the new Slovak legislation uses jointly effluent and ambient water quality standards and specifies 2005 as a target year after which criteria are tightened. Thus approximately a decade is anticipated for realization of feasible strategies, i.e., strategies which improve the state of the water environment and are affordable from a financial viewpoint. In the light of the above transition period and policy dilemma, the question is now which recommendation should be given on the basis of the analyses performed and discussed in earlier sections.

The starting point of our answer is a rough comparison of costs and impacts of different policies developed. On one side, if we excluded the uniform application of the "best available technology," BAT, costing more than 90 million US$ (see earlier), we may select the strategy based exclusively on the effluent standards of the new legislation. This would significantly improve water quality (see Table 2). However, it would also be rather expensive, requiring more than 30 million US$ (roughly equivalent to one billion Sk).

On the other side, DO-based least-cost policies show significant saving possibilities. Some of them - as pointed out - are too vulnerable and risky. However, the DO>5 mg/l strategy proved to be rather robust and acceptable. The investment cost requirement is less than half of the previous effluent based policy and the achievable water quality is nearly the same (see Figure 3). The difference in minimum DO and maximum BOD levels is less than 5%, i.e., non-significant and non-detectable (NH4-N exhibits somewhat larger deviation).

A more detailed comparison of the two policies can be seen in Table 8 including not only costs and receiving water quality but also treatment configurations (note that Level 1 indicates present treatment for most of the sites and the highest number refers to BAT). It can be clearly seen that the effluent standard based strategy nearly always leads to a higher level of wastewater treatment. An exception is formed by the middle stretch of the river where the extremely poor present quality forces the same technology for the regional least-cost strategy as given by effluent standards.

Table 8. Comparison of treatment strategies: Slovak effluent standard based strategy vs. DO 5 least-investment-cost policy.

Hand-lova	Lehota	Prie-vidza	Parti-zanske	Bano-vce	Topol-cany	Nitra	Zl.Mo-ravce	Vrable	Surany	Nove Zamky
										9
					8					8
					7	7				7
					6	6				6
		5		5	5	5				5
4	4	4		4	4	4	4			4
3	3	3	3	3	3	3	3	3	3	3
2	2	2	2	2	2	2	2	2	2	2
1	1	1	1	1	1	1	1	1	1	1
0	0	0	0	0	0	0	0	0	0	0

Legend for Table 8:

	Min. DO (mg/l)	Max. BOD (mg/l)	Max NH4-N (mg/l)
― Slovak effluent standard based policy (32.1 million USD)	5.4	11.3	2.3
▬ DO 5 mg/l least-cost policy (15 million USD)	5.2	11.4	3.2

Uniform effluent standards generally lead to uniform technologies. However, this is not the case if the upgrading of existing facilities overloaded to different extent serve also as viable alternatives. In this sense, both strategies resulted in technologies which may vary from site to site. The least-cost policy did not lead to a particularly preferred technology, although most of the upgrading would be based on chemical enhancement of existing biological treatment plant. However, more or less the same statement is also valid for the effluent standard based strategy, suggesting higher cost-effectiveness on the treatment plant level without having a regional consideration.

In summary, our recommendation is to implement the DO>5 mg/l least-cost policy as a short-term policy which can then be further expanded as financial resources become available. For the purpose of enforcement, regionally variable effluent standards can be used - like in several countries - which also belong to the results of the strategy development.

9. CONCLUSIONS

(1) Depending on the policy formulation for the emission control in the Nitra River basin (based on effluent standards, ambient standards or their combinations), the range of necessary expenditures to reach the standard is extremely broad (between 3 and 95 million US$). This indicates significant saving possibilities. The most expensive solution is to replace all the treatment plants with new ones satisfying the most stringent recommendations of the European Community. The present (and future) Slovakian effluent-quality standard system implies an investment of 32-35 million US$. A least-cost policy with investment about 15 million US$ would result in improvement from the present Class IV or Class V to Class III water (in terms of DO, BOD and NH4-N), roughly equivalent with the former solution based on effluent standards. The water quality (in terms of the dissolved oxygen) is nearly identical for all three cases.

(2) Since control actions influence several constituents simultaneously, least-cost policies developed solely on the basis of DO ambient criterion lead to significant improvement with regard to other components as well. These strategies are attractive: an investment of 4-15 million US$ improves DO from about 2 mg/l to 3-5 mg/l. An uncertainty analysis based on regret methodology showed that the policy at 15 million US$ is not too vulnerable with respect to uncertainties in parameters of the DO model used within the optimization. The possible overexpenditures are not high either. As contrasted to effluent standard based strategies, the least-cost ones are rather non-uniform. A long-term policy can be obtained by a sequence of least-cost strategies under gradually tightened ambient quality standards.

(3) The economic consequences of setting standards are strong and non-linear. The improvement of DO_{min} from 4 mg/l to 5 mg/l requires more than 10 million US$ investment, while BOD and NH4-N also play a similar role.

(4) In the present study, only municipal emissions were directly incorporated into the optimization (which represent 70% of the total BOD discharge into the river system). A sensitivity analysis demonstrated the significant role of industrial pollutant loads, which clearly calls for the development of an integrated least-cost policy covering all the emissions of various origins.

(5) The dynamic programming technique applied in this effort is well suited to handle river basin water quality management problems generically. It allows to incorporate various simulation models (linear or non-linear), expressing the impact of emissions on ambient water quality, and to consider the details of alternative waste water treatment technologies. Possible extensions to the optimization process include the incorporation of parameter uncertainty by a variance-based approach or by a scenario analysis. Different formulations of the problem, in terms of objective functions and constraints, can also be used to assess the robustness of the strategies formulated. Incorporation of quality indicators other than BOD, DO, N and P is necessary to model the various processes that are occurring in a river more accurately. Scheduling of the control activities over the planning horizon is another important problem to be solved.

REFERENCES

Biswas A.K. (1981) "Model of the Saint John river, Canada," In: Models for Water Quality Management, (A.K. Biswas, ed). McGraw-Hill Inc, New York, 68-90

Bellman, R.E. (1957) Dynamic Programming. Princeton University Press, Princeton, New Jersey

Bishop, A.B., Grenney, W.J., Narayanan, R., and Klemetson, S.L. (1974) Evaluating Water Reuse Alternatives in Water Resources Planning. PRWG123-1, Utah Water Research Laboratory, Utah State University, Logan, UT

Breithaupt and Somlyódy, L. (1994) Water Quality Modelling of the Nitra River Basin (Slovakia): A Comparison of Two Models. Working Paper WP-94-110, International Institute for Applied Systems Analysis (IIASA), Laxenburg, Austria

Burn, D.H. and Lence, B.J. (1992) "Comparison of optimization formulations for waste-load allocations," J. Envir. Engrg., **118**(4):597-612

Cardwell, H., and Ellis, H. (1993) "Stochastic dynamic programming models for water quality management," Water Resources Research, **29**(4):803-813

Hahn, H.H. and Cembrowicz, R.G. (1981) "Model of the Neckar river, Federal Republic of Germany." In: Models for Water Quality Management,(A.K. Biswas, ed). McGraw-Hill Inc, New York, 158-221

Henze, M. and Ødegaard, H. (1994) "An analysis of wastewater treatment strategies for Eastern and Central Europe." Paper presented at the 1994 Budapest Biennial Conference of the International Association on Water Quality (IAWQ) and published in Water Science and Technology, 1995

Koivusalo, H., Varis, O., and Somlyódy, L. (1992) Water quality of Nitra River, Slovakia. - Analysis of organic material pollution. WP-92-084, International Institute for Applied Systems Analysis (IIASA), Laxenburg, Austria

Loucks, D.P., ReVelle, C.S., and Lynn, W.R. (1967) "Linear programming models for water pollution control," Management Science, **14**: B-166 - B181

Masliev, I., Petrovic, P., Kuníková, M., Zajícová, H., and Somlyódy L. (1994) Longitudinal water quality profile measurements and their evaluation in the Nitra River basin (Slovakia). IIASA Working Paper WP-94-104, Laxenburg, Austria

Masliev, I., and Somlyódy, L. (1993) Uncertainty analysis and parameter estimation for a class of linear dissolved oxygen models. Working Paper WP-94-9, International Institute for Applied Systems Analysis (IIASA), Laxenburg, Austria

Morrissey, S.P. and Harlemann, D.K.F. (1990) Chemically Enhanced Wastewater Treatment. MIT Ralph M. Parsons Laboratory, Report No. R90-14

Murcott, S. and Harlemann, D.R.F. (1994) "Use of chemically upgraded treatment (CUT) in Slovakia and Hungary." Paper presented at the 1994 Budapest Biennial Conference of the International Association on Water Quality (IAWQ) and published in Water Science and Technology, 1995

Newsome, D.H. (1972) "The Trent river model - An aid to management," Proc., International Symp. on Mathematical Modelling Techniques in Water Resources Systems. Ottawa, Canada, 613 - 632.

Orlob, G.T. (1982) Mathematical Modelling of Water Quality. Wiley/IIASA, Chichester

ReVelle, C.S., Loucks, D.P., and Lynn, W.R. (1968) "Linear programming applied to water quality management," Water Resources Research, **4**(1):1-9

Somlyódy, L. (1993) Municipal Wastewater Treatment in Central and Eastern European Countries: Present Situation and Cost-Effective Development Strategies. Report submitted to the World Bank for publication, Environmental Action Program for Central and Eastern Europe (manuscript).

Somlyódy, L., and Paulsen, C.M. (1992) Cost effective water quality management in Central and Eastern Europe. WP-92-091, International Institute for Applied Systems Analysis (IIASA), Laxenburg, Austria

Somlyódy, L. , Masliev, I., and Kularathna, M. (1993) Water Quality Management of the Nitra River Basin (Slovakia): Evaluation of Various Control Strategies. Working Paper WP-93-63, International Institute for Applied Systems Analysis (IIASA), Laxenburg, Austria

Somlyódy, L., Masliev, I., Kularathna, M., and Petrovic, P. (1994) Water quality management in the Nitra River basin. IIASA Collaborative Paper CP-94-62, Laxenburg, Austria, 205 p.

Somlyódy, L., and Varis, O. (1992) Water quality modelling of rivers and lakes. WP-92-041, International Institute for Applied Systems Analysis (IIASA), Laxenburg, Austria

Thomann, R.V. and Mueller, J.A. (1987) Principles of Surface Water Quality Modelling and Control. Harper and Row, New York

Warren, J., and Bewtra, J.K. (1974) "A model to study the effects of time-variable pollutant loads on stream quality," Water Research, 8:1057-1061

CHAPTER 13

THE STATE OF THE ART IN ECONOMIC INSTRUMENTS AND INSTITUTIONS FOR WATER QUALITY MANAGEMENT

Mark Griffin Smith[1]

1. INTRODUCTION

The United States and the Western European countries control water quality using a variety of instruments and institutions. These tools range from regulatory command and control (CAC) approaches of technological, emissions and ambient standards to economic or incentive-based approaches such as charges, subsidies and transferable discharge permits (TDPs). As the economies of the CEE countries move from central planning to the free market, it is appropriate to review both the literature on and experience with economic instruments for water quality management to understand how they might be applied in that setting. The Central and Eastern European countries face serious water quality problems and the resources needed to address these problems are large (Somlyódy, 1993). The challenge of improving water quality in CEE requires finding cost-effective approaches that are appropriate to the institutional context of individual CEE countries.

The purpose of water quality control is to maintain water quality at desired levels at the lowest possible cost (Kularathna and Somlyódy, 1994). While much has been written by economists about the determination of the optimum or "desired level" of water quality (for example, Feenberg and Mills, 1980; Smith and Desvouges, 1986), the intangible nature of most water quality benefits has meant that, in practice, water quality standards have not been established on the basis of economic criteria. This notwithstanding, economics has made considerable contribution to the identification and evaluation of cost

[1] Colorado College, Colorado Springs, Colorado, U.S.A.; in 1994 -1995 on a sabbatical leave at the International Institute for Applied System Analysis, Laxenburg, Austria

NATO ASI Series, Partnership Sub-Series, 2. Environment – Vol. 3
Remediation and Management of Degraded River Basins
Edited by V. Novotny and L. Somlyódy

minimizing approaches for water quality management under an exogenously determined set of water quality objectives. This chapter reviews both what has been proposed and what has been tried toward the end of identifying appropriate water quality management policies for Central and Eastern Europe (CEE).

2. FIRST PRINCIPLES

Economists describe pollution as a "negative externality". Externalities arise when there is a non-market impact resulting from the consumption or production activity of one economic agent (a person, household, firm, state-run enterprise, etc.) that affects the welfare of another economic agent. Untreated municipal sewage is a good example of an externality as its effects can include both impacts on goods such as fish that are bought and sold in markets as well as swimming and sport fishing which are not. The important distinction is that the effect is a non-market or non-priced effect so that the market neither rewards or penalizes its producer.

When economic activity generates pollution as an externality, there are a number of significant implications for the market allocation of resources (Tietenberg, 1992):

(1) Too much output is produced.
(2) Too much pollution is produced.
(3) The prices for the pollution generating product are too low.
(4) There are no incentives to look for less polluting means of production.
(5) Recycling and reuse of polluting substances are discouraged because release into the environment is inefficiently cheap.

The misallocation of resources associated with pollution requires some means of "internalizing" the externality so that its producer faces some consequence from its pollution generating activity.

Coase (1960) observed that all externalities are essentially cases in which property rights are undefined. Where property rights are well defined, pollution problems can be resolved either through the market, negotiation or litigation between property owners. In absence of clear property rights, government intervention is necessary to correct the failure of the market to efficiently allocate resources when an externality is generated in conjunction with production or consumption[2]. The most widespread approach used to affect the control of pollution is to set standards based upon abatement technology, effluent levels and/or ambient environmental quality. These standards are then monitored and enforced using fines and penalties. Research has shown that this "Command and Control" (CAC) fails to achieve desired environmental quality improvements at minimum cost (Tietenberg, 1985). Economic instruments provide a means to meet the same objectives at lower cost.

3. ECONOMIC INSTRUMENTS: EFFICIENCY AND EQUITY PROPERTIES

The primary argument for economic instruments is based on efficiency[3], that they can be used to achieve the desired level of effluent reduction at the lowest cost. The mechanism by which this is achieved is through equalizing the marginal cost of abatement (MCA) across all pollution sources. This is illustrated in Figure 1 which compares a standard requiring uniform emissions reduction against an effluent tax. Under the standard the total cost of emissions control for firms A and B is the sum of areas BDF and EDF. Under the tax the total cost of emissions control is the sum of areas CDE and IDG. Clearly the cost of control is higher under the uniform reduction standard. This will be true any time there is a divergence in the cost of control among sources. Two other important differences between the standard and charge are apparent from this diagram. First, under the charge, levels of emissions control diverge across sources. Second, under the charge, polluters not only pay the cost of control area CDE for firm A and area IDG for firm B, but also a tax to the government, area HCE0 for firm A and area HIG0 for firm B. Both of these results have implications for the perceived fairness of a charge policy which will be discussed below.

---

[2] The common law tradition of England and the United States requires that a party have legal "standing" to seek remedy from the courts for damage from pollution. Standing requires that the affected party can demonstrate loss in the value or enjoyment of their property. Under the common law tradition it is legally impossible for anyone to sue on behalf of "the environment" or "the fish" in an attempt to affect water quality improvements. While it is conceivable that property rights could be granted to make such suits feasible, it is not clear that such a litigious system would be more effective than regulatory and other economic approaches where uncertainty, information and transactions costs are high.

[3] The term efficient is used here and throughout the paper in the sense of production efficiency, a given output is produced at the lowest cost. In this case a target level of pollution control is achieved at the lowest possible cost. Its use here should be distinguished from it's more general use in economics, allocative efficiency, which implies that certain conditions are met on both supply <u>and</u> demand sides. True allocative efficiency cannot be achieved without knowledge of both the costs and the benefits of pollution abatement.

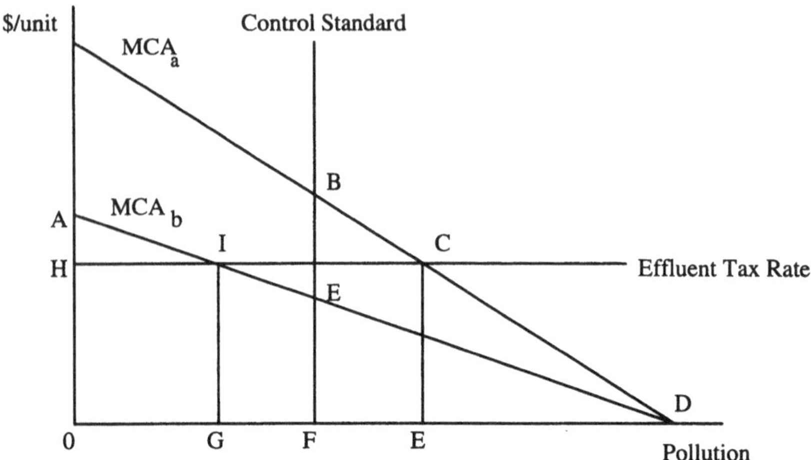

Figure 1. Comparison of the efficiency of an effluent tax versus a standard.

At a more macro level effluent charges place a price on the use of the environment for disposing of wastes thereby sending a signal to polluters that generating effluent imposes a cost on society. This signal will in turn affect production and consumption decisions throughout the economy resulting in the more efficient use of resources (von Hayek, 1945).

Economists recognize that the concept of fairness or equity lacks a unique definition. Treating everyone the same, pay as-you-go, soak the rich, help the poor are all widely understood, but often mutually exclusive, concepts of equity. While in theory it is easy to couple instruments which promote efficiency with instruments that promote "fairness", this is harder to achieve in practice. Who will bear the cost and who will reap the benefit are central questions in the design of water quality management policy. They are ultimately the key to the political acceptability of any proposed water quality management program. Thus it is important in evaluating economic instruments for water quality management to ask not only how they will promote efficiency, but also how they will distribute the benefits and costs.

4. THE INSTRUMENTS

Economic approaches to environmental quality management consist of three primary instruments: taxes, subsidies and transferable discharge permits (TDPs). This section describes each of these instruments as well as briefly discussing deposit-return systems as an alternative for controlling diffuse toxic pollutants.

4.1 Taxes/Charges

Taxes or charges on pollution producing activities can be imposed in a variety of ways - on the pollutant, on the final product or on inputs into the production process. In all cases the tax will have three main effects. First, the tax will increase the cost of polluting and therefore create an incentive to reduce emissions. The magnitude of this effect will depend upon the level of the tax and the responsiveness of the firm to a change in the cost of production.[4] Second, the tax forces the internalization of the environmental costs caused by pollution and therefore makes the polluter pay for disposing emissions into the environment. Third, taxes raise revenue.

A variation on taxing the pollutant alone is to use taxes to create a price differential between products on the basis of their contribution to pollution. Products that generate more pollution are taxed more heavily, products that generate less are taxed less, perhaps even subsidized. For example the purchase of a toilet could include a tax based upon the amount of water used per flush with low flow toilet eligible for a rebate. The objective of tax differentiation is to create an incentive for "environmentally friendly" behavior.

4.2 Subsidies

Whereas the purpose of taxes is to discourage pollution generating activities, the purpose of subsidies is to encourage pollution reduction. While in theory, polluters could be subsidized on the basis of how much emissions had been reduced, in practice it is politically unpopular to pay someone not to pollute. Subsidies usually come in the form of either grants, soft loans or tax allowances for capital expenditures undertaken to control pollution (Opschoor and Vos, 1989).

---

[4] Economists use the term elasticity to describe the responsiveness of supply or demand to a change in price. Elasticity is defined as: $\%\Delta$Quantity/$\%\Delta$Price.

4.3 Deposit-Refund Systems

In deposit-refund systems surcharge is placed on a potentially polluting product. When the product or its residual is returned, and the pollution thus avoided, the surcharge is refunded. Although the most wide-spread deposit-refund schemes are intended to reduce solid waste disposal, litter and energy use (e.g., beverage containers in most western European countries, car hulks in Norway and Sweden), similar schemes are under consideration for car batteries. Soil contamination from the mercury and cadmium in batteries can lead to groundwater contamination. Such systems might be similarly practical for other water pollutants emanating from highly dispersed sources such as household chemicals.

4.4 Transferable Discharge Permits (TDPs)

Under a system of transferable discharge permits firms are allocated permits to discharge emissions up to a certain limit. Firms that manage to reduce their level of emissions below this level can sell or trade their unused permits to other firms which can then exceed their initial pollution limit. The aim of a TDP program is to create a market in "pollution rights" which will enable firms to achieve the economic efficiency objective of equating the marginal cost of pollution abatement across all sources thus minimizing the total cost of pollution control.

Transferable discharge permit programs in their most general form allow trading amongst different firms and across a region (watershed) or nation. More limited forms of "trading pollution" have also evolved to attempt to achieve efficient control of emissions at the level of the firm or a sub-region. These are offsets, bubbles, netting and banking.

Offsets. The purpose of an offset policy is to allow new sources or the expansion of old sources in areas which have yet to achieve the targeted level of ambient environmental quality. New sources or the expansion of old sources is permitted by obtaining emission reduction credits (ERCs) from existing sources. ERCs are made available when existing sources have reduced their levels of emissions below those required by law. The primary purpose of an offset policy is to allow economic development in non-attainment areas rather than achieve economic efficiency in pollution control. Currently offsets are included as a component of air quality management under the U.S. Clean Air Act Amendments of 1977 and under the Plant Renewal Clause of the German Technical Guidelines for the Control of Air Quality of 1974 (Tietenberg, 1990; Opschoor and Vos, 1989). Most of the offsets under the Clean Air Act Amendments have been internal transactions (Opschoor and Vos, 1989).

Bubbles. The "bubble" draws its name from the conceptual notion of treating multiple sources as if they were under a bubble from which there is but a single source of pollution. Originally, the Clean Air Act required compliance by each individual source. Under this policy what matters are total emissions from the bubble, rather than the emissions from individual sources. Bubbles allow the reallocation of emissions among existing point sources under the condition that total emissions do not exceed the sum of the mandated levels for individual sources. Bubbles thus allow firms to minimize the cost of emissions control by reallocating emissions reduction to the lowest cost points of control, i.e., to equate the marginal cost of control across sources. While the Clean Air Act Amendments of 1977 allow multi-plant bubbles, in practice most bubbles cover only a single plant (Opschoor and Vos, 1989).

Netting. Netting allows existing firms to use emissions reduction credits (ERCs) earned by reducing emissions at existing sources to modify or expand other sources within the same plant. Netting allows firms to avoid the requirements of the new source review process. Its primary purpose is to provide regulatory relief rather than promote efficiency (Tietenberg, 1990).

Banking. Banking allows firms to retain credit for emissions control activities that exceed required levels. Banking allows for the discrete or "lumpy" nature of capital investment in pollution control. Banked credits or ERCs can be retained to allow for future expansion or sold under the offset or bubble schemes described previously.

5. ECONOMIC ISTRUMENTS EVALUATED: CONCEPTUAL ISSUES AND PRACTICAL EXPERIENCE

The preceding section defined the basic principles of economic or incentive-based instruments for environmental quality management. This section presents the major arguments for using economic instruments in water quality management, evaluates the experience with their application, and attempts to define specific lessons from that experience.

5.1 Effluent Charges

The Case for Effluent Charges. The basic arguments for effluent charges were outlined above: they increase the cost of polluting thus creating incentives to reduce effluent discharge, they force the polluter to pay the cost of using the services of the environment to dispose of wastes; and they raise revenue. Along with these attributes, a effluent charge system may have a number of other attractive features (Brown and Johnson, 1984):

(1) Charges create incentives for firms to look for ways to reduce pollution. This may involve input substitution, changes in production processes and changes in the character of their output as well as effluent treatment processes. For example the German chemical firm, BASF, introduced a system of internal liability for effluent within different branches of the firm in response to effluent charges imposed by the federal government. This system resulted in a 20 percent reduction in effluent discharge (Brown and Johnson, 1984).

(2) Charges increase incentives for municipalities to rationalize sewage pricing policy by establishing waste-load based charges on firms which discharge into the municipal sewage treatment system. Faced with effluent charges, municipalities will seek means to pass back these costs to indirect dischargers thus creating the same incentives for direct dischargers outlined above.

(3) Charges stimulate municipalities to improve effluent monitoring. While the incentive to keep better track of effluent follows from a self-interested motivation to reduce cost, a secondary public benefit results from the generation of more complete and precise data with which to manage water quality. Such is the case in Germany (Brown and Johnson, 1984).

(5) Charges make revenues available for financing water quality improvements. To the extent that these funds are made available to dischargers for pollution control and investment in pollution reducing industrial processes, it will mitigate the unpopularity of a charge scheme.

(6) Charges shift the burden of financing the water quality management program from the taxpayer to the polluter. This may not only make water quality management more attractive to the public, it may also make it more attractive to CEE governments seeking to fulfill a variety of obligations with limited resources.

(7) Charges make more revenues available for water quality management and therefore monitoring and enforcement. As a consequence a higher level of compliance may be achieved. Financing enforcement through effluent charges may generate excessive enthusiasm for enforcing compliance. This may require some mechanism to prevent abuse.

Experience. The following briefly summarizes the most notable examples of the use of effluent charges in water quality management. No attempt has been made to provide a complete description of these programs, but to offer sufficient background as to glean the key lessons from their experience. Brown and Johnson (1984) review the German program, Andersen (1991) the Dutch as well as the Danish programs and Opshoor and Vos (1989) assess the German, French, Italian and Dutch charge systems.

Table 1 highlights the salient features of the French, German and Dutch effluent charge programs. There are notable similarities and some significant differences. Most notably, all combine effluent charges with the use of standards; none rely on effluent charges alone to provide sufficient incentive to achieve the desired water quality objectives. Revenues generated are used in all to both fund the administration of the program and finance public and private investments in pollution abatement. Except in the case of large plants in Holland and France, charges are based upon average or expected loads rather than actual loads. Household charges are based on a flat rate in the two countries in which they are applied, France and Holland. Charges vary by region in both France and Holland reflecting differences in each regional pollution control construction program rather than differences in assimilative capacity.

Each program also has distinctive features. The distinguishing feature of the German program is the schedule of charge reductions associated with the degree of compliance. No charge reduction is earned for simply meeting the minimum standard, however firms which reduce discharges beyond this standard can receive charge reductions up to 100%.

A striking feature of the Dutch system is that charges can apply to the water boards themselves when discharging into the waters of the state, i.e., large rivers, channels and reservoirs. The state may impose charges upon a water board if the waters under the board's authority are of unacceptable quality. An adjacent water board receiving unacceptable water may also request that the offending water board is charged.

Table 1. Cross-National Comparison of Effluent Charge Programs

Country	Pollutants	Administration	Notes
France	- Suspended Matter - Oxidizable Matter - Soluble Salts - Inhibitory Matter - Organic/Ammonia Nitrogen - Total Phosphorus	- Six river basin authorities which are financially independent - Charges levied on firms and households - Revenues raised fund administration and public and private abatement activities - Charges not related to abatement costs - Charges vary by authority - Combined standard/charge system	- Revenues recycled directly back for abatement activities. - Low incentive effect due to low charge rates - Overall impact on improving water quality unclear
Germany	- Settling Substances - Oxidizing Substances - Mercury - Cadmium - Toxicity to Fish	- State (Lander) based - Charges levied on direct dischargers only - Revenues fund administration and public and private abatement activates - Administration costs are high - Charges set by federal government - Combined standard/charge system	- Charges based on compliance levels with discounts for exceeding standards - Exemptions possible for hardship - Notable improvement in water quality - Has possibly promoted technological innovation
Holland	- Biodegradable Matter - Suspendable Solids - Toxic Substances - Heavy Metals	- Administered jointly by the national government and 140 local water boards - Variable rates based on loads - Charge levied on firms and households - Administration costs low - Charges set by water boards - Charges apply to local water boards as well as households and firms - Combined standard/charge system	- Primary intent is financial, but appears to have incentive effect as well - 80% decrease in pollutant load - Aggressive water quality program has not impeded industrial growth - Program may have promoted over expansion of treatment capacity

Sources: Anderson (1991), Brown and Johnson (1984), Opshoor and Vos (1989).

The French program points to both strengths and weaknesses of self-financing local or regional authorities. On the one hand the direct recycling of charges back into pollution abatement efforts has lessened the political opposition to water quality management by industry. On the other it is recognized that, at some point, industry may effectively block higher water quality standards by refusing to pay higher charges.

Work by Pethig (1989) is especially relevant given the models of the French and German effluent charge systems whose principal purpose is to finance the activities of the water authority. Pethig shows, under a general set of theoretical conditions, that water quality management will be inefficient when firms using publicly provided wastewater treatment only pay for the costs of building and operating the plant, i.e., to finance the system. Industrial wastewater abatement activity will be too low and the public treatment facility will operate at an inefficiently high level. Efficiency requires that firms are charged not only for the cost of treatment but also for the "free service" provided by the assimilative capacity of the receiving water body. Both French and German programs implicitly recognize this fact by imposing standards in conjunction with their charges.

Lessons. The effluent charge systems now in use bear little resemblance to an optimal system of charges reflecting the strength, content, location and timing of the discharge as well as the flow, temperature and water quality goals for the receiving water body. Moreover, they are set too low to induce the necessary level of control to achieve the desired level of water quality alone. Nevertheless, experience has shown that they effectively raise revenue for water quality management and have generated some incentive for pollution abatement and innovation in control technology. Their political acceptability can be promoted by recycling revenues back into investments in abatement projects which clearly demonstrate water quality improvements.

5.2 Subsidies

If the purpose of effluent taxes is to raise revenues and discourage pollution generating activities, the purpose of subsidies is to promote environmentally desirable behavior. However environmentally desirable water pollution control might be, the public opposes subsidies to polluters for two reasons - paying someone to stop doing something bad seems at best unfair, at worst immoral. Second, they cost money. Furthermore, granting a subsidy for pollution abatement violates the widely-held environmental quality management objective, the polluter-pays-principle (Opshoor and Vos, 1989).

The main arguments for subsidies are borne of political expediency. Both industry and municipalities argue that they need help to meet new requirements created by environmental legislation, i.e., to make the transition from a lower to a higher level of abatement activity. Many governments accept this position and have instituted subsidy programs for capital investment in pollution control but not the cost of operating the plant once built (Opshoor and Vos, 1989).

It is also recognized that certain dischargers will have difficulty complying with environmental standards. Where uniform standards or charges are applied, subsidies are a means of redistributing the costs and facilitating compliance. Evidence from the U.S. experience suggests that the true hardship cases are small in number and that the regulatory "stick" has been more important than the subsidy "carrot" in achieving compliance with the Clean Water Act (Freeman, 1990).

Subsidies have been critical in buying political acceptance for water quality management. Industry is more willing to agree to controlling pollution if the cost of control is subsidized. Opshoor and Vos (1989) conclude in their survey of the use of economic instruments in the OECD countries that subsidies have contributed little to enhancing environmental quality, they may have been necessary for establishing environmental programs in the first place.

Subsidies may hasten investment in pollution control for two reasons. One, because they lower the cost of compliance and two, those eligible for the subsidy may adopt a "get it while you can" attitude if it is uncertain how much money will be available or how long the program will last. Again, limited empirical evidence from both the United States (Freeman, 1990) and Germany (Opshoor and Vos, 1989) suggests compliance deadlines are more important than subsidies in achieving rapid compliance.

Subsidies may be the only means of achieving pollution reduction from those who are not required to do so. In North Carolina agricultural best management practices (BMPs) to reduce nutrient loadings are voluntary. However, reducing nonpoint source loadings may be more cost effective than tertiary treatment. Municipalities in the Tar-Pamlico watershed in North Carolina have joined together in a river basin association to fund a cost sharing program for agricultural BMPs (Swanek, 1994). Since they cannot impose BMPs on farmers they must buy their cooperation with the program.

Subsidies appear to have achieved little improvement in environmental quality that would not have been achieved by strict enforcement of water quality standards. Subsidies have also contributed to the general level of uncertainty in water quality management. In the U.S. the government's cost sharing contribution varies from 30 to 85 to 55 percent. Congress has in turn funded different levels of support than were initially requested. The subsidy program may have in fact delayed some pollution abatement investment because municipalities expected the EPA's cost sharing component to rise (Freeman, 1990).

The most frequently cited criticism of the U.S. construction grants program for municipal water treatment is that it has led to excessive capital investment in sewage treatment. Treatment plant are built with excessive capacity and a bias toward expensive capital intensive processes and away from equally effective, but lower capital cost methods of treatment such as sewage lagoons (U. S. Congress, 1985).

While subsidy programs may be necessary in the short-run to gain political support for water quality management and address the needs of true hardship cases, once established they are hard to terminate. All subsidy programs breed dependency by narrow interests and help finance the growth of their political clout which in turn is used to push for the continuance of the program. Revenues that could be more effectively spent to address new problems are tied to problems or approaches that no longer require support. Opshoor and Vos (1989) report that France and Germany have begun to shift the emphasis of their spending programs away from effluent treatment to implementation of new, cleaner process technologies, a promising sign.

5.3 Effluent Trading

The primary argument for effluent trading systems is cost minimization. Allowing dischargers to trade permits among themselves will, under competitive market conditions, achieve the cost minimizing condition that the marginal cost of pollution control is equated across every firm. In addition to this argument TDPs potentially have a number of other attractive qualities:

(1) Effluent trading systems are, in theory, administratively simple. Once permits have been distributed, interaction occurs among dischargers rather than between individual dischargers and the government. The water authority continues to monitor and enforce water quality standards, but the permit system obviates the need to administer compliance with technological standards for specific dischargers or the system of effluent charges.

(2) TDPs allow for flexibility. Some firms will choose to significantly reduce their effluent discharge

and sell permits, others will choose to increase it and buy them. A TDP market allows for the separation of who pays for pollution control from who installs it thus creating greater flexibility in meeting water quality standards.

(3) TDPs allow for the development of leasing markets wherein firms may acquired permits to meet short-run needs. This feature is particularly attractive in transition periods in which a firm would be better off to lease permits than to invest in new pollution control equipment to be used with outdated process technology. Leased TDPs allow firms the flexibility to forestall investment in pollution control until this investment can be coordinated with new investment in the production process itself, potentially at a lower cost than end-of-the-pipe measures (Tietenburg, 1990).

(4) The initial allocation can be used to achieve distributional objectives at no expense to cost effectiveness. TDPs can be auctioned off, given away, given to some and sold to others. From a cost efficiency perspective the method of initial distribution is irrelevant (Coase, 1960). The efficiency of the system is not driven by the initial distribution of the permits but by the trading activity that occurs among dischargers which drives marginal abatement costs to equality.

The method of initial distribution is not, however, inconsequential to polluters. TDPs have value which is either retained by the firm if they are given away or transferred to the government if they are sold. The fact that the government creates this value by establishing a market for TDPs allows them the additional policy flexibility to differentially favor municipalities over industry, one industry over another, growing versus decaying sectors, etc..

(5) Correctly administered, TDPs create a secure property right, that is an entitlement to discharge a certain amount of pollutant over a specified period of time. In doing so, they reduce the uncertainty associated with standards and charges both of which can be changed at any time. If under a TDP system the government decides to achieve higher water quality standards it must purchase and retire a share of the outstanding permits. The reduction in regulatory uncertainty engendered by a TDP systems allows polluters to make more rational long term decisions about pollution control.

(6) TDPs eliminate the need to continuously revise the water quality management program in response to economic growth. Under a static program of either effluent charges or standards,

the level of water quality will be negatively related to growth in population and economic activity. A stable level of water quality can be achieved only by increasing the stringency of the standard or the amount of the charge. Such changes raise the level of regulatory uncertainty against which dischargers must make long-term capital investments in pollution control. Because the ultimate level of water quality is established by the number of permits initially allocated there is no need to administratively adjust the system in response to increased economic activity. Reallocation of permits will occur within the TDP market to accommodate new economic activity. The price of permits will increase with demand, however this price will equal the opportunity cost of pollution control rather than the government's attempt to estimate the value with new charges or standards.

(7) TDPs cost the polluter less than effluent charges. This issue was previously discussed above as an equity concern. It may be a greater concern in CEE countries where there is a pressing need for capital investment in production technology as well as pollution control. TDPs allow firms to retain more resources which can be invested in either pollution control or process technology.

Experience. The experience with transferable discharge permits systems is very limited. The only example of a true permit trading system for water quality management exists on the Wisconsin's Fox River in the United States. The other programs might more appropriately be called "offset" schemes since they facilitate offsets between different sources; they do not, however, involve an initial distribution and subsequent trades in permits.

Table 2 outlines the principal features of the existing programs. One other program in Colorado at Cherry Creek Reservoir is not included because of its similarity to the program at Dillon Reservoir. The SO_2 program has also been included to highlight some of the differences between air and water quality management.

The table reveals a number of interesting aspects of emissions trading in practice versus optimistic theoretical results. While there has been speculation in the literature about optimal means of controlling multiple pollutants (Lence et al., 1988; Lence, 1991), existing programs target only one pollutant. While the Tar-Pamlico program includes both phosphorus and nitrogen, that program does not involve permit trading. In all cases the target pollutant is the water quality limiting parameter for that particular water body.

Table 2. Comparison of Effluent Trading Programs

Case Study	Pollutant	# of Trades	Geographic Extent of Market	# of Discharges	Nature of Discharger
Fox River	BOD	One	Fox River 62 km in length	6 Municipalities & 13 Pulp/Paper Plants	Point - Municipalities and Pulp/Paper Mills
Dillon Reservoir	Phosphorus	One - Some internal offsets have occurred, credits from several nonpoint source control projects have been "banked" for future use.	Reservoir 65 sq. km.	5 Towns & Various Real Estate Developments	Point & Nonpoint - Municipalities and Real Estate Developments
Tar-Pamlico River Basin	Nitrogen & Phosphorus	Creation of a Basin Association, Cost Sharing Program for Agricultural BMPs, Minor Capital and Operational Upgrades of Existing Treatment Plants	Tar-Pamlico River Basin - 13,000 sq. km.	12 Municipalities, Numerous Farms, 2 Industrial Plants	Point & Nonpoint - 12 Large and Small Towns, Farms, One Phosphate Mining Operation
SO_2	SO_2	Many - well organized market, permits tradeable on Chicago Board of Trade and through other brokers	21 States in the East and Midwest	110 Largest Power Plants	Point - Coal-fired Power Plants

Sources: Letson (1992), Novotny (1986), Swanek (1994), Wyatt (1994)

The second observation is that trades are almost non-existent. It has been more than a decade since the Fox River and Dillon reservoir programs began. Both have seen one trade. If the efficiency of TDP programs results from the reallocation of abatement activity that tradeable permits allow, how can the program effect any cost savings if no trades occur. Novotny (1986) assesses the Fox River program and concludes that there are six reasons why more trading has not occurred: (1) the program as implemented exacerbates rather than elevates transactions cost for both traders and the State; (2) the market is thin, so the transactions cost for traders trying to find each other is high; (3) trades cannot be made *solely* to reduce costs (emphasis added); (4) trades must be made for a least one year, but not more than five years (the life of the permit). It is not clear to anyone if the State will allow those who have accumulated permits to extend them after five years, therefore increasing uncertainty about the value of the permit in the long-run. (5) The program was established on top of existing standards and dischargers were still required to comply with these standards thus reducing the scope of trading activity. Finally, (6) water quality control costs are less than one percent of product cost for industries involved thus providing little incentive to trade.

John Palmisano, the architect of the SO_2 trading program, has observed that the number of trades is not necessarily the best indicator of the cost effectiveness of a trading program (Barr, 1991). Trading may create opportunities within a firm to control effluent at a lower cost simply because the burden of regulatory compliance is less within the new system than in the old. The new policy allows them to exploit new ways of reducing discharges without regulatory review. This may be true for both the Dillon and Tar-Pamlico cases, where activities have been undertaken to offset discharges that have not resulted in actual "trades".

The third observation is that all programs involve relatively homogenous sets of dischargers. This may limit trading activity where there is little opportunity to exploit differences in abatement cost functions. Nevertheless, it makes it easier to identify uniform trading rules and administrative procedures which lower the transactions costs of trading.

Finally, it should be noted that even the much celebrated SO_2 trading program, as practiced, does not successfully resolve the problem of the spatial variation of sources. While the EPA has retained the right to approve trades it has permitted trades that, while maintaining the overall level of SO_2 emissions, result in a reduction in air quality over the region it was largely designed to protect - New England and the mid-Atlantic States.

Lessons. The necessary conditions for a well functioning effluent market in TDPs are for the control authority to be able to define for each pollutant and each emitter a vector of transfer coefficients which links emissions at location X with concentrations at each pre-defined receptor location. Under this condition specific trades can be identified (Tietenberg, 1992). The TDP market must also be competitive. Examining these conditions reveals the potential shortcomings of TDP systems for water quality management.

In most water quality management problems, in contrast to numerous important air quality management problems, the pollutant is not well mixed, spatial variability of the pollutant characterize the system. Different emitters have differential impacts on distinct receptors. This fact makes it impossible to make trades on the basis of a uniform trading ratio amongst all emitters as each emitter's impact on the receptors is different.

There are three reasons why this is problematic. First, it makes trades complicated. Parties wishing to engage in a trade have no straightforward way to estimate whether or not their trade will comply with ambient water quality standards. The complexity of trades serves as a barrier to trading activity. Second, because trades are complicated and both emitters and the control authority are concerned that trades comply with the standards, the authority must approve each trade (Novotny, 1986). Third, the non-uniformity of emitters impacts on receptors necessitates grouping emitters into submarkets in which their impacts are similar. This reduces the number of players in the market thereby increasing the likelihood that there will be too few actors and too few trades to insure a competitive market.

There are additional issues. If permits are initially distributed free of charge to all existing dischargers, these existing sources are favored at the expense of new sources. If these new sources are firms they will incur the additional expense of acquiring TDPs in the market from other sources (potentially competitors) that initially received their permits for free. At one level it is simply unfair, at another it may discourage investment in new production capacity which has lower operating costs except for the cost of the permits. This problem will be exacerbated to the extent that existing firms have market power. In theory the control authority could withhold permits in the initial allocation to make available to new sources in the future, in practice all existing sources are grandfathered into the system, permits are distributed gratis and none are withheld from the initial distribution (Tietenberg, 1990).

Finally, firms may be unwilling to participate in the TDP market for a variety of reasons. Selling permits may foreclose future options. If the asset value of the TDPs is not large the firm may prefer to retain the flexibility of using its permits later. A municipality may not wish to limit the potential for future growth or face the uncertainty of trying to buy the necessary future permits. Unused TDPs not only represent the right to pollute, but the option to pollute more in the future. Where the current value of the permit is low, dischargers may prefer to hold them.

6. CROSS CUTTING THEMES IN INCENTIVE-BASED APPROACHES TO WATER QUALITY MANAGEMENT

Not all economic analysis of water quality management has focused on the efficacy of alternative instruments. Other important studies have examined enforcement, the political economy of environmental quality management, capital turnover and the spatial variability problem. This literature is reviewed here.

Magat and Viscusi (1990) performed an empirical study on the regulation of the pulp and paper industry in the U.S. under the Clean Water Act. Their objective was to analyze the relationship between inspections and compliance. They conclude that enforcement of water quality standards in the pulp and paper industry are an "unusual success story." They identify the basis of this success as the coupling of *feasible* standards with *stringent* enforcement (emphasis added) where enforcement is measured as the frequency of inspections. In addition they found that increasing inspections reduced non-reporting of pollutant discharge levels. Their conclusions differ from their own previous work on health in safety regulations in which stringent standards are coupled with weak enforcement. That policy does not work.

Enforcement is similarly the focus of Russell's (1990) study of monitoring and enforcement of pollution control laws in Europe. Russell surveys monitoring and enforcement practices in six countries (Belgium, France, German, Italy, Spain and the United Kingdom) and concludes that:

> *A general characteristic of the European monitoring and enforcement systems might fairly be drawn as follows: Infrequent, often pre-arranged, visits are made to measure discharges. Defining what constitutes a violation is likely to some large extent to be within the discretion of the inspector who makes the visit. When a violation is discovered, the penalty for it is likely to be fairly small, at least when measured against aggregate corporate profits.*

Russell notes that both the probability of inspection and the maximum fine limit are an order of magnitude greater in Germany than in the other countries surveyed but does not present any evidence as to whether this achieves greater compliance. In concluding, he suggests that economists might do well to focus more attention on the problem of motivating compliance rather than attending only to the problem of policy design under the erroneous assumption of perfect compliance.

In an analysis of the distributional impacts of alternative pollution control measures Dewees (1990) asserts that economists have failed to understand the political effects of economic instruments because their work has focused on either the efficiency of alternative instruments or the *diffuse* distributional impacts on the general public, taxpayers, regions, product consumers, etc. (emphasis added). Dewees' analysis focuses upon the impacts of charge and effluent programs on capital (shareholders) and labor (employees,) those interests on which there are large impacts on a small number of people. He finds that shareholders and employees are more negatively impacted by either of the two economic instruments than by standards, and that in fact they may prefer standards to no regulation at all if they are tougher on new firms, i.e., they create barriers to entry for new competitors. He concludes that charge and effluent trading policies can be made politically acceptable to capital and labor if they are compensated for their losses.

In a study with potentially significant implications for transitional economies Maloney and Brady (1988) analyze the impact of environmental quality regulation on capital turnover in the electric power industry. At issue is the policy under the U.S. Clean Air Act requiring new sources to meet more stringent standards than existing sources. They find that this policy creates significant incentives to continue operation of (dirty) existing plants with a concomitant decrease in environmental quality. Ironically those states with the most stringent new source performance standards had SO_2 emissions rates which were 27% higher as a result of delayed investment in new plant and equipment induced by the tougher standards. Their work suggests that policies which differentiate between new and old sources by requiring new sources to meet tougher standards will delay desired environmental quality improvements.

One of the most significant problems of applying either charges or TDPs for water quality management is separating out the impacts of dischargers on receptors. The problem of applying economic instruments in water quality management can be greatly simplified if the impacts of different dischargers do not overlap receptors, or dischargers can be grouped according to their impacts on specific receptors. Eheart (1990) and Eheart et al. (1990) provide two useful techniques for addressing these problems.

Eheart (1990) describes a simplified technique for identifying when the impacts of nonconservative pollutants from one discharger can be considered independently of other dischargers. Eheart et al. (1990) present a method for defining groups of dischargers whose impact on water quality is relatively homogenous. While previous studies have used groupings of dischargers to examine the impacts of group differentiated charge or permit trading schemes (Brill et al., 1984; Kshirsagar and Eheart, 1982), these studies identified groupings on an ad hoc basis. Where grouping is possible the task of administrating either permit trading or differentiated effluent charges will be greatly simplified.

7. THE EVOLVING INSTITUTIONAL CONTEXT

Instruments are but one part of water quality management. Just as the experience of the last two decades has demonstrated shortcomings in policy tools, so has this experience shown that the institutions that develop and use these tools fall short of their mission. Environmental policy has been criticized as being arbitrary, centralized, narrowly focused and sometimes ineffectual. In response to these criticisms the institutions are evolving to meet the challenge of more effectively managing environmental quality.

In what follows, several of the most recent trends in U.S. environmental policy are briefly described. They are negotiated rule making, ecosystem or watershed approaches to water quality management, decentralization and national expert programs. The bias towards the American experience comes with the apologies of the author.

Negotiated Rule Making. Negotiated rule making has evolved as a response to the criticism that environmental regulations are too arbitrary and the process in which regulations are developed is too adversarial. Ordinarily, environmental legislation as enacted by Congress sets only very broad environmental objectives, such as the "fishable, swimmable" standard of the Clean Water Act. It is the responsibility of the U.S. Environmental Protection Agency (EPA) to promulgate specific regulations through which these objectives will be achieved. In doing this, the EPA develops draft regulations, publishes them for public comment then incorporates these comments as it sees fit into its final regulations. Public participation in this process, whether from industry, environmental groups or state enforcement agencies, comes as criticism of the draft regulations. The alternatives open to anyone who is unhappy with the regulation are to either apply political pressure or find grounds to sue EPA over the proposed regulation.

In negotiated rule making, the EPA agrees to involve representatives of all interested parties in the process of drafting regulations from the start. The development of regulations on discharges from pulp mills, for example, might involve representatives of the pulp and paper industry, environmental groups, state enforcement agencies as well as the EPA. The motivation is that the regulations developed out of this participatory process will meet industry's desire for cost effectiveness, the environmentalists' desire for improved environmental quality and the enforcing agencies desire for administrative efficiency. Successful examples of negotiated rule making include drafting regulations on underground injection, asbestos in schools and pesticide standards (US EPA, 1992). Potential pitfalls of the process include the absence of goodwill on the part of the participants and the fact that the remedies of political pressure and litigation are still open to everyone if the process breaks down.

Policy dialogues are a similar participatory process used to establish consensus on broader policy issues such as reauthorization of Superfund and use of plant genetic resources.

Ecosystem or Watershed Approaches. From a systems analysis perspective ecosystem or watershed approaches are not new, they are simply untried. The fundamental idea is to evaluate each ecosystem or watershed individually, establish water quality objectives for the watershed, then manage for the water quality limiting parameters rather than a set of pre-established criteria. For example, the State of Colorado has a stringent water quality standard for silver. Cities in Colorado are currently developing programs to control silver pollution mainly generated by home and commercial photo processing. However, there is no evidence which shows that silver is a water quality limiting pollutant any where in the State except for high mountain streams contaminated by mine drainage. Sediment loads create a more significant problem but there are no sediment standards. An ecosystem approach would address the sediment problem first before engaging in an expensive program to reduce silver concentrations. Variants of this idea have been articulated in Somlyódy (1993) and Water Environment Federation (1992).

Decentralization. Over the past ten years the EPA has attempted to delegate more responsibility to the state and local level. While responsibility for implementing the EPA's programs has always been with the State's, there have been greater efforts to assign greater financial and decision-making authority to them as well. The Reagan Administration crafted this policy of "New Federalism" for three reasons. First, the Administration's political philosophy was that the federal government was too big and that programs that could be run at the state level should be. Second, the Administration sought to reduce the financial burden on an over-extended federal treasury by charging the states with more fiscal authority and responsibility. Third, the policy was consistent with grassroots sentiment that the federal government was out-of-touch with the people.

While the original rational for the policy was largely philosophical and financial, the devolution of authority to the state and local level coincides with the current trends towards negotiated rule making and ecosystem management. Both negotiated rule making and ecosystem management include involving the affected parties in the process of program design and taking local concerns into account. Although much of this may be positive, it is worth remembering that the federal government became involved in environmental protection because of a combination of, lack of will at the state and local level, lack of technical expertise and fear that some states might use low environmental standards to attract investment and promote economic development. To the extent that these factors are relevant in the CEE countries we would do well to closely consider the balance between local and national control.

National Expert Programs. The EPA has historically organized itself and addressed problems on the basis of media specific programs. There are branches for water, air, hazardous waste, etc.. Essentially these branches carry out the programs mandated by the various major environmental laws, the Clean Water Act, the Clean Air Act, Superfund which are, in general, media specific laws. While this organizational structure is for the most part logical, it has created two problems. First, reducing pollution in one media has at times increased pollution in another, i.e., an air pollution problem is transformed into a water pollution problem. Second, firms generating multiple pollutants find themselves dealing with not one, but many different offices at EPA. This makes the task of complying with environmental regulations more difficult, costly and frustrating.

In response to these problems the EPA has recently created a small number of "National Expert" programs organized by industry. Two examples are programs in mining waste and pulp and paper. The intent of these programs is to create a single office within EPA with which the industry has to deal to both reduce their regulatory burden and to achieve a coordinated approach to emissions reduction across media.

At this point these programs are new and have yet to establish a record to evaluate the efficacy of this approach. One problem noted by the director of the mining waste program is that these programs have been established on top of existing programs thereby creating conflicts amongst offices within EPA over jurisdiction and resources.

8. CONCLUSIONS AND RESEARCH IMPLICATIONS

The European Economics Community (EEC) *Task Force Report on the Environment and the Internal Market* (Bonn, Economic Verlag, 1990) recommends five basic principles for environmental policy in the Single Market (as cited in Howe, 1993):

(1) the prevention principle;

(2) the polluter pays principle;

(3) the "subsidiarity" principle, i.e., placing program responsibilities at the lowest (most local) level consistent with effective overall system performance;

(4) the economic efficiency/cost effectiveness principle; and

(5) "legal efficiency", i.e. enforceability.

These principles are no less relevant for the CEE, thus providing criteria against which to judge the applicability of economic instruments in CEE countries.

It is clear from the literature that economic or incentive-based instruments are conceptually consistent with all five criteria and potentially the most effective means of achieving (2), making the polluter pay and (4), economic efficiency. It is also clear that: (a) few of the existing programs using economic instruments were designed to achieve an incentive effect on polluter behavior; (b) there are no pure incentive based program for water quality management; and (c) the limited experience with economic instruments has produced little convincing evidence of significant cost savings. Does this imply that economic instruments should be abandoned altogether? No, not yet. First, while economic instruments may have failed to live up to their promise, there is substantial evidence that standard based approaches have been excessively expensive (Tietenburg, 1985). Second, twenty years experience with water quality management has yielded important lessons with which more effective approaches can be designed. The task is to use this experience to identify the most effective mix of strategies to meet water quality objectives at minimum cost. This survey of Econo-Alice in Water Quality Land suggests the following lessons:

Lesson 1: Make it simple. The successful German effluent charge program began by controlling only five pollutants, a strictly limited set of threshold values and an uncomplicated rate structure. Modifications of the law to increase the number of pollutants (to 10) and revise the charge system came after more than 10 years of operational experience. The program is administratively simple for both regulators and the regulated.

Effluent trading programs for water quality appear to be moving in the same direction. The complexity of trades involving different impact coefficients was one of the impediments to trading in the Fox River case. The Dillon and Cherry Creek Reservoir programs are based upon trades between point and non-point sources at a fixed ratio. While situations in which trades at a fixed ratio are consistent with the dynamics of the receiving water body may be limited, it may be worthwhile to identify where such opportunities exist. Hughes (1991) identifies saline water emissions from coal mines in both Poland and the Czech Republic as one such opportunity.

What the "make it simple" edict suggests for research is that we focus our efforts on identifying the best simple program rather than the program that is simply the best.

Lesson Two: Clean Water Costs Money. Even a cost minimizing approach to improving water quality in the CEE will require substantial capital expenditures. France and Germany generate these funds with effluent charges. In the United States funding comes in part from the federal treasury, in part from combined water and sewerage charges and in part from the authority of municipalities to issue tax-exempt bonds. The design of a successful water quality management program for CEE countries requires that we ask, at the outset, where the money will come from.

The priority for economic development and the existing debt burden in CEE countries make it unlikely that their governments will be willing to commit substantial resources to improving water quality. The money must come either from effluent charges or the water quality managements authority to attract capital from the public or the private sectors.

Effluent charges have already been discussed. Other than charges, what is needed to attract capital investment in water quality improvement? The answer is the creation of municipal or regional water quality authorities with the power to issue debt and guarantee repayment. This may not be as unrealistic as it might sound. Water and sewerage service are provided by monopolies to captive markets. Water and sewerage users are highly insensitive to price changes therefore increased prices will result in higher revenues rather than a decline in demand. Thus it is highly likely that investments in water quality improvements financed by charges on water and sewerage customers can and will be paid back. Evidence of the potential for attracting investment in sewage treatment is witnessed by the substantial interest by American investors in financing wastewater treatment in Mexico after the NAFTA agreement.

The research task is to work with national and regional environmental authorities as well as municipalities to identify financing schemes that are consistent with the existing institutional structure and impediments to the flow of capital into water quality improvements.

Lesson Three: The solution will not be pure. The German, French and Dutch effluent charge systems are used in conjunction with standards. The American TDP programs have not replaced previous standards but rather have been applied on top of them. All countries' have means for enforcing noncompliance by issuing fines, revoking permits or both. While a sufficiently large charge will, in theory, induce polluters to reduce discharges to the desired level, no one has yet applied a charge that is large enough to obtain this result. TDPs cannot be applied without an enforcement mechanism otherwise there will be no incentive to acquire the necessary permits.

Nor can we assume away the existing institutional framework in the CEE countries. Whether effective or not standards exist, monitoring and enforcement programs are already in place. While it is possible that some countries will be willing to accept revolutionary change in water quality management, it is more likely that most countries will retain significant elements of the existing system.

The research task is then to identify the incremental steps from the existing institutional framework in each country that will result in more cost effective pollution control.

Lesson Four: We're not smarter than they are. It is the presumption of the traditional rule making process in the United States that neither those who will benefit from a proposed regulation nor those who will be harmed by it have much to contribute to the process of developing the regulation itself. The German success with implementing their effluent charge program contradicts this view. In contrast to the implementation of the Clean Water Act in the United States where much interpretation, litigation and political maneuvering occurred after specific regulations had been promulgated, implementation of the German law was easier because effected parties had been involved in the development of the policy from the start (Brown and Johnson, 1984). The analysis of Magat and Viscusi (1990) of standard setting in the pulp and paper industry also supports the use of broad-based participation in policy design.

CEE governments can learn from both the German experience and the emerging trend toward negotiated rule making in the United States. The result will be a water quality policy that is more likely to be politically acceptable to administer, economically feasible for industry and consistent with the aspirations of the people for improved water quality.

The research implication is that we must work with both those who are affected by water quality management policy and those who administer the policy to understand the current institutional framework, understand the objectives of water quality management *as they see them*, identify the feasible policy options and provide the necessary technical support to help them evaluate alternative policy options.

Lesson Five. There's something out there bigger than us. The transition of the CEE countries towards market economies has unleashed economic forces that extend far beyond individual sectors, regions or markets. Relative prices are changing and have yet to achieve a stable equilibrium. These changes in relative prices will affect both the ways in which goods are produced and consumers' choice of goods themselves. These effects will in turn have an impact on water quality. Such impacts have already been observed where water quality has increased as a result of the decrease in aggregate output in the CEE countries over the last several years.

If market economics fulfills its promise in the CEE countries and per capita GNP rises, what is the implication for water quality? Two forces will be at work. Higher incomes generate higher levels of consumption and their associated residuals. Higher incomes also generate greater demand for environmental quality and the ability to pay for it.

The fundamental and often hard lesson of economics is - there is no free lunch. While relative prices are still in flux it would seem an propitious time to end the free lunch at the expense of water quality. By placing a price on water pollution now, CEE governments have the opportunity to send a powerful signal into the market - that the services of the nation's rivers and lakes are not free, that pollution imposes a cost on society, a cost that must be accounted for. In doing so at this time, before substantial new investment has been made in restructuring the productive base of the economy, firms will make different decisions about industrial processes, the use of inputs and the composition of outputs. They will be forced to take the cost of pollution into account. The result will be a productive base that is fundamentally less polluting. Because the capital investment that is made now will last thirty to fifty years, it is the single most effective action that can be taken.

REFERECENS

Andersen, M. S. (1991) "Green Taxes and Regulatory Reform. Dutch and Danish Experiences in Curbing with Surface Water Pollution." Science Center Berlin

Barr, S. (1991) "Psst...Wanna buy a license to pollute." Management Review (November 1991): 50-53.

Brill, E.D.; Eheart, J.W.; Kshirsagar, S.R.; Lence, B.J.(1984) "Water-quality impacts of biochemical oxygen-demand under transferable discharge permit programs." Water Resources Research **20**(4): 445-455.

Brown, G., Jr., and R. Johnson (1984) "Pollution control by effluent charges: It works in the Federal Republic of Germany, Why not in the U.S.?" Natural Resources Journal **24**: 929-966.

Coase, R. "The problem of social cost (1960)" The Journal of Law and Economics 3(October 1960): 1-44.

Dewees, D. N. (1990) "Instrument choice in Environmental policy." Economic Inquiry, 53-71.

Eheart, J. W. (1990) "Methods for distinguishing between single and multiple discharger situations." Journal of Water Resources Planning and Management **116**(3): 335-348.

Eheart, J.W.; Brill, E.; and J. Leibman (1990) "Discharger grouping for Wwter quality control." Journal of Water Resources Planning and Management **116**(1): 21-37.

European Community. (1990) Task Force Report on the Environment and the Internal Market, 1992, The Environmental Dimension. Economic Verlag, Bonn.

Feenberg, Daniel and Edwin S. Mills (1990) Measuring the Benefits of Water Pollution Abatement. Academic Press, New York

Freeman, A. M. (1990) "Water Pollution Policy," in Public Policies for Environmental Protection (P. R. Portnery, ed.), Resources for the Future, Washington, DC

Herzog, H. W., Jr. (1976) "Economic efficiency and equity in water quality control: Effluent taxes and information requirement," Journal of Environmental Economics and Management pp. 170-83.

Howe, C.W. (1993) " The U.S. environmental policy experience: A critique with suggestions for the European Community," Environmental and Resource Economics, **3**:359-379

Hughes, G. (1992) " Are the costs of cleaning up Eastern Europe exaggerated? Economic reform and the environment, Oxford Review of Economic Policy, 7(4):106-136

Kularathna, M. and Somlyódy, L. (1994) River Basin Water Quality Management Models: A State-of-the-Art Review. Working Paper, WP-94-3, Institute for Applied Systems Analysis (IIASA), Laxenburg, Austria.

Kshirsagar, S.R. and Eheart, J. W. (1982). "Grouped markets for transferable discharge permits for water quality management." Working Paper 4, National Scinece Foundation Award PRA 79-13131, National Science Foundation, Washington, D.C..

Lence, B. J., J. W. Eheart, and E. Brill (1988) "Cost efficiency of transferable discharge permit markets for control of multiple pollutants," Water Resources Research **24**(7): 897-905.

Lence, B.J. (1991) " Weighted sum transferable discharge permit program for control of multiple pollutants," Water Resources Res. **27**(12):3019-3027

Letson, D.. (1992) "Point/nonpoint source pollution reduction trading: An interpretive survey," Natural Resources Journal 32, (Spring 1992): 219-232.

Magat, W.A., and W.K. Viscusi (1990) "Effectiveness of the EPA's regulatory enforcement: The case of industrial effluent standards," The Journal of Law and Economics **33**(2): 331-60.

Maloney, M. T. and G. L. Brady (1988) "Capital turnover and marketable pollution rights," The Journal of Law and Economics **31**(1): 203-26.

Novotny, G. (1986) "Transferable Discharge Permits for Water Pollution Control in Wisconsin." Wisconsin Department of Natural Resources, Madison, WI

Opschoor, J. B. and Vos., H. (1989) Economic Instruments for Environmental Protection, Organization for Economic Cooperation and Development, Paris, France.

Pethig, R. (1989) "Efficiency vs. self-financing in water-quality management," Journal of Public Economics **38**(1): 75-93.

Russell, C. S. (1990) "A preliminary view of monitoring and enforcement of pollution control laws in Europe." A paper delivered at the Symposium on Conflicts and Cooperation in Managing Environmental Resources, Freudenberg, Germany, 15-16 November 1990.

Smith, V.K. and William Desvousges (1986) Measuring Water Quality Benefits. Kluwer Academic Publishers, Norwell, MA

Somlyódy, L. (1993) Quo Vadis Water Quality Managment in Central and Eastern Europe?. Working Paper, WP-93-68, International Institute for Applied Systems Analysis (IIASA), Laxenburg, Austria.

Swanek, R. (1994) North Carolina Department of Natural Resources, Water Quality Division, personal communication, 7 June 1994.

Tietenberg, T.H. (1985) Emissions Trading: An Excercise in Reforming Pollution Policy. Resources for the Future, Washington, DC

Tietenberg, T.H. (1990) "Economic instruments for environmental regulation." Oxford R. Econ Policy, **6** (Spring 1990): 17-33.

Tietenberg, T. (1992) Environmental and Natural Resource Economics. 3rd. ed., HarperCollins, New York.

von Hayek, F.A. (1945) "The use of knowledge in society." American Economic Review 35(September 1945): 519-30.

U.S. Congress, Congressional Budget Office (1985) Efficient Investment in Wastewater Treatment Plants, U.S. Government Printing Office, Washington, DC

U.S. Environmental Protection Agency (1992) Fact Sheet: negotiated Rulemaking/Regulatory Negotiation, Washington, DC

Water Environment Federation (1992) A National Water Agenda for the 21st Century Final Report. Water Quality 2000, Water Environment Federation, Alexandria, Virginia.

Wyatt, L. (1994) Northwest Colorado Council of Governments, personal communication, 2 June 1994.

CHAPTER 14

USE OF ECONOMIC INSTRUMENTS TO ENHANCE CEE WATER QUALITY: INSTITUTIONAL CHANGES AND RESEARCH CHALLENGES

Charles M. Paulsen[1]

1. INTRODUCTION

The preceding chapter by Mark Smith amply demonstrates that much research has been directed at the "efficiency properties" of economic instruments for pollution control. Beginning in the early 1970's, a large body of both theoretical and empirical research has demonstrated that the use of transferable discharge permits or uniform emission charges is the least expensive means to meet a limit on total discharges (c.f. Tietenberg, 1985). This work has been extended to deposition or ambient quality impact permits (for example, Spofford and Paulsen, 1988). One can show from theory that the least costly policy to meet limits on ambient quality is through the use or tradable permits in ambient impacts or ambient quality decrement charges (Tietenberg, 1985, among others).

Economists are often dismayed, however, when surveying the field of real-world applications of economic instruments. With a few significant exceptions, including US and EC systems for trades in sulfur dioxide, there are almost no large-scale applications of the theory to real-world situations. Some policy analysts hope that, as Central and Eastern European economies transform from central planning into free-market systems, they will adapt economic instruments to control discharges and improve ambient quality (for example, Zylicz, 1993). This hope seems to be based on three conjectures or arguments. The first is that since CEE countries are both financially strapped and facing serious environmental problems, the efficiency of economic instruments will be more attractive to CEE governments than was the case in Western Europe and North America, whose economies were far more prosperous when environmental problems first received wide-spread attention in the

1 Paulsen Environmental Research, Portland, Oregon, USA

NATO ASI Series, Partnership Sub-Series, 2. Environment - Vol. 3
Remediation and Management of Degraded River Basins
Edited by V. Novotny and L. Somlyódy

early 70's. The second is that economic instruments are an obvious extension of free-market reforms: as supplies and prices for marketable products are de-regulated and opened up to market forces, it follows that supplies and prices for non-marketed "goods" such as BOD or nitrates should be allowed to float as well, rather than being controlled by a central authority. The third is that with societies and economies in transition, it should be easier to implement fundamental reforms in environmental policy, since these reforms will be seen as accompanying other radical institutional changes.

Note that only the first of these foundations is primarily an economic concern, based on efficiency or cost savings. Even this efficiency argument has a non-economic component, since it is obvious that just because cost savings are possible does not necessarily mean that a society must take advantage of them. The other arguments are more closely related to institutional changes than to traditional micro-economic concerns about efficiency. This is an area that has received relatively little attention from resource economists who have been the primary proponents of the use of economic instruments.

This chapter first offers a brief review of recent analyses on CEE water quality and cost-effective policies to improve it. Next, it expands on why cost-effective pollution control has been so rarely applied in OECD countries in the past, and some speculation why it may have become somewhat more popular in recent years. It then examines the institutional trends that explain the economists' hope that cost-effectiveness may prove to be more popular in CEE countries. It concludes with suggestions for potential applied research topics that seem to follow from these issues.

2. POLICY RESEARCH TO DATE

When one considers the fact that post-socialist research on CEE water quality is at most five years old, there has been a remarkable amount of work done on the topic. In addition to work done by Hughes (1992) and others at the World Bank, most of the empirical research has centered on work done at IIASA with numerous collaborators from other institutions (e.g., Chapter 1 of the present volume, Somlyódy & Paulsen, 1993; Paulsen, 1993a). Much of this work has focused on method development and broad policy prescriptions (of which, more below). In addition, work done under the PHARE program and similar studies (e.g., USAID, 1993) has begun the tedious but necessary tasks of compiling discharge inventories, making first-cut estimates of discharge reduction costs, and the like. What follows is an attempt to summarize the broad policy implications that can be drawn from this body of work. The purpose of the summary is not to compile a bibliography of past work, but to set the stage for what I see as some important gaps in the existing knowledge base.

The first point is that the cost of meeting EC standards for water-borne, point-source discharges will be extremely high relative to CEE countries' economies. Table 1 shows estimates, based on Paulsen (1993b). Using 1992 gross domestic product (GDP) as a point of comparison for the costs, one can see that meeting EC standards would cost from 0.6 to 4.5 times total domestic production, depending on the country. By contrast, most estimates of annual spending suggest that OECD countries spend anywhere from one to three percent of GDP on all environmental protection activities combined, far below the (admittedly imprecise) estimates for CEE countries for water pollution reduction alone.

One option, of course, would be for CEE governments to "buy now and pay later" by borrowing the required capital from international lending institutions. Most CEE countries already carry heavy debt loads, and their prospects for strong economic growth in the near term are limited at best (see Table 1). Of course, there are many other priorities for public investment, ranging from industrial privatization and improvements in transportation and communication infrastructure to mitigating the long-term effects of improper disposal of toxic substances. This means that it will probably not be possible for central governments to finance more than a modest portion of water-borne discharge reduction from general tax revenues.

Table 1. Resources and Potential Costs for Improving Water Quality in Central and Eastern Europe.

County	1992 Population, Millions	GDP, Millions of US$, 1992	1992 Per-Capita GDP, US$	Per-Capita Cost to Meet EC Water Quality Standards, 1992 US$	Total Debt as % of GDP, 1991	% Change in Industrial Production, 1990-1992
Bulgaria	8.47	6,903	815	3,755	N/A	-54
Former CSFR	15.66	36,093	2,305	4,927	27	-40
Hungary	10.3	35,494	3,446	2,116	78	-32
Poland	38.3	72,579	1,895	1,230	61	-32
Romania	23.2	14,152	610	1,422	N/A	-54

The disparity between environmental demand, on the one hand, and financial supply, on the other, generally leads analysts to two broad policy prescriptions. The first is that improvement in any environmental arena, whether air-borne emissions, water-borne discharges, or toxic waste disposal, will necessarily be a gradual process. Different writers obviously have different conceptions of the amount of time required and what the end point for improvement might be, but the consensus seems to be that one to three decades will be needed for CEE countries to meet standards similar to those presently attained in OECD countries (Somlyódy, 1993). Note the use of the term "presently attained," as opposed to standards that are on the books but not actually met in practice. This is not a trivial distinction, since in many areas (e.g., US air quality) many cities have never attained prevailing ambient quality standards.

The second prescription is that policies should be cost-effective, least-cost, or some variant on these terms. As with the gradualism, the use of the term differs somewhat among authors, but what is usually meant is that CEE governments should put policies into place that either reduce discharges or improve ambient quality without spending more than is necessary to do so. A wide variety of methods have been suggested to do this (see previous chapter by Smith or Kularathna and Somlyódy, 1994 for recent surveys). Although the mathematical methods and specific policy prescriptions vary considerably, the overall purpose of these efficiency analyses is to meet a set of discharge limits (in mass units per day) or a set of ambient water quality standards, at the lowest possible cost. The research methods used generally involve a combination of optimization methods, to find the least-cost set of discharge reduction techniques, and a simulation model to compute the water quality effects of different control technologies. The results, not surprisingly, usually show that discharge policies that try to minimize costs have lower costs than policies that do not take expenditures into account. The cost savings depend on two broad phenomena: that some dischargers can control emissions more cheaply than others, and that different dischargers have a different impact on water quality, due to their location relative to critical reaches. One example, from previous work on the Nitra, is shown in Table 2.

The efficiency arena has been an extremely fruitful research area, in both analyzing the relative costs, different policies and in developing methods that gracefully incorporate non-linearities, uncertainty, and other complex phenomena (see, for example, previous chapters by Smith or Kularathna and Somlyódy, 1994).

Table 2. Results of Control Policies for Nitra River

Dissolved Oxygen Standard	Control Policy	Capital Cost, 10^6 US$	Annual Cost, 10^6 US$	Ratio-Annual Cost to BAT Cost	Minimum Dissolved Oxygen
None	Base Case	0.0	0.0	-	0.7
-	Minimum discharges (BAT)	64.7	14.4	1.00	6.9
DO ≥ 4	Uniform % reduction	23.4	5.7	0.40	4.3
DO ≥ 4	Limit on regional discharges	18.3	3.9	0.27	4.6
DO ≥ 4	Regional least-cost	9.2	1.9	0.14	4.0
DO ≥ 6	Uniform % reduction	40.7	10.4	0.72	6.9
DO ≥ 6	Limit on regional discharges	33.0	6.9	0.48	6.5
DO ≥ 6	Regional least-cost	23.6	5.2	0.36	6.0

Source: Somlyódy and Paulsen (1993).

We are not aware of any applications of the above methods on a national scale for water-borne discharges. Simpler comparative-static methods have been applied as research exercises to both OECD and CEE countries. An example that is specific to CEE countries is the impact of economic restructuring on environmental quality. While empirical applications are still fairly limited (e.g., Hughes, 1992; Csermely et al., 1994), many other writers have remarked on the potential importance of this phenomenon (e.g., Somlyódy and Paulsen, 1993; Zylicz, 1993). Although the details of empirical estimation of the effects of restructuring can be complex, the basic idea is straight-forward. The notion is that as subsidies for raw materials are removed and money-losing industries close, plants will become more efficient and generate less pollution. One example is the removal of subsidies for fertilizer purchase in Hungary. Since these subsidies ended in 1990, concentrations of phosphorus have declined markedly in several Hungarian rivers (Somlyódy and Paulsen, 1993). Another is the well-known effect of rising water prices in many CEE countries. This has markedly reduced consumption. If consumption remains low in the future, this should result in cost savings for municipal water treatment plants, since most of the investment cost for these plants is determined by required hydraulic capacity. This is an area where additional empirical work may identify additional potential cost savings.

3. APPLICATION IN THE OECD

As noted in the introduction, the actual application of cost-effective methods to control water-borne discharges has been extremely rare. Recent surveys by the OECD (1992) and by Opschoor (1993) show that although waste dischargers are subjected to all manner of fees and fines, almost none of these is high enough to actually have any direct effect on their behavior. That is, it is almost invariably the case that economic incentives that are applied to either air or water-borne discharges are not large enough to have any incentive effects. Instead of reduction in discharges, the intended effect of the fees is almost exclusively to raise revenue for environmental improvement, such as wastewater treatment works, paying the administrative costs of inspection and monitoring programs, and so forth.

Given that this is the case, it seems useful to examine two aspects of the OECD experience with cost-effective environmental policies: why they have generally been unpopular, and why some of the exceptions have actually worked. The aim in both cases is not to do a comprehensive review of such policies, since this can be found in the research noted above. Instead, by surveying the OECD experience, we hope to discover some potential lessons for CEE governments, where the case for cost-effectiveness is more pressing than in the OECD. The discussion uses the definitions found in Smith (previous chapter) for the various specific economic instruments, such as transferable discharge permits, transferable ambient quality decrement permits, and emission charges.

The first class of problems includes difficulties for individual plants, firms, or dischargers, including municipal wastewater treatment plants. Other things being equal, their costs will increase, since besides direct control costs, they must also pay fees or pay for permits under any sort of incentive scheme. In cases where permits are issued free to existing firms, this difficulty is alleviated to some degree, but new firms will always have to buy permits before they can begin operation. In addition, most schemes require that discharges be monitored much more accurately under a fee or permit plan than would be the case with command-and-control regulation, and the cost and responsibility for monitoring usually falls most heavily on the dischargers themselves. Another problem for sources is the fact that some dischargers will usually need to reduce emissions more than others (under most ambient quality permit policies). This is often regarded as inequitable. Finally, if multiple pollutants are controlled using economic instruments (e.g., phosphorus, nitrogen, and BOD), or if ambient quality decrement markets exist at more than one receptor, sources will need to hold multiple permits in pollutants whose production is controlled jointly. This greatly complicates matters for potential traders in pollution markets.

From the viewpoint of environmental groups, the fact that dischargers can purchase more permits if they choose to spend the money is often regarded as a "license to pollute." While on the face of it, this view seems to ignore the fact that existing command-and-control systems are also "licenses," the existing systems often have the built-in assumption that discharges are a necessary evil that will eventually, via technological progress, tighter regulations, and more effective enforcement, be made to vanish entirely (Sagoff, 1993).

For environmental authorities, charged with implementing and enforcing a fee or transferable permit scheme, numerous potential difficulties may arise. From an analytical or scientific viewpoint, perhaps the most serious is that the discharge and ambient effects of economic instruments are regarded as very uncertain. There seem to be two root causes for this skepticism. The first is the (well-founded) belief that water quality modeling is equal parts art and science, and that one cannot place enough confidence in discharge-water quality models to use them as a guide for setting fees, permit prices, and so forth. The second is the belief that the assumptions which underlie the economic models are unrealistic: that firms do not really minimize costs, have perfect information regarding the effectiveness of treatment technologies, and so forth.

A second type of problem is that the use of economic instruments may require more accurate, defensible estimates of discharges than some other policies. In particular, if a policy based on economic instruments replaces one that specifies treatment technologies (e.g., biological treatment with nitrogen removal) the authority's task may become considerably more complex. Under a policy that specifies a particular treatment technology, a regulator may need only inspect the facility to verify that a treatment plant is installed and operating. In contrast, when using economic incentives, the actual amounts of different substances discharged becomes critical, since this is the basis for fees collected and the "portfolio" of permits held by each source.

A third type of problem (at least in the US) is the "regulatory bias" of legislators and lawyers who write the regulations. It is often believed that it is easier to simply issue a decree, such as "all sources will minimize discharges," than to carefully analyze the probable results of a variety of different policies. The individuals responsible for creating policies in the West often have little or no training in the analytical methods required to simulate the results from different types of policies. In addition, many details of different policies, such as the required treatment technology for a particular industry, are often decided based on what regulators, industry groups, and environmental groups decide that they can live with, rather than what might be best for society as a whole.

4. RECENT DEVELOPMENTS IN THE OECD

Taken in combination, problems with economic instruments, whether real or imagined, have greatly reduced the scope of environmental problems to which economic instruments have been applied. Indeed, the few arenas where economic instruments have been applied, they are always layered on top of extensive conventional regulations. Nevertheless, some notable applications have recently been instituted, including the SO_2 trading program in the US, the extensive use of "environmental taxes" in Sweden, and others (OECD, 1992). There seem to be several reasons for these developments. The first is economic: as environmental regulations have become stricter in OECD countries, they have also become more expensive for private and public organizations that must implement them, and economic instruments are viewed as having considerable cost-saving potential. The second is that environmental groups, particularly in the US, are interested in further reductions in emissions or discharges; the potential cost-savings from economic instruments are seen as a selling point for both industry and regulators. Finally, at a more conceptual level, the Bruntland report (1987) emphasizes the need to link environmental improvement and economic development, and economic instruments may be viewed as one way to achieve this. All of this gives some grounds for optimism among resource economists working on the subject.

5. CEE INSTITUTIONAL TRENDS

Most CEE governments have stated that they intend to apply for EC membership when their economies stabilize sufficiently. As part of this process, they will need to develop policies to substantially reduce discharges. While timetables and other details of these policies are still under development, it is obvious that very substantial discharge reductions will be required to come close to meeting EC standards. It is apparent from work by Hughes (1992) and others that although market reforms can, by themselves, reduce discharges by inducing increases in manufacturing efficiency, they cannot meet the very large discharge reductions that would be needed to meet EC standards. Neither government tax revenues nor industrial profits can hope to meet this demand for discharge reduction any time soon (see Table 1). There is also an understandable desire to use EC effluent standards as the basis for water-borne discharge control, rather economic instruments or other more innovative methods. Indeed, there is anecdotal evidence that these standards are often advocated by OECD-based consultants, since they are both familiar and profitable for consulting firms.

All of this is set against a background of substantial institutional instability. Most CEE governments are composed of coalitions of minority parties, making passage of any environmental legislation difficult. Once legislation is in place, uncertainty surrounding how (and even whether) it will be enforced is pervasive. Many industrial sectors may vanish completely within the foreseeable future, while the course of privatization for many countries remains problematic. Most countries in the region have had only one parliamentary election since 1989, and policies may change substantially following the next round of elections (e.g., Poland's economic reforms following the 1992 elections). Even if one sets aside the states of the former Yugoslavia, civil wars of varying intensity are ongoing in Georgia, Tajigistan, Armenia and Azerbaijan, and the potential for similar events in other countries of the former USRR, including Russia, cannot be discounted out of hand.

This instability is simultaneously a problem and an opportunity. One problem, of course, is that instability in discharge-control policies makes investment in capital-intensive sewage control technology problematic even for profitable firms and solvent municipalities.[2] Broader instability problems of the sort faced by citizens of Ukraine, for example, are far more serious. The opportunities are perhaps more difficult to discern, and relate to the economists' optimism noted in the introduction. Basically, it amounts to assuming (or at least hoping) that since things are changing, they can just as well change for the better as for the worse. In this case, this means that environmental policies could be formulated that would be less costly than those of the OECD, while substantially improving water quality.

There is certainly some evidence to support this hope. Hungary is seriously considering implementing an effluent charge system that would reduce discharges at substantially lower cost than a Best Available Technology (BAT) or uniform percentage reduction policy (see Paulsen and

---

[2] This is especially problematic given high inflation rates and even higher real interest rates.

Lehoczki, 1994). Poland has formulated a very flexible control policy for air pollution control that promises considerable saving (Zylicz, 1993). In most cases, however, these policies are still under consideration in their respective parliaments, and changes seem likely before actual implementation.

6. RESEARCH POSSIBILITIES

Although the problems with economic incentives and other efficient policies are obviously serious, the pay-off from their application in CEE countries is potentially high. I believe that the problems are primarily questions of acceptable, practical application of existing tools and techniques, rather than technical questions that would require the development of entirely new methodologies. Three areas seem to be especially likely to be fruitful for future work: efficiency analyses, methods to make cost-minimizing solutions acceptable, and careful treatment of institutional uncertainty.

6.1 Efficiency

As already noted, analyses of the technical efficiency of various water-quality improvement policies have received extensive attention in the technical journals (Kularathna and Somlyódy, 1994). Their review rightly emphasizes the need to handle uncertainty in water quality impacts in a systematic way. This emphasis, of course, is based on the fact that prediction of water quality impacts of any particular array of discharges is at best an inexact science. Formulating and calibrating appropriate water quality models for any given river basin and policy scenario are obviously difficult problems, and further work is clearly required in this area.

Another problem for the use of marketable permits to enhance water quality appears to have received much less attention. The problem with "markets" in this context is that there are usually only a small number of participants. For example, in the Nitra (Kularathna and Somlyódy, 1994) there would be at most 20-30 participants in potential markets, even if one included relatively small industrial sources. Since there are only a few potential buyers and sellers, markets may not "clear," because at any given time a potential purchaser of permits may not be able to locate a potential seller. This problem has received more attention for air-borne emissions, where researchers have advocated the use of simulation and laboratory experiments to project how small-scale markets might react (e.g., Ferreira dos Santos, 1993). It would be useful to apply these methods to small-scale markets for water-borne discharges as well.

A third efficiency-related research possibility concerns the rules used for permit trading. For example, some trading schemes require that ambient quality constraints not be violated as a result of a trade (Klaassen and Amann, 1992). That is, as trades in emissions over time occur, constraints on ambient (water) quality should be maintained. This is another area that has received more attention from air-pollution specialists. It would be useful to see how trading rules affect water quality, efficiency, and distribution of costs and discharges for applied water quality problems.

6.2 Acceptability

Acceptability of cost-minimizing solutions has obviously been a problem in the OECD, and it seems likely to be problematic in CEE countries as well. As the old economists' maxim says, "there is no constituency for efficiency," and without a constituency, efficiency will be overlooked in setting water-quality control policies. Clearly, this problem is partly one of education and persuasion, which are not traditional research topics among water quality analysts or economists[3]. In addition, however, there are several related problems that are amenable to modeling. First, a key question that is sometimes overlooked in cost-minimization analysis is how costs are distributed, as opposed to what total costs are. For individual firms and municipalities who must install and operate sewage treatment plants, this question is crucial to obtaining their support for any proposed policy. A closely related question is what sort of financial mechanisms might be used to distribute costs more acceptably, such as fees, taxes, and subsidies. Note that it should in principle be possible to design instruments that are *both* efficient *and* that distribute costs equitably, but this is obviously more difficult than doing only one or the other.

A second research topic related to acceptability is how an efficient policy will fit with existing practice in individual countries, river basin authorities, and other jurisdictions[4]. In practice, economic instruments are invariably only one stratum in a multi-layered regulatory cake. Analyses that take close account of existing rules and regulations are obviously more likely to be persuasive to environmental authorities than those that assume a "pure" market system for analytical convenience. Along the same lines, models that are based as closely as possible on existing data already being

---

[3] An obvious area of investigation regarding the OECD would be how successful efficiency-enhancing policies were conceived and implemented.

[4] A related problem is that fact that many countries have very strict standards that are not enforced. This compromises the authority of responsible agencies.

collected for other purposes are far more likely to be accepted than those that require systematic collection of new discharge inventories or other information.

Finally, my impression is that environmental authorities charged with implementing marketable permits, fees, and other efficiency-based policies are often overwhelmed by the complexity of the models that underlie the policies (Opschoor, 1993). I think that one could easily construct suites of models for several river basins, ranging from extremely simple to state-of-the-art. The next step would be to see if there is any necessary correlation between model complexity and policy relevance. Simply because a model produces more accurate predictions or has a better scientific basis does not necessarily mean that these factors will make any difference in setting pollution-control policies. For example, work done in the early 1970's with industrial process models (Russell, 1973) showed that whether linear programming models had hundreds or thousands of decision variables was relatively unimportant for predicting the responses of plants to a wide variety of emission charges and discharge constraints. I suspect that this would prove true for river-basin planning as well, especially for policies designed to make first-order improvements to severely degraded river basins.

6.3 Institutional Uncertainty

As indicated earlier, institutional uncertainty is a pervasive feature of environmental policy in CEE countries. Even if one concentrates on the more stable governments (Poland, Hungary, the Czech and Slovak republics, Romania), there are still many uncertainties for which no ready OECD parallels exist. Performing credible water quality policy analyses under these circumstances will require an unprecedented sensitivity to a wide range of uncertainties. While a detailed treatment of how one might approach these is beyond the scope of this paper, some examples may at least give a taste of what would be required.

The obvious place to begin is with the possible fate of many industrial sectors. As subsidies are withdrawn from many industrial sectors, their managers usually have three choices: lobby for continued subsidies, modify their operations and product mix, or close. Setting aside continued subsidies, these options will inevitably change their water-borne discharges, and may markedly alter the cost-effectiveness of particular control policies (see Paulsen, 1993, for a didactic example). Environmental affects are rarely given the same weight as employment and income in decisions regarding industrial policy. However, it is clear that CEE industrial policy may have important effects on environmental policy.[5] Two examples are found in Hughes (1992) and Csermely et al. (1994), in which the authors estimate the effects of increases in energy prices and changes in industrial structure on CEE and Hungarian air quality. Additional work along these lines could usefully be applied to water quality problems. Such analysis could serve two purposes. The first would be to estimate how much water quality improvement can be achieved "free of charge" as a result of closure of uneconomical industrial plants and sectors. The second would be to discover how sensitive cost-effective policies may be to such changes.

A second area of uncertainty concerns government environmental policies. Although analysts generally investigate a variety of source control policies in any applied analysis (e.g., BAT, minimum discharges, uniform percentage reduction, etc.) there are much broader uncertainties that should be investigated in transitional societies like those in CEE. Again, several examples may help illustrate the breadth of the problem. First, whatever polices are chosen initially, they may well change over time. For example, what might happen if a government first tries to implement BAT policies, then decides to switch to something less costly before some sources have actually constructed treatment plants? What if discharge sources believe that some years will elapse between establishing a policy and effective enforcement? Second, funding sources are very much up in the air, at least at present, which leads to a host of additional uncertainties. Will funding levels and levels of policy control be the same or different? For example, if funding for municipal treatment plants is obtained by taxes or fees on users of the system, will this involve complete local control over the type of treatment plant installed? If (as occurred in the US) the central government supplies most of the funding for a treatment plant, does this imply much greater central control over plant design?

Obviously, only time will tell how these uncertainties will actually be resolved. However, all of them are amenable to applied research, at least in the form of sensitivity analysis. When one contemplates a comprehensive analysis that would examine scientific uncertainty, acceptability, and institutional uncertainty in a systemic fashion, it is readily apparent that this will be a substantial undertaking. Nevertheless, the research pay-offs from such a study are likely to be substantial. In addition, the rate of return from a real-world application will probably be quite high.

---

[5] It was past economic policies and industrial activity that is largely responsible for the present-day state of the environment.

REFERENCES

Bruntland Commission (1987) Our Common Future: Report of the World Commission on Environment and Development. Oxford University Press, Oxford, Engl.

Csermely, Á., P. Kaderjak, and Z. Lehoczki (1994) "Direst Impacts of Industrial Restructuring and Air Pollution in Hungary," Fifth Annual Conference of Environmental and Resource Economists, June, 1994, Dublin. Draft, February

Ferreira dos Santos, R. (1993) "Transferable Discharge Permits for Air Pollution Control: A Methodology for the Allocation of Joint Benefits," IIASA Conference Proceedings, Economic Instruments for Air Pollution Control, Laxenburg Austria, October 1993.

Hughes, G. (1992) "Are the Costs of Cleaning Up Eastern Europe Exaggerated?" Economic Reform and the Environment, Oxford Review of Economic Policy, vol. 7, no. 4.

Klaassen, G., and M. Amann (1992) "Trading of Emission Reduction Commitments for Sulphur Dioxide in Europe," Working Paper No. SR-92-03, International Institute for Applied Systems Analysis, Laxenburg, Austria, May.

Kularathna, M. and L. Somlyódy (1994) "River Basin Water Quality Management Models: A State-of-the-Art Review," Working Paper No. WP-94-3, International Institute for Applied Systems Analysis, Laxenburg, Austria.

OECD (1992) "Environment and Economics: A Survey of OECD Work," Organization for Economic Cooperation and Development, Paris, January, 1992.

Opschoor, J. B. (1993) "Trends in the Use of Economic Instruments in OECD Member Countries," IIASA Conference Proceedings, Economic Instruments for Air Pollution Control, Laxenburg Austria, October 1993.

Paulsen, C. M. (1993a) "Policies for Water management in Central and Eastern Europe," Energy and Natural Resources Division Discussion Paper ENR 93-20, September, 1993. Resources for the Future, Washington, DC

Paulsen, C. M. (1993b) "Cost-Effective Control of Water Pollution in Central and Eastern Europe," Resources, Fall, 1993, pp. 28-31

Paulsen C. M. and Z. Lehoczki (1994) "Emission Charges for Reducing Hungarian Industrial Discharges," in press.

Russell, C. S. (1973) Residuals Management in Industry, Resources for the Future, Washington, DC.

Sagoff, M. (1993) "Environmental Economics: An Epitaph," Resources, Spring, 1993, pp. 2-7.

Somlyódy, L. (1993) "Quo Vadis Water Quality Management in Central and Eastern Europe?" Working Paper No. WP-93-68, International Institute for Applied Systems Analysis, Laxenburg, Austria.

Somlyódy, L., and C. M. Paulsen (1993) "Cost-Effective Water Quality Management Strategies in Central and Eastern Europe", Resources for the Future Discussion Paper 94-05, December, 1993.

Spofford, W.O., Jr., and C.M. Paulsen (1988) "Efficiency Properties of Source Control Policies for Air Pollution Control: An Empirical Application to the Lower Delaware Valley," Quality of the Environment Division Discussion Paper QE88-13, Resources for the Future, Washington, D.C.

Tietenberg, T. H. (1985) *Emission Trading:* An Exercise in Reforming Pollution Policy, Resources for the Future, Washington, DC.

USAID (1993) "Water Quality Pre-Investment Studies in Four Danube River Tributary Basins: Summary Report," US Agency for International Development, Water and Sanitation for Health Project, Washington, DC, July, 1993.

Zylicz, T. "Cost-Effectiveness of Air pollution Abatement in Poland," IIASA Conference Proceedings, Economic Instruments for Air Pollution Control, Laxenburg Austria, October 1993.

CHAPTER 15

SUMMARY AND CONCLUSIONS

1. SUMMARY OF WORKSHOP PRESENTATIONS

The degradation and poor quality of surface and groundwater resources in the CEE countries has been extensively documented since the political changes that lead to restoration of democracy in the region. However, in most CEE countries, it has become clear that the cost of meeting the standards based on the European Community criteria may be beyond their economic means.

The magnitude and the reasons for the adverse situation were outlined in the presentations by the participants of the ARW. Professor Valentina Priazhinskaya from the Russian Academy of Sciences, pointed out that, at present, 75 percent of rivers and lakes were unsuitable as sources of potable water and, consequently, 50 percent of tap water provided by water supply utilities violates health standards. Also, in spite of the immense area of Russia, 30 percent of all groundwater is polluted. Only 30 percent of municipal and industrial sewage is inadequately treated while the rest is discharged untreated. A similar situation was documented by Professor Jan Suschka for the Upper Silesia region in Poland where over 90 percent of all surface water bodies are unfit for drinking water supply. In this region, 50 percent of all sewage discharged into the receiving waters receives no treatment. In Slovakia (according to Professor Juraj Namer of the Technical University in Bratislava), only 51 percent of sewage is collected by sewers and even a smaller fraction of sewage receives any treatment. In general, the level of wastewater treatment throughout the CEE region is relatively low and, with the exception of the Czech Republic, does not exceed 40 percent even in more advanced CEE countries. High industrial pollution, reliance on outdated production technologies, the lack of pretreatment, and the absence of proper economic instruments and enforcement represents the second group of adverse factors. In addition to water quality degradation, excessive discharges of untreated or partially treated discharges have also caused in-situ soil and sediment contamination. The third cause of the adverse situation in water quality is intensive agriculture that contributes sediment, nutrients from fertilizers, and organic chemicals.

NATO ASI Series, Partnership Sub-Series, 2. Environment – Vol. 3
Remediation and Management of Degraded River Basins
Edited by V. Novotny and L. Somlyódy

Professor Suschka presented River Klodnica in Poland as a typical example of a severely stressed water body. This river drains approximately 50 percent of the Upper Silesia Region. The total average low flow of the river is 8.3 m^3 /s from which about 5.3 m^3 /s are wastewater discharges (80 percent municipal sewage and 20 percent industrial discharges, including saline mine drainage). Only 40 percent of sewage and industrial discharges receive some treatment, which is mostly limited to primary mechanical treatment. As shown on Figure 1, the concentrations of BOD_5 in the river are on the level of untreated sewage, resulting in long anoxic stretches of the river.

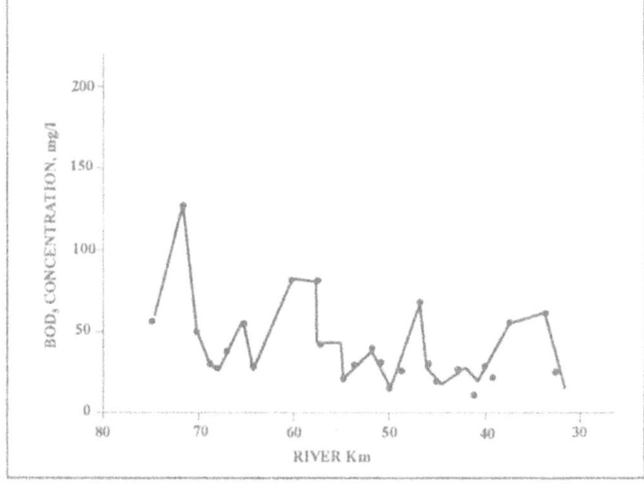

Figure 1

BOD profile of the Klodnica River in Upper Silesia (after Prof. Suschka)

Professor Priazhinskaya also pointed out that diffuse pollution in Russia as well as in other CEE countries, is a major threat to population. Airborne lead, sulphur dioxide and other aerosols affect the health and economy of Russia. High levels of nitrates in ground- and surface water are widely spread throughout the CEE countries. Biodiversity is also threatened.

Several presenters reported ongoing river basin studies and water quality monitoring. Monitoring has been generally adequately funded in most CEE countries and research studies as well as routine monitoring conducted by state research institutes and river basin agencies is well established and advanced. There is also a large water quality/quantity data base available at these institutions. For example, Professor Pavel Petrovič of Slovakia described the hydrological data of the Nitra River basin (see Chapter 12 by Somlyódy et al.). It is a well known fact that the previous regimes often kept the results of monitoring secret and did not inform the public about the state of deterioration of the environment. Today, on the other hand, the governments and population in all CEE countries are aware of the environmental hazards caused by pollution and degradation of water bodies and watersheds. However, in most countries environmental remediation does not seem to be the top priority. Although the data collection in the CEE countries has been carried out, in some cases extensively, the data were not used for solving real life problems. Consequently, no integrated data bases (including, e.g., hydrology, water quality, water supply, wastewater collection and treatment) exist today and the access to the data bases is not easy because of the collapse of the earlier institutional system and the appearance of many follow-up institutes/companies struggling for survival and/or interested in profit making.

In almost every CEE country, water quality planning studies have been and are being prepared, often partially funded by Western financing (e.g., the World Bank financed international study of the Danube River or the European Community project dealing with the remediation of the River Elbe). Participants from Czech Republic (Ing D. Beránková and Dr J. Zdařil) described briefly a program for improvement of water quality in the Morava River basin (the largest tributary of the Danube River originating almost solely from the Czech Republic). In the Czech Republic, the problems of pollution of the river basins of Elbe and Morava has received attention by the authorities and an ambitious clean-up programs have been proposed for both water bodies, sponsored, in part, by international organizations such as the European Community and the World Bank. Dr Feher of the Hungarian Water Resources Research Center (VITUKI) outlined the Hungarian national water quality research program which is aimed at development of water quality objectives for selected Hungarian watersheds.

In some CEE countries, integrated river basin management has been implemented. River basin agencies are in place (e.g., Czech Republic, Poland), with most financing derived from the government. Dr Dariusz Stanislawski from the Polish Ministry of Environmental Protection outlined the problems which are slowing implementation of integrated management systems. Polish authorities, through the new Water Law, instituted the River Basin Water Management Authorities. The law and management system it has created utilizes the mechanisms of the market economy and has created a basin-wide system of financing water quality management. The funds for management were to be derived from payments for licenses to use water and fees for discharging wastewater into the receiving water body system.

However, many delays and obstructions typical for emerging democracies have occurred. Discussions and disagreements among conflicting lobbying groups and the public have caused that the initially declared popular, almost unanimous support for the new, market economy oriented water resource management in Poland to vanish. Dr Stanislawski pointed out that the situation in Poland and other CEE countries differs from that in many advanced Western countries which initiated new water basin management agencies years ago. The Western countries had relatively stable, prospering economies with relatively small pressures for immediate solutions of local conflicts, consequently, implementation was gradual and occurred in a favorable economic climate. In Poland and other CEE countries, solutions have to be implemented at a much faster pace because the deterioration of water resources has reached alarmingly high levels and, at the same time, the economies of these countries are in a very volatile state.

2. CONCLUSIONS AND RECOMMENDATIONS

The workshop concluded that during the transitional period which may last between fifteen to thirty years (depending on the country) realistic and implementable standards must be developed, state by state, implemented and enforced. New economic realities in the CEE countries provide favorable conditions for significant improvement of water quality which would be accomplished using affordable technologies, integrated basin wide-water quality management and by implementing economic instruments and incentives that would lead to a transition to cleaner industrial technologies and promote further treatment of municipal and industrial discharges. The same opportunities have arisen for agriculture. On the other hand, the economic means of these countries are limited, therefore, a proper balance between efficient and equitable abatement must be developed and implemented during the transitional period.

Even though the present status of the environment in the CEE countries may resemble the situation in Western countries twenty to thirty years ago, it has become quite apparent that the process of environmental management may not exactly repeat the past developments in the West. It was concluded by the participants that the past Western experience should be disseminated and applied under the very specific political, economic and social conditions of the CEE countries.

The conclusions and recommendations of the workshop which follow, were prepared by a sub-committee which included all key speakers and one representative from each participating CEE country. The draft was then presented to and discussed at the plenary session.

1. WATER POLLUTION PROBLEMS SPECIFIC TO THE CEE REGION AND RELATED ISSUES

- Unlike in the West, where pollution problems have been recognized sequentially during the past 20 to 30 years, water pollution problems throughout the CEE region are not new and had been identified by scientists of the region but were not recognized or were even suppressed by the previous CEE authorities. Also unlike the West, where the recognition and solution of the problems sequentially progressed from traditional point sources and pollutants to nonpoint sources and priority pollutants, the sudden emergence of the problems in the CEE countries presents a managerial and priority establishment dilemma.

- Water, sediment and soil contamination in the CEE region is significant and extensive. Therefore remediation and abatement may take a longer time. Priorities must be established, beginning first with the protection of human health and then followed by protection of the well being of aquatic biota and other water body uses.

- Economic transition and constraints on environmental remediation, together with previous problems related to acquisition of western pollution control technologies, generate a special situation in the CEE region.

- Some management and planning practices in the CEE typical for the previous central planning system persist still today and the change is difficult. Certain important tools of environmental management (e.g., lobbying) have no tradition. Non-governmental environmental organizations (NGOs) are not yet strong enough to exert needed pressure on the government.

- The involvement of scientists in formulation of environmental policies is often not satisfactory and significant communication gaps exist between the science, practice and decision-making.

- When addressing environmental (water pollution) problems, the general trends and site-specific features should be considered simultaneously.

- Monitoring of conventional pollutants is satisfactory in most CEE countries, however, monitoring of priority pollutants which requires specific and costly instrumentation is insufficient. Consequently, adequate water quality and hydrological data bases are available only for flow and traditional pollutants in most CEE countries but their accessibility is sometimes not easy.

2. METHODS, TECHNOLOGIES AND ENGINEERING

- While the necessary skills and knowledge are available in the CEE countries the ownership of methods, technologies, and tools is crucial. Methodologies and technologies can be transferred relatively quickly although in this aspect there is a large diversity within the CEE region.

- Careful evaluation of different methods and technologies is needed, respecting local conditions and focusing on efficiency. Adoption of Western criteria must be made with caution and always considering implementability. For the transition period special realistically achievable CEE water quality standards must be developed rather than uncritically imported from the West. The new environmental restrictions should consider the risk to the population and aquatic ecology.

- Imitation of Western expensive high technologies may have to be avoided in the transitional period. There is a strong need for innovative, less expensive low-technology solutions which seem to be appropriate in many parts of the CEE region, both in urban and rural areas. This focus on the less expensive technologies is applicable to industrial and municipal as well as agricultural sources or, in a broader sense, to watershed management. The CEE region has, therefore, specific research and development needs.

- Prevention, re-use, re-cycling, pretreatment, closing material cycles and related economic incentives should receive a higher priority.

3. POLICY IMPLICATIONS

- Scheduling of goals and standards, phasing in as well as implementation and enforcement of environmental policies for the upcoming twenty to thirty year period are crucial. Definition of overall receiving water quality goals set by central legislative bodies should serve as a starting point. Implementation (action plans) should be basinwide, keeping in mind the natural unit of water management.

- Integrated river basin management is considered as the only feasible management process. The relative advantage of the CEE countries is that the related management agencies have already been established in the past and the known failures of the West can be avoided in the future.

- It is recommended to develop step-wise planning and management procedures to identify the goals, objectives, and alternatives similarly to many Western countries. Although basin-wide water quality planning and management is an established and only feasible process used widely also in the West, central command type planning has been discredited and must be avoided. On the other hand, pure market approaches to water quality are not possible and generally cannot be implemented.

- Water quality impact and economic implications should be evaluated jointly when formulating new environmental control legislation.

- It is recommended to begin implementation of pollution control as early as possible and not to wait for external instruction and assistance. For this reason, responsibilities should be directed to local authorities.

- It is proposed to utilize institutional changes presently under way in the CEE countries for introduction of economic instruments which should be applied "holistically."

- Develop efficient mechanisms for the transfer of costs and funds for pollution abatement and environmental remediation among jurisdictions. It has been found that the "polluter pays" principle is the most "equitable" and should be followed wherever possible. The use of subsidies for pollution control, which has been the prevalent practice in the CEE countries, should be critically reevaluated and they should be applied with caution during a specific transitional period.

- It is recommended to involve key actors, including ministries, water quality inspectorates, dischargers, local governments, scientists, the general public and NGO's into the decision-making process. Policy workshops should be organized for a broader audience including members of these groups, in addition to more narrowly focused technical and scientific meetings and workshops.

4. EDUCATION, COMMUNICATION AND INFORMATION EXCHANGE

- Improved education is needed on all educational levels in order to realize the importance and impacts of consumption habits, the overall attitude towards environment, nature and ecology, and prevention of pollution.

- Improved interactions are needed among scientists, engineers and practitioners on one side and decision-makers on the other side. Closer involvement of the research community is essential in forming and analyzing legislation and providing impetus for other policy debates.

- Efforts to integrate the CEE professionals with their Western counterparts are desirable and needed. In this respect, the present international professional organizations may serve as a means and forum of communication and knowledge exchange.

- Technology and transfer of know-how from the Western developed countries to the CEE region is important to enhance the likelihood of appropriate technology applications therein. On the other hand incentives are needed to avoid the "brain drain."

- Access to basic scientific publications, good library resources, and international and regional networking is crucial for several CEE countries.

- Dissemination of models, methods, and results of their application to a broader community (managers, legislators, etc.) is essential for successful applications to real world problems.

- There is a need for summarizing Western standards, environmental laws, etc., and their comparative assessment with those presently in force in the CEE region which should be disseminated in the CEE region to key persons.

SUBJECT INDEX

LIST OF PARTICIPANTS AND SPEAKERS

ARW TITLE: Remediation and Management of Degraded River Basins with Emphasis on Central and Eastern Europe

DIRECTORS:

Professor Vladimir Novotny
Department of Civil and Environmental Engineering
Marquette University
1515 West Wisconsin Ave.
Milwaukee, Wiscosnin 53233
UNITED STATES

Professor Lászlò Somlyódy
Department of Water and Wastewater Engineering
Budapest University of Technology
Müegyetem rpk. 3
H-1111 Budapest
HUNGARY

LOCATION: Laxenburg, Austria

DATES: 13. - 16. 6. 1994

1. **SPEAKERS**

Dr. Luis Veiga da Cunha
Director, Priority Area on Environment
North Atlantic Treaty Organization (NATO)
Scientific and Environmental Affairs Division
B-1110 Brussels
BELGIUM

Professor Petr Grau
AquaNova International a.s.
Pod vilami 22
CR-140 00 Praha 4 - Nusle
CZECH REPUBLIC

Dr. Milan Straškraba
Biomathematical Laboratrory
Czech Academy of Sciences
Branišovská 31
CR-370 05 České Budějovice
CZECH REPUBLIC

Professor Mogens Henze
Technical University of Denmark, Building 115
DK-2800 Lyngby
DENMARK

Professor Sven Erik Jϕrgensen
DFH, Institute A
Miljϕkemi (Environmental Chemistry)
University Park
DK-2100 Copenhagen ϕ
DENMARK

Professor Ing-Dr. Ulrich Förstner
Department of Environmental Engineering
Technical University of Hamburg
Eissendorfer Str. 40
D-2100 Hamburg 90
GERMANY

Professor Dr. Hermann H. Hahn
Institut für Siedlungswasserwirtschaft
Universität Fridericiana zu Karlsruhe
Kaiserstraße 12
D-76131 Karlsruhe
GERMANY

Dr. Neithart Müller
Universität Fridericiana zu Karlsruhe
Kaiserstraße 12
D-76131 Karlsruhe
GERMANY

Dr. Wim Salomons
Institute for Agrobiology and Soil Fertility
P.O. Box 129
9750 Haren
THE NETHERLANDS

Mr. Ilja Masliev
Institute for Water and Environmental problems
Russian Academy of Sciences
Papanizev Str. 105
656099 Banaul-99
RUSSIA

Professor Dominic DiToro
Department of Environmental Engineering & Science
Manhattan College
Riverdale NY 10471
UNITED STATES

Dr. Charles M. Paulsen
Environmental Management Program
Resources for the Future
18640 Wood Duck Way
Lake Oswego, OR 97035
UNITED STATES

Dr. Peter Shanahan
HydroAnalysis Inc.
481 Great Road No.3, P.O. Box 631
Acton, Massachussetts 01720
UNITED STATES

Professor Mark Smith
Colorado College
Colorado Springs, Colorado
UNITED STATES

2. OTHER PARTICIPANTS

Dipl-Ing Dr. Tech Ludwig Csépai
Zivilingenieur für Bauwesen
Haupstraße 25
A-2340 Mödling
AUSTRIA

Professor Helmut Fleckseder
Wien TU/Wassergüte
Karlsplatz 13
A-2040 Wien (Viena)
AUSTRIA

Dr. Ivanka Dimitrova
Bulgarian Academy of Sciences, Institute of Water Problems
'Akad. G. Bontchev' Street, Bl
BG-1113 Sofia
BULGARIA

Professor Valentin Nenov
Bourgas University of Technology
Department of Water Technology
BG-8010 Bourgas
BULGARIA

Ing Danuše Beránková
Water Management Institute T.G. Masaryk
Dřevařská 12
657 57 Brno
CZECH REPUBLIC

Dr. Václav Eliáš
Director of the Institute of Hydrodynamics
Czech Academy of Sceinces
Podbabská 13
Praha 6
CZECH REPUBLIC

Ing Jaroslav Zdařil
Director
Water Research Institute (VUV) T.G. Masaryk
Dřevařská 12
65 757 Brno
CZECH REPUBLIC

Professor Irina Blinova
Water Protection Laboratory
Tallin Technical University
Jarvenana 5
EE0001 Tallin
ESTONIA

Dr. Olli Varis
Laboratory of Hydrology and Water Resources Management
Helsinki University of Technology
FIN - 02150 Espoo
FINLAND

Dr. Valentina Krysanova
Institute of Climate Impact Research
Department of Integrated Systems
Postfach 601203
Telegrafenberg
D-14412 Potsdam
GERMANY

Professor Kálmán Buzás
Technical University of Budapest
Dept. of water and Wastewater Engineering
Müegyetem rpk. 3
H 1111 Budapest
HUNGARY

Mr János Fehér
Water Resources research Center (VITUKI)
Kvaswsay u. 1
H-1095 Budapest
HUNGARY

Mr T. Kőszegi
Ministry of Transport. Communication and Water Management
Dob ut. 75-81
H - 1077 Budapest
HUNGARY

Dr. Miklós Pannonhalmi
North Transdanubian Water Authority
Postacim 9002 Györ, Pf: 471
Árpád u. 28-32
H-9021 Györ
HUNGARY

Dr. Pál Varga
National Environment Authority
Fö ut. 44-50
H-1011 Budapest V
HUNGARY

Professor Dr. Giuseppe Bendoricchio
Instituto di Chimica Industriale
Facoltà di Ingegneria
Univeristà di Padova
Via Marzolo 9
35131 Padova
ITALY

Professor Dr-Ing Andrea Capodaglio
Department of Hydraulic and Environmental Engineering
Università degli Studii di Pavia
Via Abbiategrasso, 213
I-2700 Pavia
ITALY

Dr. Eva Skarbovik (NORWAY)
presently JPO, Freshwater Unit
United Nations Environment Programme
P.O. Box 30552
Nairobi
KENYA

Dr. Dariusz Jan Stanislawski
Adviser to the Minister
Ministry of Environmental Protection, Natural Resources and Forestry
Department of Water Economy
Wawelska 52/54
PL-00 922 Warsaw
POLAND

Professor Jan Suschka
Technical University of Lodz
Filial Bielsko-Biala
Textile and Environmental Protection
Willowa Str. 2
PL-43 300 Bielsko-Biala
POLAND

Professor Angheluta Vadineanu
Department of Ecology
University of Bucharest
Splaiul Independentei 91-95
76201 Bucharest
ROMANIA

Professor Valentina G. Priazhinskaya
Water Problems Institute
Russian Academy of Sciences
10 Novaya Basmannaya St.
107078 Moscow
RUSSIA

Academican Oleg F. Vasiliev
Institute for Water and Environmental Problems
Russian Academy of Sciences - Siberian Division
Papninzev Str. 105
656099 Barnaul
RUSSIA

Professor Dr. Juraj Námer
Slovak Technical University
Department of Sanitary Engineering
Radlinského ul. 11
SR-813 68 Bratislava
SLOVAK REPUBLIC

Ing-Dr Pavel Petrovic
Water Research Institute (VUVH)
Nábr. arm. gen. L. Svobodu 5
SR-812 49 Bratislava
SLOVAK REPUBLIC

Dr. Alojz Bitenc
University of Ljubljana
'Jozef Stefan ' Institute
Department of Computer Automation and Control
Jamova 39
SL-61111 Ljubljana
SLOVENIA

Professor Mitja Brilly
FAGG-Hydraulics Division
University of Ljubljana
Hajdrihova 28
61 000 Ljubljana
SLOVENIA

Tjaša Bulc
Water Management Institute
Hajdrinova 28
Ljubljana 61 000
SLOVENIA

Dr. Vladimir Vanek
VBB VIAK
Gijersgatan 8
S - 216 18 Malmö
SWEDEN

Professor G.A. Sukhorukov
Ukrainian Science Centre for Water Protection
6 Bakulin Str.
310888 Kharkov
UKRAINE

Professor Keneth M. Strzepek
Center for Advanced Decision Support for Water & Environmental Systems
University of Colorado
Applied Science Suite D
2945 Center Green Courth South
Boulder, Colorado 80301
UNITED STATES

Mr Denver Stutler
Camp, Dresser & McKee, Inc.
1950 Summit Park Drive, Suite 300
Orlando, Florida 32810
UNITED STATES

Dr. David Yates
Center for Advanced Decision Support for Water & Environmental Systems
University of Colorado
Applied Science Suite D
2945 Center Green Courth South
Boulder, Colorado 80301
UNITED STATES

The Partnership Sub-Series incorporates activities undertaken in collaboration with NATO's Cooperation Partners, the countries of the CIS and Central and Eastern Europe, in Priority Areas of concern to those countries.

The volumes published as a result of these activities are:

Vol. 1: Clean-up of Former Soviet Military Installations. Edited by R. C. Herndon, P. I. Richter, J. E. Moerlins, J. M. Kuperberg, and I. L. Biczó. 1995

Vol. 2: Cleaner Technologies and Cleaner Products for Sustainable Development. Edited by H. M. Freeman, Z. Puskas, and R. Olbina. 1995

Vol. 3: Remediation and Management of Degraded River Basins. Edited by V. Novotny and L. Somlyódy. 1995

GPSR Compliance
The European Union's (EU) General Product Safety Regulation (GPSR) is a set of rules that requires consumer products to be safe and our obligations to ensure this.

If you have any concerns about our products, you can contact us on

ProductSafety@springernature.com

In case Publisher is established outside the EU, the EU authorized representative is:

Springer Nature Customer Service Center GmbH
Europaplatz 3
69115 Heidelberg, Germany

www.ingramcontent.com/pod-product-compliance
Lightning Source LLC
LaVergne TN
LVHW012111170826
845678LV00001BA/3
* 9 7 8 3 6 4 2 6 3 3 4 6 1 *